Water Diplomacy in Action

ANTHEM WATER DIPLOMACY SERIES

More effective resolution of our increasingly complex, boundary-crossing water problems demands integration of scientific knowledge of water in both natural and human systems along with the politics of real-world problem solving. Water professionals struggle to translate ideas that emerge from science and technology into the messy context of the real world. We need to find more effective ways to bridge the divide between theory and practice and to resolve complex water management problems when natural, societal and political elements cross multiple sectors and interact in unpredictable ways. The **Anthem Water Diplomacy Series** is a step in that direction. Contributions in this series diagnose water governance and management problems, identify intervention points and possible policy changes, and propose sustainable solutions that are sensitive to diverse viewpoints as well as conflicting values, ambiguities and uncertainties.

Water Diplomacy in Action

Contingent Approaches to Managing Complex Water Problems

Edited by Shafiqul Islam and Kaveh Madani

ANTHEM PRESS

Anthem Press
An imprint of Wimbledon Publishing Company
www.anthempress.com

This edition first published in UK and USA 2017
by ANTHEM PRESS
75–76 Blackfriars Road, London SE1 8HA, UK
or PO Box 9779, London SW19 7ZG, UK
and
244 Madison Ave #116, New York, NY 10016, USA

British Library Cataloguing-in-Publication Data
A catalogue record for this book is available from the British Library.

Library of Congress Cataloging-in-Publication Data
Names: Islam, Shafiqul, 1960– editor. | Madani, Kaveh, editor.
Title: Water diplomacy in action : contingent approaches to managing complex
water problems / editors, Shafiqul Islam and Kaveh Madani.
Description: London, UK; New York, NY: Anthem Press, 2017. |
Includes bibliographical references and index.
Identifiers: LCCN 2016052841 | ISBN 9781783084906 (hardback : alk. paper)
Subjects: LCSH: Water-supply–International cooperation. |
Water-supply–Political aspects. | Water-supply–Government policy. |
Water-supply–Management. | Water resources development.
Classification: LCC TD345 .W26155 2017 | DDC 333.91–dc23
LC record available at https://lccn.loc.gov/2016052841

ISBN-13: 978-1-78308-490-6 (Hbk)
ISBN-10: 1-78308-490-1 (Hbk)

ISBN-13: 978-1-78308-493-7 (Pbk)
ISBN-10: 1-78308-493-6 (Pbk)

This title is also available as an e-book.

CONTENTS

FIGURES

TABLES

THE BLIND MEN, THE ELEPHANT AND THE WELL: A PARABLE FOR COMPLEXITY AND CONTINGENCY

Maimuna Majumder

In the early 1970s, the leading cause of childhood mortality in Bangladesh was diarrheal disease (Smith, Lingas and Rahman 2000; Anthamatten and Hazen 2012). Due to inadequate sanitation infrastructure, the untreated surface water that the majority of the population used for their daily needs was contaminated with a variety of disease-causing agents including *Vibrio cholerae*, rotavirus and *Escherichia coli*, among others (Hoque et al. 1996; Alam et al. 2006; Begum et al. 2007; Hashizume et al. 2008).

The underdeveloped immune systems of children below five years make them particularly susceptible to death from diarrheal disease. To address this issue, the United Nations Children's Fund (UNICEF) initiated a project in 1972 to build shallow tube wells that tapped into local groundwater aquifers (Smith, Lingas and Rahman 2000). Such tube wells are not as vulnerable to contamination as are surface water sources, which made them a seemingly ideal solution for the problem at hand.

By 1980, one million tube wells had been installed—a number that then grew tenfold in the decade that followed, bringing coverage to nearly the entirety of the Bangladeshi population in the process (Jones et al. 2008; Smith, Lingas and Rahman 2000; Chappell, Abernathy and Calderon 2001). As under-five mortality began to fall as a result of UNICEF's shallow tube well project, the wells themselves became status symbols in the communities they served (Anthamatten and Hazen 2012; Sultana 2007, 2013). In fact, they became so intertwined with Bangladeshi cultural practices that many families even began to include them in their daughters' dowries. At the time, UNICEF did not know that the same wells that prevented diarrheal disease so effectively in these rural Bangladeshi communities were also the cause of arsenic poisoning.

Shallow tube wells, which draw water from within a depth of 150 meters from the surface, are prone to arsenic contamination from metal deposits in the surrounding soils (Smith, Lingas and Rahman 2000). Unfortunately, arsenic poisoning—or arsenicosis—has a latency period that can be as long as 20 years (Martinez et al. 2011). By the time symptoms started to appear in the 1980s, tube wells had already become the primary water source for millions of Bangladeshis—resulting in what the WHO has since called "the largest mass poisoning of a population in history" (Smith, Lingas and Rahman 2000).

Red Well, Green Well

In 1999, UNICEF initiated a multimillion dollar campaign to screen wells for arsenic and educate communities about the dangers of consuming arsenic-contaminated water (Chappell, Abernathy and Calderon 2001). Within five years, the majority of shallow tube wells in rural Bangladesh were painted either red or green; red rims indicated high levels of arsenic contamination, while wells with green rims were safe for use (Sultana 2009; Hanchett et al. 2002). More than 20 percent of the shallow tube wells were painted red (Chappell, Abernathy and Calderon 2001). Suddenly, one in five families no longer had a safe and reliable source of water at their disposal.

But limited access to clean water was not the only consequence that emerged from the 1999 UNICEF campaign. With the red-rimmed tube wells came tremendous stigmatization, and to this day, individuals with arsenicosis face discrimination both financially and socially (Ahmad et al. 2007; Brinkel, Khan and Kraemer 2009).

Rural Bangladeshi women—who largely rely on their husbands and fathers for monetary support—suffer disproportionately (Sultana 2007; Ahmad et al. 2007; Brinkel, Khan and Kraemer 2009). The same wells that had once made it easier for women to wed now leave them with diminished prospects for marriage because of suspected arsenicosis.

The Blind Men and the Elephant

The story of arsenic poisoning in rural Bangladesh—and the events that both preceded and succeeded it—is in many ways a modern-day manifestation of "The Blind Men and the Elephant," a timeless parable that originated centuries ago on the Indian subcontinent and has since diffused worldwide.

In the tale, a group of blind men touch various parts of an elephant to learn what it is like. Each man feels only one part, such as the tusk or the tail, and upon comparing notes, they quickly discover that their individual experiences are in complete disagreement with one another.

The moral of the parable is simple: the elephant is a complex beast; thus, one blind man cannot fully comprehend it through a single touch. To understand the complete picture, a blind man must rely not only on his own experience, but also on the experiences of others.

The Elephant and the Well

The UNICEF shallow tube well project, as well as the testing and education campaign that followed nearly three decades later, serve as ill-fated examples of a metaphorical blind man's simple solutions to a complex problem.

In both 1972 and 1999, UNICEF failed to consider the experiences of the local population. Both the initial well installation project and the well painting campaign were designed *for* rural Bangladeshi communities instead of *with* them. As a result, UNICEF did not fully recognize the potential social and economic ramifications of either intervention—or how those ramifications would compound over time and marginalize women in the process.

Today, these communities still live with the negative implications of the interventions and, like the blind men seeking to understand the nature of an elephant, providing access to safe water in rural Bangladesh has proven to be complex. Each intervention outlined above addressed whatever issue was most evident—first, under-five mortality and then, arsenic poisoning. However, by focusing solely on the immediate, context was disregarded and complexity was dismissed—and the Bangladeshi people continue to feel the repercussions of this neglect. After all, just because a blind man only feels a tusk does not mean that the tail ceases to exist.

Our challenge is to acknowledge that we have discovered the tusk, but also to be aware that we may have missed the tail. We must be willing to consider and adopt interventions appropriate for particular contexts because it is nearly impossible to find a simple and permanent solution to complex problems. We need to seek and find contingent approaches to resolve these water problems.

References

Ahmad, Sheikh A., Muhammad H. S. Sayed, Manzurul H. Khan, Muhammad N. Karim, Muhammad A. Haque, Mohammad S. A. Bhuiyan, Muhammad S. Rahman and Mahmud H. Faruquee. 2007. "Sociocultural Aspects of Arsenicosis in Bangladesh: Community Perspective." *Journal of Environmental Science and Health, Part A* 42, no. 12: 1945–58. doi:10.1080/10934520701567247.

Alam, Munirul, Marzia Sultana, G. Balakrish Nair, R. Bradley Sack, David A. Sack, A. K. Siddique, Afsar Ali, Anwar Huq and Rita R. Colwell. 2006. "Toxigenic *Vibrio cholerae* in the Aquatic Environment of Mathbaria, Bangladesh." *Applied and Environmental Microbiology* 72, no. 4: 2849–55. doi:10.1128/AEM.72.4.2849-2855.2006.

Anthamatten, Peter, and Helen Hazen. 2012. *An Introduction to the Geography of Health*. New York: Routledge.

Begum, Y. A., K. A. Talukder, G. B. Nair, S. I. Khan, A.-M. Svennerholm, R. B. Sack and F. Qadri. 2007. "Comparison of Enterotoxigenic *Escherichia coli* Isolated from Surface Water and Diarrhoeal Stool Samples in Bangladesh." *Canadian Journal of Microbiology* 53, no. 1: 19–26. doi:10.1139/w06-098.

Brinkel, Johanna, Mobarak H. Khan and Alexander Kraemer. 2009. "A Systematic Review of Arsenic Exposure and Its Social and Mental Health Effects with Special Reference to Bangladesh." *International Journal of Environmental Research and Public Health* 6, no. 5: 1609–19. doi:10.3390/ijerph6051609.

Chappell, W. R., C. O. Abernathy and R. L. Calderon. 2001. *Arsenic Exposure and Health Effects IV*. Oxford: Elsevier.

Hanchett, Suzanne, Qumrun Nahar, Astrid Van Agthoven, Cindy Geers and Ferdous Jamil Rezvi. 2002. "Increasing Awareness of Arsenic in Bangladesh: Lessons from a Public Education Programme." *Health Policy and Planning* 17, no. 4: 393–401. doi:10.1093/heapol/17.4.393.

Hashizume, M., B. Armstrong, Y. Wagatsuma, A. S. G. Faruque, T. Hayashi and D. A. Sack. 2008. "Rotavirus Infections and Climate Variability in Dhaka, Bangladesh: A Time-Series Analysis." *Epidemiology and Infection* 136, no. 9: 1281–89. doi:10.1017/S0950268807009776.

Hoque, B. A., T. Juncker, R. B. Sack, M. Ali and K. M. Aziz. 1996. "Sustainability of a Water, Sanitation and Hygiene Education Project in Rural Bangladesh: A 5-Year Follow-Up." *Bulletin of the World Health Organization* 74, no. 4: 431–37.

Jones, Huw, Pornsawan Visoottiviseth, M. Khoda Bux, Rita Födényi, Nora Kováts, Gábor Borbély and Zoltán Galbács. 2008. "Case Reports: Arsenic Pollution in Thailand, Bangladesh, and Hungary." *Reviews of Environmental Contamination and Toxicology* 197: 163–87.

Martinez, Victor D., Emily A. Vucic, Daiana D. Becker-Santos, Lionel Gil, Wan L. Lam,. 2011. "Arsenic Exposure and the Induction of Human Cancers." *Journal of Toxicology* 2011 (November): 431287. doi:10.1155/2011/431287, 10.1155/2011/431287.

Smith, Allan H., Elena O. Lingas and Mahfuzar Rahman. 2000. "Contamination of Drinking-Water by Arsenic in Bangladesh: A Public Health Emergency." *Bulletin of the World Health Organization* 78, no. 9: 1093–1103. doi:10.1590/S0042-96862000000900005.

Sultana, Farhana. 2007. "Water, Water Everywhere, but Not a Drop to Drink: Pani Politics (Water Politics) in Rural Bangladesh." *International Feminist Journal of Politics* 9, no. 4: 494–502. doi:10.1080/14616740701607994.

———. 2009. "Community and Participation in Water Resources Management: Gendering and Naturing Development Debates from Bangladesh." *Transactions of the Institute of British Geographers* 34, no. 3: 346–63. doi:10.1111/j.1475-5661.2009.00345.x.

———. 2013. "Water, Technology, and Development: Transformations of Development Technonatures," in *Changing Waterscapes Environment and Planning D: Society and Space* 31, no. 2: 337–53. doi:10.1068/d20010.

PREFACE

Today, a billion people lack access to clean water. Water scarcity affects one in three people on our planet. These complex water problems are sobering—but not new. A hundred years ago, in the first issue of the *American Economic Review*, Katharine Coman published "Some Unsettled Problems of Irrigation."[1] She used irrigation systems in the United States west of the one hundredth meridian to describe intricate collective-action problems. This was almost half a century before Garett Hardin identified the so-called tragedy of the commons and associated theoretical and implementation problems related to common pool resource management.[2]

The process of converting the desert into farmland demonstrated the challenges of achieving a collective good (in this case, building and running an irrigation system). In her reflections on Coman's article, Elinor Ostrom summarized that "changing the formal governance structure of irrigation is not sufficient to ensure efficient investment in facilities or that farmers are able to acquire property and make a reasonable living. Building knowledge and trust are, however, essential for solving collective-action problems."[3]

Today, in our highly interconnected and globalized world, water is an integral part of any discussion on energy, agriculture, public health, transportation, environment and the future. In the past, we have tried to solve our water problems with reservoirs, dams and treatment facilities. Now, we know that these responses often did not balance the value-laden needs of individuals and communities with those of industry and agriculture, nor did they sufficiently protect our natural resources. We have discovered that many of our intended solutions are not sustainable, especially considering global population growth, consumption ethic and the changing global climate. We must find ways to resolve water problems, account for legacy effects from past allocation and infrastructure development decisions, and meet demands for industrial and economic growth while ensuring a more sustainable future.

Our challenge is how to translate solutions that emerge from science and technology into the messy context of the real world. In the present volume, we hope to address this challenge by synthesizing two emerging ideas—complexity and water diplomacy—to understand and manage risks and opportunities for an uncertain water future. We argue that the origin of many complex water problems is a dynamic consequence of competition, interconnections, and feedback among variables, processes, actors and institutions. Consequently, science alone cannot solve complex water problems, nor can policy operating without input from science and contextual politics. More effective resolutions to our increasingly complex water problems demand integration of scientific knowledge of water in both natural and human systems with the politics of real-world problem solving to develop sustainable solutions that address issues of equity as well as the individual and group rationality criteria.

We need to define what makes a class of water problems complex and how to resolve competing—and often conflicting—needs inherent to these complex water problems. This volume synthesizes insights from theory and practice to address complex water problems through contingent and adaptive management using the water diplomacy framework.[4] Today, we face an incredibly complex array of interconnected water issues across multiple boundaries that share two key defining characteristics: competing values, interests and tools to frame the problem, and differing views on how to resolve the problem, which is plagued by uncertainty and ambiguity. The root cause of many water problems lies at the intersection of multiple causal forces buried in observational signatures with often conflicting views and values related to *who decides, who gets water and how*. In such situations, neither numbers nor narratives will resolve the dilemma.

In an attempt to address these complex water issues, the Water Diplomacy Framework (WDF) diagnoses the sources of complexity for a given water problem, identifies intervention points, and proposes sustainable resolutions that are sensitive to diverse viewpoints and uncertainty as well as to changing and competing needs. The WDF actively seeks value-creation opportunities by blending science, engineering, policy and politics through a contingent negotiated approach. This framework is based on the following key ideas and assumptions:

- Conflicts over water resources occur due to lack of resilience in management systems when the natural, societal, and political processes and variables interact to create complex and competing water demands. Water-related problems are complex not only because they involve various stakeholders (e.g., farmers, industrial users, urban developers, environmental activists) who are competing for a limited resource, but also because water problems cross multiple boundaries (e.g., natural, societal and political), scales (spatial, temporal, jurisdictional and institutional) and levels (local, regional and global) that are difficult to incorporate into a traditional modeling approach.
- We view water problems as a network of variables, processes, actors and institutions. The characterization and management of complex water networks use methodological approaches that blend knowledge from various domains including water science and engineering, environmental and ecological economics, water resources planning and management, systems analysis, political science, social science, law and decision analysis.
- We posit that the use of a network perspective allows one to treat water as a flexible resource, where fixed quantities can be expanded through appropriate application of technology, management and policy interventions. Water management and operations networks are open and continuously changing. Uncertainty, variability and ambiguity are an integral part of these networks and must be acknowledged and discussed in problem framing and formulation.

This book starts with the premise that neither numbers nor narratives are adequate to understand and manage complex water problems. We need a synthesis of numbers and narratives, theory and practice, and objectivity and interpretation to resolve complex water issues. It begins with the age-old metaphor of the blind men and the elephant as a way to introduce ideas of complexity science and contingent approaches needed to resolve water problems. The book is broadly structured in three parts: the first two parts focus on theory—why, what and how of complexity—of complex water problems (Chapters 1 through 9) while the third part describes and analyzes case studies on how to resolve complex water problems (Chapters 10 through 15).

The introductory chapter provides an overview of complexity science from multiple domains of knowledge. A cursory look at this vast body of literature in complexity sciences shows an incredible diversity in terms of ontological and epistemological assumptions, foundational concepts, levels of analysis, research methods and so on. It focuses on two broad ways of thinking about the different faces of complexity: numbers and narratives; models and meanings; objective and interpretive. The breadth and depth of scholarship within each domain suggest that a comprehensive synthesis may not be attainable. The intent is not to provide a complete description of multifaceted reality but to develop an interdisciplinary understanding of these two broad ways of thinking about complexity. It is hoped that an interdisciplinary understanding of different schools of thought from complexity science can provide a pragmatic way to diagnose sources of complexity, identify intervention points, and develop equitable and sustainable solutions for complex water management problems. A case example of water sharing between Israel and Jordan demonstrates how the creation of an actionable space with a commitment to collaborative adaptive management can produce a relatively sustainable water agreement. It is argued that instead of searching for *optimal solutions* to address complex problems one needs to look for an *optimal space* where certain solutions are actionable given the constraints the context imposes.

The shared waters of transboundary basins are a source of both conflict and opportunity for cooperative behavior. Increased variability and uncertainty related to water quantity and quality may multiply both water-related risk and opportunity for riparian states sharing transboundary waters. Pohl and Swain (Chapter 2) argue that cooperative arrangements for transboundary basins can support sustainable development programs. However, shaping the agreements and tools for transboundary water requires both technical and diplomatic expertise, necessitating a hydro-diplomacy approach that includes technical experts with skill in negotiation and mediation as well as policy makers and diplomats who have knowledge of water management and the development of water infrastructure. The authors suggest that a strong international institutional platform dedicated to supporting transboundary water cooperation is needed to address the shortcomings of current efforts and create avenues for improved cooperative water management for a range of riparian interests, especially in areas where enduring hegemons introduce significant power asymmetries between basin stakeholders.

Water resources allocation and satisfying the needs of stakeholders is one of the complexities associated with water resources management. Zarezadeh et al. (Chapter 3) introduce several bankruptcy rules that can be used for water and natural resource allocation. In a world with growing demand for resources, the situations in which the demand for a resource is more than the available supply are not rare. Thus, the authors suggest considering such situations as 'natural resource bankruptcy' and applying commonly used financial bankruptcy rules to allocate the available resource among the stakeholders whose total claim cannot be fully satisfied. The authors apply the suggested bankruptcy rules to allocate energy resources in the Caspian Sea conflict, which is considered to be one of the longest and largest transboundary natural resource conflicts in recent decades.

Uncertainty is an essential element of planning and managing complex water systems—an element that has been overlooked by traditional predict-and-design approaches. Wong Turlington et al. (Chapter 4) suggest Flexible Design as an adaptive approach to planning and designing water infrastructure that recognizes a wide range of uncertainties, both hydrological and human. Flexible Design can be considered as a novel approach to design that acknowledges uncertainty and attempts to learn from experiences by adjusting our decisions over time. Using a case analysis inspired by Singapore, the authors argue how their proposed approach can help reduce downside risks, increase upside opportunities, and facilitate long-term management and design of complex water infrastructure systems.

In water resources decision making (e.g., allocation, design, operations) analyses, assumptions and decisions are normally made in stationary settings. However, stationary assumptions and formulations fail to accurately capture the inherent characteristics of complex water systems. The water resources community now appreciates the nonstationarity of water systems and is in search of new methods to solve problems in nonstationary settings. Robust communication of hydrologic risk and the determination of risks of extreme events such as floods and droughts are gaining more interest among those who appreciate nonstationarity as an essential characteristic of complex water resource systems. In Chapter 5, Mondal and Mujumdar overview some methods for characterizing nonstationary behavior in hydrologic extremes within the statistical extreme value theory (EVT) framework. Using a flood-prone river basin in India as an example, they show how these methods can be used to estimate design flood quantiles and their uncertainties under transient conditions.

The application of complexity theory to study water resource problems calls for the expansion of traditional, arbitrary water resource system boundaries. Sustainable water management appreciates the complex nexus of water and other systems such as energy, food, climate, ecosystem, economy, society and politics. Without careful investigation of such nexus effective solutions cannot be developed and unintended consequences on other sectors are unavoidable. Carr and D'Odorico (Chapter 6) analyze the global patterns of virtual water transfer associated with international food trade. By examining how international virtual water trades affect the ethical and physical link between societies and the water resources that sustain them, the authors conclude that globalization of water through trades prevents famine in water-scarce regions, increases equality in access to water and allows for a more water-efficient food production. On the negative side, however, water globalization can lead to loss of environmental stewardship, increased trade dependency and reduced societal resilience to drought.

In complex management problems, crossing of boundaries, interconnectedness, nonlinearity, feedback, and sensitivity to timing and small changes can profoundly affect system performance. The water resources community needs analytical tools to critically examine such complexities. In Chapter 7, Ibrahim and Islam couple network theory with agent-based modeling to develop a network model that can help in analyzing the coevolutionary interactions and complex dynamics of agents involved in a hypothetical transboundary water management problem. Their modeling exercise suggests that sustainable solutions to complex water management problems are available even in the presence of competing interests. However, a well-designed negotiation process formation of an initial coalition of cooperating players and a search for creative solutions with mutual gains would be essential for developing an adaptive path to achieve such resolutions.

Another vital step to achieve sustainable solutions to complex water management problems is changing the way individuals, institutions and societies must invest in their capacity to address the complexities of water problems. Marshall et al. (Chapter 8) call on their readers to reach beyond the standard water-management approaches and invest in capacity development to strengthen the ability of individuals, institutions and societies to navigate the complexities of transboundary water conflicts as an exercise in water diplomacy. In their view, capacity development can occur at the individual level, institutional level or societal level, and may include initiatives to improve one's technical capabilities in water resources management, international water law, data collection and monitoring, and interpersonal communications and conflict resolution. They argue that each member of the water-resources community has a role to play and a responsibility to fulfill. To play their roles, the members of the community should not just think outside the box, but act as if there were no box at all.

Analysis of human-water systems can benefit and provide actionable insight from the use of carefully crafted models and computational tools. Computational models and tools are continuously developed and refined to help us better understand the interrelated dynamics of human-water systems. Significant improvements in computational technologies have incredibly increased our modeling capacity. Nevertheless, the use of more computational complexity does not necessarily result in a better understanding of complex systems. These models might not be useful for designing effective intervention when they fail to reasonably model the causal and physical relationships of different variables, even if their results may reasonably explain historical records. In Chapter 9, Bhatia et al. propose a new method for adding physics to data sciences, using ideas and tools from statistics and signal processing to machine learning and nonlinear dynamics. Using three case studies on precipitation, they show how their proposed physics-guided data mining framework can fill some of the crucial gaps in complex water-climate models to improve the reliability of such models.

Transboundary water management issues are highly contextual; while scientific and engineering solutions can address some aspects of management problems, the organizational and political context of water problems is integral to developing sustainable and adaptable solutions to complex transboundary water disputes. In Chapter 10, Choudhury presents two normative anchors to address these problems: sustainability of the water quality, and supply and equity of allocation and use. This chapter examines the contingent nature of interaction and discusses three enabling conditions that allow for negotiation of an enduring interaction for sharing transboundary waters. The author uses the case of the 1960 Indus Treaty between India and Pakistan and the 1994 Peace

Treaty between Jordan and Israel to illustrate the efficacy of the proposed enabling conditions. These three conditions—(a) active recognition of interdependence among contending stakeholders; (b) framing mutual interests through joint fact finding and creating mutual benefits; and (c) monitoring agreements through a joint authority and building capacity to manage emergent problems—are described in detail and illustrated through the two example cases.

Huntjens (Chapter 11) takes a practitioner's viewpoint to explore water cooperation efforts between Israel and Palestine, reflecting on observations from his participation in two water cooperation efforts between stakeholders in these two countries facilitated by third parties. In his opinion, "cooperation on water by the Israelis and Palestinians is a golden opportunity" but zero-sum approaches have stifled past efforts. He uses these two recent cases to explore how multitrack efforts that support stakeholder engagement and collaborative approaches with expert support can help participants move toward a mutual gains framing and help support social learning. The first case, the production of a master plan for the Lower Jordan River Basin, occurred between 2012 and 2014, with the goal of integrating the separate plans of Jordan, Israel and Palestine. The second case focuses on an effort to create an addendum to the 2009 Geneva Initiative Water Annex that addresses issues not included in the 2009 Annex.

Uncertainty permeates many aspects of water problems, including planning, design and construction for water infrastructure projects, such as desalination plants. AlMisnad et al. (Chapter 12) explore Flexible Design for infrastructure projects, an approach that allows system managers to adjust the project design to better reflect actual needs and conditions that unfold during project implementation. Because infrastructure projects that involve the construction and operation of a system often represent a long-term public-private partnership between a public authority and contractors, it is important to design contracts that are suitable for sharing risks and maximizing the value of the flexibly designed infrastructure.

The authors use a case study of a desalination facility in Qatar to explore contract arrangements and risk preferences of participants to provide some advice on how authorities may consider and choose between alternate arrangements in a Flexible Design process. The facility in this case study is under a Build-Own-Operate-Transfer (BOOT) contract, in which the private firm constructs the facility and the public authority is contracted to purchase water services. The case details the risks and uncertainties inherent in this type of problem and provides an analysis of how different contract options can encourage or discourage parties from adopting flexibility through changes in the distribution of gains and risks in the contract arrangements.

Infrastructure projects developed for transboundary rivers have upstream and downstream impacts that cross borders. The construction of the Grand Ethiopian Renaissance Dam (GERD) on the Blue Nile will impact Ethiopia and the greater Nile basin, in terms of both how the dam impacts the flow of the Nile and also cascading outcomes to political relationships, economic outcomes and development goals, including hydro-power production. In Chapter 13, Berndtsson et al. discuss the Nile within its MENA context and impacts of climate change and changing population and demographics on water-related resources and risks. They specifically focus on how the GERD has evolved from an idea to a completed dam that will soon be in operation, and the implications for efficient water-management efforts and collaborative management between riparian states as well as the changes in the hydro-political relationships between riparian states due to GERD.

Stakeholder engagement and participatory decision making provide a "bottom-up" approach to water resources management. Many of these processes seek to create better arrangements than strict top-down hierarchal approaches. Thoadeniya and Maheshwari (Chapter 14) describe features of stakeholder engagement processes in relation to their suitability for incorporation in IWRM, as well as examine past stakeholder engagement examples to extract lessons for stakeholder engagement in the context of IWRM and water diplomacy. This analysis takes a broad view of participatory processes and the descriptions distinguish between levels of involvement and the tools and techniques employed in the stakeholder engagement or participatory decision-making process. They highlight that participatory approaches are useful when data are too incomplete or a situation is otherwise inappropriate for technical approaches (such as computer-based models) to be used to address a water problem. Additionally, if the outcomes from a decision must be accepted by the stakeholders prior to a successful implementation, then a stakeholder engagement process that builds trust between parties can contribute to success. Advantages of successfully implemented programs improve the acceptability of difficult decisions, can improve stakeholder relationships, and may have cascading benefits such as capacity building and opportunities for addressing adaptability and resilience for future decisions. The authors use case examples from Australia, Europe, Africa and North America to briefly illustrate the concepts presented.

Water management involves managing conflicts between stakeholders. Game theory can be used to provide strategic insights into complex water management problems and to develop management institutions that are less vulnerable to conflicts. In Chapter 15, Moazezi et al. use the Graph Model for Conflict Resolution (GMCR) to model one of the most complex and ongoing water resource conflicts in California—the Sacramento–San Joaquin Delta conflict. By employing a range of non-cooperative game theory solution methods (concepts) they try to capture the effects of the options, preferences and behavioral characteristics of the major decision makers of the delta conflict on its evolution over time. They show how these concepts can be used to identify possible resolutions to the conflict and possible strategic moves by different players to avoid some undesirable outcomes.

There is great diversity in how water problems manifest across knowledge and political domains. The diversity of problems and cases, tools and methods explored within this volume speak to the many challenges for understanding how complexity shapes problems and their paths to resolution and the contention encountered when applying and adopting tools and methods to specific cases.

While a comprehensive synthesis of an interdisciplinary understanding of complexity science as applied to real-world water challenges remains illusive and likely unattainable, our hope is that the diverse discussion and exposures to methods and cases exploring these challenges and pathways to successfully addressing them will support further development and refinement of perspectives and tools to better address existing and emerging complex problems involving shared water resources.

Notes

1 K. Coman. "Some Unsettled Problems of Irrigation." *American Economic Review* 1, no. 1 (1911): 1–19.
2 The opening paragraphs of this piece originally appeared on our blog: "Water Diplomacy Network: Translating Ideas and Insights into Actions for a Sustainable Water Future." http://blog.waterdiplomacy.org/2013/01/water-diplomacy-network/
3 E. Ostrom. "Reflections on 'Some Unsettled Problems of Irrigation.'" *American Economic Review* 101, no. 1 (2011): 49–63.
4 S. Islam and L. Susskind. *Water Diplomacy: A Negotiated Approach to Managing Complex Water Networks* (New York: Routledge, 2012).

Part I

ROOTS AND CAUSES OF COMPLEXITY AND CONTINGENCY IN WATER PROBLEMS

Chapter One

COMPLEXITY AND CONTINGENCY: UNDERSTANDING AND MANAGING COMPLEX WATER PROBLEMS

Shafiqul Islam

Abstract

While classical scientific methods seek generalizable and predictable solutions, these methods do not work for complex problems where solution spaces are neither well bounded nor predictable. A key challenge for many water problems—such as allocation between competing uses, providing access to water in urban slums or creating water-sharing arrangements between riparian countries—is that outcomes from interventions are not predictable due to the dynamic nature of interactions and interdependencies of complex water systems. An interdisciplinary understanding of different schools of thought from complexity science can provide a pragmatic way to diagnose sources of complexity, identify intervention points and develop equitable and sustainable solutions for complex water-management problems. Here we provide an overview of key concepts from complexity science and use a case example to illustrate how to operationalize complexity thinking to address a complex water allocation problem. This case example of water sharing between Israel and Jordan demonstrates how the creation of an actionable space with a commitment to collaborative adaptive management can produce a relatively sustainable water agreement. To manage complex water problems, we need to reframe traditional *specify-design-implement* to a *try-select-adapt* management approach. Instead of searching for *optimal solutions* to address complex problems we need to look for *optimal space* where certain solutions are actionable given the constraints the context imposes.

Introduction

This introductory chapter is intended to provide a broad overview of complexity science from multiple domains of knowledge (e.g., Anderson et al. 1988; Nicolis and Prigogine 1989; Lewin 1992; Kauffman 1993; Mainzer 1994; Bar-Yam 1997; Cilliers 1998; Maguire et al. 2006; Maguire 2011). A cursory look around this vast body of literature in complexity sciences shows incredible diversity in terms of ontological and epistemological assumptions, foundational concepts, levels of analysis, research methods and so on.

Clearly, there are important differences among the approaches followed by disciplinary groups but, among these examples, the *within-group* differences are much smaller compared to the differences *between* groups. In this chapter we focus on broad ways of thinking about the different faces of complexity: numbers and narratives; models and meanings; objective and interpretive. Clearly, the boundary between these domains (e.g., numbers and narratives) is not as sharp as implied; in addition, the breadth and depth of scholarship within each domain suggest that a comprehensive synthesis may not be attainable. Our goal is not to provide a complete description of the multifaceted reality but to develop an interdisciplinary understanding of these two broad ways of thinking about complexity. We hope an interdisciplinary understanding of different schools of thought from complexity science can provide a pragmatic way to diagnose sources of complexity, identify intervention points and develop equitable and sustainable solutions for complex water-management problems. This chapter is organized in three sections: Understanding Complexity; Understanding Complexity of Water; and Managing Complexity of Water Resources.

Understanding Complexity

From Clock to Confusion: Origin of Complexity

We desire clocklike precision, clear certainty, deterministic prediction, and explicit knowledge to guide us in making objective choices and decisions about the future. Yet, the world that we live in is filled with confusing signals; uncertainty and randomness are everywhere, and many of our choices and decisions are guided by tacit knowledge and subjective interpretations where numbers and narratives often collide to reinforce perceptions and create meaning.

> *Is there one or multiple world(s) and worldview(s)?*
> *Is there a difference between world(s) and worldview(s)?*

These are age-old philosophical questions without definitive answers; yet, they have practical implications in our everyday life. Do we live in one world or multiple worlds? Differences in thinking about and acting in these singular and multiple worlds with competing—and often conflicting—worldviews create complexity. These are not new problems; in fact, many of the important problems were (are), indeed, complex, and wise people knew (know) how to deal with them. For centuries, stories and metaphors have crystallized wisdom of complexity through cultures and continents using narratives— think about the story of blind men and the elephant from the Indian subcontinent, the fable of the eighteenth camel from the Middle East, or the parable of the tortoise and the hare from multiple continents. Difficulties arise when we try to translate the wisdom of complexity to prescriptive advice using tools and techniques we have developed to address simple and complicated problems.

Complexity arises because (a) the world is a system that can be put together in an infinite number of ways because its components are intricately connected and interdependent; (b) it is impossible to enumerate all the interactions and interdependencies of this system explicitly to ensure predictable outcomes; (c) such an enumeration demands our explicit as well as tacit interpretation of multiple worldviews to define the nature of system components and their interactions; and (d) our attempt to analyze this system objectively often fails because of our multiple worldviews. Casual pathways for such systems—known as complex systems—are intractable and open to a plurality of representation and interpretation. The solution space for these complex systems is neither bounded nor predictable. Yet, a common presumption is to seek certainty in our ability to plan and act in a way that will yield predictable outcomes. This notion of a predictable outcome may be theoretically achievable for well-defined scientific problems where causal connections are well understood. We can tell, for example, with absolute certainty where the sun will be when it is 7:37 a.m. in Boston one hundred years from today; yet, we cannot predict with any reasonable certainty how much will it rain at 7:37 a.m. in Boston a week from today. Neither can we predict—with reasonable certainty—the water use and availability in Boston one hundred years from now. Our classical scientific methods seek generalizable and predictable solutions while complexity arises because of interactions, interdependencies and feedback as well as a multiplicity of interpretations and representations of uncertainty. How, then, can one use traditional scientific methods to address complex water problems?

Classical scientific knowledge can be broadly grouped into two categories: "hard" sciences where generalizable mathematical formalism and numbers are used to express the laws of nature (think about the position of sun as a function of time and spatial location) with predictive certainty, while "soft" sciences explain events in terms of narratives. For example, completion of Activity A will result in Output B, leading to Outcome C and Impact D. This linear idea of cause and effect has profoundly affected our understanding of classical science for centuries. Classical (or reductionist) scientific methods tend to use a machine metaphor: a machine is built from distinct elements that can be reduced to its elemental parts without losing its machine-like character. Newton gave us three laws of motion that could be generalized from particle motion to galaxies. This broad generalization has been the foundation of the classical scientific method that we refer to here as the Newtonian Paradigm.

This Newtonian view of the world has influenced our education system, our culture, our institutions and our management practices so effectively that it has become the reality. This view of reality assumes that: (a) a system responds because something causes it to respond (cause and effect); (b) system components are linearly related, implying that responses are proportional to forces and causes are proportional to effects (linearity); (c) linear systems can be broken down into their components to analyze each component and its response separately, and all the separate responses can be added to get the response of the whole system (reductionism); (d) our world is orderly and it works like a machine with known and predictable laws (machine metaphor); and (e) the optimal way to manage this world is to organize institutions with hierarchical governance structure with clear prescriptions (command and control).

At the beginning of the twentieth century, the certainty of this Newtonian view of a mechanical world was questioned by quantum mechanics and the Heisenberg Uncertainty Principle. Einstein told us that time is relative, space is curved, matter and energy are interchangeable—and other new challenges to this Newtonian view of reality began to emerge. For example, leading biologists started questioning the idea that living organisms can be reduced to mechanical operations, and argued that even a single cell has more complexity than a typical car. Scientists in other disciplines started to explore issues of nonlinear dynamics and limits of predictability for well-structured mathematical systems. Lorenz's butterfly effect—how a butterfly flapping its wings in Beijing (a small difference at the beginning of an experiment) may create a thunderstorm in Boston (a massive difference at a later stage at a different location)—is perhaps one of the most vivid and influential metaphors to illustrate the effects of nonlinearity on predictable outcomes for well-structured and bounded systems. Clocklike precision and predictability of a mechanical Newtonian world was challenged by these discoveries in atmospheric sciences (Lorenz 1963), biology (von Bertalanffy 1968) and other disciplines. Linear and reductionist modes of thinking—in which the future world is seen as a straightforward extrapolation of the past—started losing credibility and relevance. A new domain of knowledge—known as complexity science—began to emerge to address this non-Newtonian world.

Complexity Science: Foundational Ideas and Concepts

Complexity science challenges not only the foundation of our knowledge but also the effectiveness of institutions built on that knowledge. Adopting a complexity perspective has significant ontological, epistemological and axiological implications to understand and manage water-resources systems. When relatively simple rules collectively give rise to very complex behavior—for example, when the micro-scale behavior of individuals generates an unpredictable macro-scale economy; and when three wetlands in the same region evolve differently because of small contextual differences in natural and societal processes (Narayanan and Venot 2009)—we need a different approach to understand and manage these systems. The key ideas of the complexity science perspective are summarized below from two different domains of knowledge:

- "We have learned that nature is not a well-designed puzzle with only one way to put it back together. In complex systems the components can fit in so many different ways that it would take billions of years for us to try them all" (Barabási 2003).
- "The option of optimal design is not available to mere mortals. The number of combinations of specific rules that are used to create action situations is far larger than any set that analysts could ever analyze even with space-age computer assistance" (Ostrom 2005).

The implications of these two assertions are profound. In short, we cannot prespecify the state space for complex systems—where variables, processes, actors and institutions are interconnected and interdependent with nonlinearity and feedback—and the possible solution space is theoretically infinite dimensional. That means we may not know the entire sample space with any degree of certainty. Of course, knowing the sample space does not tell us what *will* happen but allows us to theoretically determine what *can* happen. Since we cannot prestate which variables, processes, actors and institutions will become relevant for complex systems, we will not know what can happen. This appreciation for the unknowable nature of complex systems is critical to understand and manage complex water problems. We begin with a brief overview of key foundational ideas and concepts related to complexity science.

Complicated and Complex Systems

Everyone would agree that a car, with its thousands of electronic elements, is an extremely complicated system. The same can be said of the stock market. Both the car and the stock market are human constructions, but there is a significant difference between them. A complicated car was carefully designed and tested by a team of engineers who positioned every mechanical and electronic element in its place with precision, and it works in a reliable and predictable way. But no one designed the stock market, and no one claims to fully understand or control it, and yet it works—perhaps not as reliably and predictably as a car—without explicit control by anyone. While the effective operation of a car is highly dependent on the successful operation of every one of its core elements, the functioning of the stock market is much more robust in the face of perturbations and failures at the level of its elements and agents.

Looking around, one can see systems with similar characteristics from the water world: drought in California, water allocation in transboundary rivers such as the Nile, or water access in megacity slums. What is it that unites these systems and makes them different from a car? And what can we learn from these systems to help us develop more equitable and sustainable water management systems? More generally, can we understand the relationship of structure to function in nature for the benefit of science and society? These systems—stock markets, transboundary water management, urban water supply and treatment—have come to be known as *complex systems*, not to be confused with *complicated* systems such as a car. This distinction is important because although these systems arguably share similar characteristics, they are fundamentally different. A municipal water supply system—similar to a car—is complicated, while providing equitable and sustainable access of water to urban slums is complex.

Here, elements and agents are used interchangeably and consist of four broad groups: variables (e.g., quantity and quality of water) and processes (e.g., hydrologic, climatic, ecological) from the natural system, while actors (e.g., stakeholders) and institutions (e.g., markets, values, governance) are from the societal system. In complicated systems, relationships, interactions and feedback among elements are known and can be controlled. In complex systems, relationships, interactions and feedback are not precisely known; more importantly, interactions and feedback among agents usually change dynamically, and such changes cannot be reliably predicted or controlled.

Different Faces of Complexity Science

Over the last several decades, complexity science has received increased attention from several disciplines (e.g., Anderson et al. 1988; Nicolis and Prigogine 1989; Lewin 1992; Kauffman 1993; Mainzer 1994; Bar-Yam 1997; Cilliers 1998; Maguire et al. 2006;

Maguire 2011). It is important, however, to recognize that what is usually referred to as complexity science actually is a collection of frameworks, theories, models, and tools from a number of disciplines, including systems engineering, chaos theory, cybernetics and complex adaptive systems. A cursory look at the growing body of literature in complexity science reveals incredible diversity in terms of ontological and epistemological assumptions, foundational concepts, levels of analysis, research methods and so on. Complexity science has its origin in many disciplines (Table 1.5.2 in Maguire et al. 2006 provides an excellent summary of different key concepts and their disciplinary origins); two increasingly overlapping threads of intellectual development may be broadly classified as European and North American approaches to complexity (Maguire et al. 2008); yet, understanding of emergent phenomena—a key attribute of a complex system—requires both schools of thought. Another broad grouping of literature is dominated by scholars from natural and social sciences (Furtado and Sakowski 2014; Hidalgo 2015 and references therein). Clearly, there are important differences among the approaches followed by disciplinary groups, but since the within-group differences are much smaller compared to the differences between groups, we will here focus on these two broad ways of thinking about the different faces of complexity—exploring both numbers and narratives; both models and meanings; both objective and interpretive—for two reasons. First, any discussion of the nature and origin of complexity that transcends the boundary between the social and natural sciences is rare. Second, the diverging sets of literature on complexity will be enriched by translating the disciplinary jargon of one body of literature into the language of the other to enrich mutual understanding of each other's goals and objectives. Admittedly, the boundary between these domains (e.g., numbers and narratives) is not as clearly delineated as implied; in addition, the breadth and depth of scholarship within each domain suggest that a comprehensive synthesis may not be attainable. Of course, the breadth of the effort resultant from these domains implies that discussion here is destined to be narrow—may even be viewed by some domain experts as somewhat generic—and incomplete. Our goal is not to provide a complete description of multifaceted reality, but to develop an interdisciplinary understanding of complexity to find a pragmatic way of diagnosing sources of complexity, identifying intervention points and developing equitable and sustainable solutions.

Diagnosis of Sources of Complexity

We hear the term *complexity* used routinely to describe everything from the California drought to the "complexities" of dealing with impacts of climate change or providing access to clean water for millions of people. Yet, people from different domains—natural, societal, and political—have very different understandings of what the term really means and what the new science of complex systems might contribute to our understanding of and responses to water issues. Water allocation between agriculture and urban development, water access in urban slums and water sharing between riparian countries have come to be known as *complex systems*, which should not be confused with *complicated* systems such as a car. This distinction is important because these systems may share some similar characteristics; yet, they are fundamentally different. An accurate diagnosis of the type of system—complicated versus complex—is essential to design and implement intervention to achieve measurable outcome. Complex systems by their very nature resist simple diagnosis. And this diagnosis is further convoluted by disciplinary jargons. For example, social scientists often focus on how context-specific interactions of actors and institutions create complexity while natural scientists explore how nonlinearity and feedback among variables and processes—often independent of context—generate complexity. Yet, the ubiquity of complex problems has resulted in a theory, model and prescription to address these problems, and these three are fragmented and disconnected.

Case examples (such as the one presented in the opening section of this book) can provide a sense of how complexity evolves in the real world: In the 1970s, Bangladesh found that contaminated water significantly contributed to rampant water-borne disease and a high infant-mortality rate. UNICEF initiated a tube well installation program that brought higher quality water to millions of people, and infant mortality dropped sharply over several decades. In the 1990s, doctors started seeing arsenic poisoning among populations using the tube wells and, by 2000, international health experts labeled it as the largest mass poisoning in human history (Smith et al. 2000). In 2000, a multi-million-dollar program was initiated to identify and label "safe" and "dangerous" wells; approximately 20 percent of wells were identified as dangerous and marked with red paint to indicate unsafe levels of arsenic. However, no viable solutions to communities with dangerous levels of arsenic were provided. In addition, with the red-rimmed tube wells came tremendous stigmatization; individuals with arsenicosis face discrimination both financially and socially (Brinkel et al. 2009). The remarkable success of the shallow tube well project to reduce infant mortality, the largest arsenic poisoning in history, and the subsequent testing and education campaign to address the arsenic crisis, serve as ill-fated examples of a metaphorical blind man's simple solution to a complex problem. Unfortunately, the same wells that had once made it easier for women to wed—with the well included in a dowry—now leave them with diminished prospects for marriage because of suspected arsenicosis (Majumder 2016).

Complex systems by their very nature cannot be explained by simple cause-and-effect relationships. Yet, they share some distinguishing characteristics. For the sake of providing concreteness, we will use the well story to illustrate a few dominant characteristics of complex systems. At the story's heart is a collection of interdependent elements or agents—variables, processes, actors, and institutions—that interact in a nonlinear way, with feedback to create an emergent outcome. The interactions, nonlinearity and feedback among these elements can theoretically make the resulting state space infinite dimensional. We provide two examples to illustrate this point. First, high arsenic levels in Bangladesh may not be directly linked to the shallow drinking-water wells; instead, they appear to be related to changes in groundwater use for dry-season irrigation. Biogeochemical data analyses suggest that arsenic mobilization may be associated with a recent inflow of carbon due to irrigation pumping related to changes in dry-season cultivation of a new variety of rice (Harvey et al. 2002). These interdependencies and interactions were not known at the initiation of installation of shallow drinking water wells or the shift to dry-season rice cultivation. Second, labeling unsafe wells and the related arsenic-awareness campaign were intended to minimize the impact of arsenic poisoning; yet, they produced significant social stigmatization and discrimination.

This intricate coupling—between natural and societal processes with their interactions, nonlinearity and feedback—suggests that we may never know the entire sample space with any degree of certainty. From many individual interactions, new and often surprising phenomena may emerge—such as the high level of arsenic in shallow wells and the diminished prospect for marriage because of suspected arsenicosis—and the emergent behavior of the whole cannot be predicted by examining the individual elements of the system: the whole is more than the sum of its parts. Since we cannot prestate which variables, processes, actors and institutions will become relevant for complex systems, we do not know what can happen.

This appreciation for the unknowable nature of complex systems is critical to understand and manage complex water problems. Complexities of arsenic levels in drinking water wells exist within their own environment and are part of that environment. As the environment changes, elements and their interactions will also change. Addressing the impact of high arsenic levels is not merely a question of understanding the mobilization of solid-phase arsenic to groundwater; but is related, for example, to providing access to arsenic-free water as well as to addressing social stigmatization. We need to recognize that any intervention to address a problem will change aspects of the interactions and interdependencies of elements, and the elements of the system will adapt to their environment. In other words, solution space cannot be prestated with certainty; consequently, our search for diagnosis, intervention and implementation needs to be contingent.

Systems Engineering and Complex Systems

Broadly speaking, in its origin and scope, complexity science is close to systems thinking. While systems thinking provides a useful perspective and toolbox for the systems engineer, there is a clear and large distinction between systems engineering and complexity science. Some systems thinkers argue that they have always worked with complex systems, while some complexity scientists argue that complexity science is built on a fundamentally different set of assumptions than systems engineering. The following features—adapted and revised from Ramalingam et al. (2008)—highlight some of the key differences between systems engineering and complexity science:

- Systems engineering assumes that systems are composed of elements with prespecified rules that can be used to calculate potential equilibrium, whereas complexity science emphasizes situations usually far away from equilibrium;
- Systems engineering assumes that systems are composed of machine-like rational agents with predictable dynamics, whereas complexity science recognizes that dynamics of agents are contingent and often unpredictable.
- Systems engineering asserts that the functioning of a system can be understood as an aggregated and predictable response of isolated elements, whereas complexity science views interdependence and interconnectedness of individual elements within a context in which emergent simplicity and complexity are possible system-level outcomes;
- Systems engineering assumes that systems change their structures in accordance with prespecified, rule-based algorithms, whereas complexity science acknowledges that change

is dynamic, context-specific, and cannot be prespecified; consequently, understanding and learning need to be adaptive and perpetual.

Adopting a complexity perspective—to understand and manage complex water systems—has significant ontological, epistemological and axiological implications. Maguire et al. (2011) assert complexity science challenges not only the foundations of our knowledge but also the institutions we have engineered to manage complex systems. The root cause of many complex water problems lies at the intersection of multiple causal forces buried in observational signatures with often-conflicting views and values related to: *Who decides, who gets water and how?* In such situations, neither numbers nor narratives will resolve the dilemma. To understand the complexity of water, we begin by making the distinction of water as an object or a resource.

Understanding the Complexity of Water

Water: Object or a Resource?

The complexity of many of our water problems stems from our fragmented understanding of water as an object (e.g., hydrogen and oxygen makes water) or a resource (e.g., who decides who gets water). Let us begin by exploring this distinction and related implications.

Water as an Object

A water molecule (chemical formula: H_2O) contains one oxygen atom and two hydrogen atoms that are connected by covalent bonds. Water is a liquid at standard temperature and pressure, but it can exist in solid (ice) and gaseous states (water vapor). Water has the property of being liquid, but none of the atoms out of which water is constituted has this property; both hydrogen and oxygen exist in gaseous states at standard temperatures and pressure. The liquidity of water is determined by the way hydrogen and oxygen interact and by the patterns resulting from this interaction. These patterns are different for ice and vapor. Can these patterns be predicted by understanding atomic structure and the properties of hydrogen and oxygen independently? "No." The answer to this question teaches us an important lesson: We need to challenge our notion of reductionism, based on which the whole must have properties that can be reduced to the properties of the parts. It raises questions about the idea that we could understand the world by examining smaller and smaller pieces of it and, when assembled, the small pieces would explain the whole.

Can, we then, consider water as a complex object? Consider a cubic centimeter of water: Elementary physical chemistry tells us that at 760 mm pressure and 0° C temperature, 10^{19} molecules are crowded into a space of one cubic centimeter. That is, 10 billion billion molecules are interacting and moving in all directions. Does this sheer number of interactions of water molecules qualify as a complex system? No; the laws of physical chemistry and thermodynamic theory tell us that these billions of randomly colliding molecules are not "doing anything" intentionally, and their apparently complex behavior is predictable with a few generalizable physical laws. This lack of intentionality of water as an object and the generalizability of physical laws make an object sharply different than water as a resource.

Water as a Resource

Water—as an object—usually follows gravity and flows downhill. Yet, water flows uphill to money where water is scarce—such as the American West—and it finds its own economic level. In other words, "if you've got the money, you can get the water", even in the Great American Desert. Consider the Colorado River, which supplies water to many large cities, including Denver, Los Angeles and Phoenix. None of these cities is in the Colorado River watershed. The physical laws governing the flow of water appear to be irrelevant when confronted by the political and financial capacity of a metropolis to access and control water as a resource.

Can water flow downhill by gravity—a generalizable physical law for an object— be substituted by water flows uphill to money when treating water as a resource? Apparently not; it is one of those intellectual traps that is somewhat uncritically accepted. Sometimes this is accurate; however, across a range of water allocation decision making, it may be inaccurate and misleading. It is important to think in what situations (and why) water flows uphill toward money; yet, at other times it does not. For example, we have water-allocation rules that frequently create barriers to move water as a resource, even if the move might be hydrologically possible and economically preferable. Such a move may not be implementable for political and legal reasons. As Fleck (2015) pointed out, annual agricultural production in the southern part of New Mexico is roughly $7,000 per acre, while in the middle valley it is about $3,500 per acre; yet, the southern part faced a serious water shortage in 2015. The Rio Grande Compact sets a boundary between these two parts of the river across which it is physically possible but legally impossible for a farmer in the middle valley to sell water to a willing buyer uphill toward pecan and chili money in the south. Although the ag–urban contrast is more common; this ag-to-ag comparison highlights the importance of an accurate diagnosis of the sources of complexity in water-allocation and access. Thinking through these contextual differences is critical to design and implement effective intervention. In other words, we need a reframing of our understanding of water: Water, as a resource, flows neither by gravity nor by money, but by the complexity created by interactions of the knowledge and political communities.

Sources of Complexity in Water: Interactions of Knowledge and Political Communities

Water access, demand, usage and management become complex because of the crossing of multiple boundaries: political, social and jurisdictional, as well as physical, ecological and biogeochemical. The origin of many complex water problems lies in the dynamic consequences of nonlinearity of interconnections and feedback among elements—variables, processes, actors and institutions—of the system. Not all water problems are complex; a key source of complexity is the uncertainty of interactions and feedback among elements that function in the political and knowledge domains of water management.

Framing the complexity of water problems—rooted in the interactions and feedback within and between the knowledge community and the political community—shifts our focus from seeking efficient tradeoffs from primarily rational perspectives to identifying and negotiating the needs of multiple stakeholders with competing—and often conflicting—values and interests. Thus, a purely economic framework that balances supply with demand and promotes only technocratic interventions fails to account for other values important in managing water as a resource. However, these other values affect boundary-crossing water policy debates and are generally framed within two communities engaged in water resource management—the knowledge community and the political community (Choudhury and Islam 2015).[1]

The knowledge community includes actors and activities from the technical sphere of water research, water management and water policy. One key contribution from this community is the formulation and widespread adoption of the integrated water resources management (IWRM). Alongside the knowledge community, the political community also affects water resource management. Decision making processes within the political community involve not only input from the knowledge community, but the perceived legitimacy, necessity and effectiveness of proposed actions that are aligned with community values. The political community includes both actors and activities that are engaged in governmental policy, and the non-governmental activism of civil-society groups that seek to frame the underlying values of governance that affects water allocation priorities.

For example, the nature of complexity introduced by the political community can be understood in terms of negotiation on the meaning of a value such as equity. The meaning varies from seeking equitable allocation among different sectors of the economy (for example, agriculture, energy, and urban), to providing poor communities access to affordable water or establishing sustainable management of water to affect generational equity.

The two globally understood guiding principles to addressing complex water issues—sustainability and equity—are evident in the values and activities of the knowledge and political communities as described above. As a result, we can no longer consider them as dispensable options in an economic tradeoff—say, to achieve efficient pricing of water or efficient allocation of water—but as embedded in the complexity of deliberation and decision making that arise from the dynamic interactions of the knowledge community and political community.

Complexity Introduced through Growth in Know-How in the Knowledge Community

The ongoing growth and refinement of natural, societal and technological knowledge continues to provide solutions to water problems. This enhancement in knowledge and informed action creates efficient and different use of water, for example: mapping and accessing aquifer water; desalinizing water; and addressing ecological water needs. In this context, effective use of knowledge is crucial for creating new options for competing stakeholders and resolving water disputes (Islam and Susskind 2013). In doing so, however, this growth in knowledge also introduces uncertainty about problems (such as climate change) and gives rise to the identification of new problems (such as ecosystem vulnerability). Consequently, the efficacy of this knowledge has also become contingent upon the uncertainties associated with the generation, appreciation, perception and implementation of this knowledge. For example, what are the impacts of climate change on water availability? At what space–time scales are these impacts important for decision making?

Complexity Introduced through Competing Values and Interests in the Political Community

The political community adds new problems and solutions as well, thereby increasing the ambiguity of understanding and resolving water problems. The nature of water conflicts is dependent on a constellation of interacting societal factors, adding uncertainty to the issues of water use, access, and equity. For example, the conflicting developmental needs of riparian states in terms of economic growth, pattern of urbanization or industrialization and agricultural productivity all lead to competing water needs (Agnew and Woodhouse 2011). Two other factors further expand the scope of the conflicts. One is the emergent conventions on water management; for example, how "integrated use of water" conflicts with the established convention of "prior use." The second is the growing commodification of water (e.g., creation of water markets) conflicting with the equity of access and cost of water (Prescoli and Wolf 2009).

Managing Complexity of Water Resources

Complexity Creates Contingency

There is an increased awareness that the water world we live in is filled with confusing signals; uncertainty and randomness are everywhere; many of our choices and decisions are guided by tacit knowledge and subjective interpretations where numbers and narratives often collide to reinforce perceptions and create meaning. A common catchphrase used by both scholars and practitioners to describe the seemingly intractable challenges of water management is that "the problems are complex." For example, Schnurr (2006) asserts that "water policy is a field of high ecological, social, and economical complexity"; Wallis et al. (2011) describe the complexity of water governance to be related to uncertainty and interconnectivity with other issues that are "highly resistant to traditional problem-solving methodologies, especially technological fixes"; Moore (2013) alerts that "those responsible for water governance face great complexity"; and Kumar (2015) emphasizes that hydrocomplexity arises from "evolving interdependencies between complex systems." Yet, careful scrutiny of why these challenges are complex, what creates this complexity, and what can be done to resolve this complexity is rare.

    The complexity of water governance has many faces; many of our water problems stem from our fragmented thinking of water as an object or as a resource. This thinking is further compounded by the interaction of the knowledge and political communities. We assert that multifaceted nature of these problems can neither be understood nor explained with the degree of certitude we desire. Addressing these problems will require contingent understanding and pragmatic approaches rooted in pluralistic interpretations and methodological diversity.

There is an increased awareness that solutions that work for simple and complicated water problems usually will not work for these complex problems and may contribute to the complexity itself (Islam and Repella 2015). Given the contextual conditions that operate within the political and knowledge communities and the complexities that arise from their interactions, an integrated and generalizable approach—although highly appealing and desirable—may not be realistically achievable. This recognition that "the relationship between the system and its parts is intellectually a one-way street" (Anderson 1972) or that the "search for universal laws of social development has been inadequate" (Rittel and Weber 1973; Itzkowitz 1996) is not new and has been and continues to be debated in natural and social sciences (Biswas 2004; Saravanan et al. 2009; Moore 2013).

Dichotomies exists between micro- and macro-scales, between the usefulness of numbers versus narrative, and between the objective versus the interpretive. Syntheses across these have been elusive and may not be achievable (Arrow 1951; Anderson 1972; Rittel and Weber 1973; Itzkowitz 1996). A reframing of understanding, explaining and acting is needed for effective water resources planning and management. If the primary goal is understanding for prediction, then the associated analysis needs to consider all dominant elements (variables, processes, actors and institutions) and their interactions, and it needs to satisfy the necessary and jointly sufficient conditions for a reliable predictive outcome. On the other hand, if the primary goal is an intervention to affect an outcome, one needs to identify enabling conditions and select a few contextually relevant elements that are likely to produce a measurable outcome with the explicit recognition that this identification and selection is contingent and adaptive to changes (Choudhury and Islam 2015). Therefore, effective resolution of complex water problems necessitates a contingent approach that addresses competing interests spanning both the political and knowledge communities.

In our increasingly interconnected world, anything is possible. However, in reality, not everything happens. Of all possible events, only a few will actually happen with significant consequences. How do we know which few important things will happen and when? How do we plan and prepare so that we can act accordingly when those few significant things happen? More importantly, can we identify and influence what may happen, given the resources and constraints we have? How do we decide what to do when options are many, resources are limited, uncertainty is large and consensus is difficult to achieve? Many of these decisions are complex because variables, processes, actors and institutions are interconnected and interdependent, making a range of solutions possible. But not all possible decisions are actionable.

Contingent Management of Water Resources: Search for Actionable Space

A key goal for the classical water management approach is to produce efficient and reliable methods that meet prespecified constraints and prespecified standards of performance in prespecified situations. Usually, it is a goal-oriented process seeking to achieve *known* specific ends using *well-understood* means. The dominant attributes for such a management approach are predictability, reliability, stability, controllability and transferability. These attributes are hallmarks of excellent engineering design: given a problem to solve, finding out how to do it once and then repeating that over and over again—and this works extremely well for simple and complicated problems. Unfortunately, for complex problems these dominant attributes are not helpful to provide replicable management outcomes because the solution space cannot be prespecified.

Jordan and Israel's 1994 peace treaty includes a historic bilateral agreement on sharing the water of the Jordan River, an agreement that exemplifies a contingent approach that is centered around an actionable solution space—the flexibility of water storage in an existing reservoir and the ability to shape future coordination and collaboration on water management. The water agreement was the product of direct negotiations between the two parties, with some facilitation by additional players. When political and technical constraints threatened to create a stalemate; the negotiators used what is referred to by some observers as "constructive" or "strategic" ambiguity to close the gaps between them. By not specifying certain details, the teams of negotiators shaped an agreement that they could both sell to their home constituencies. The two sides also allotted time to conduct further studies that would allow them to adjust the details of the agreement during implementation. The agreement included a number of options that created value.

For instance, the parties were able to use water-storage technology to "enlarge the pie" of water available in the basin. The treaty specifies allocations of the Yarmouk River that Israel may extract during the summer and winter periods for its needs. In exchange, Jordan is allowed to store some of its water allocation in Lake Tiberias during the winter; Israel is to release this water back to Jordan each year during the dry season. The treaty specifies that, on the Jordan River, Israel may maintain extraction levels equivalent to its level of use in 1994, and Jordan may withdraw an equal amount when there is sufficient supply. The treaty allocates specific amounts of groundwater to Israel south of the Dead Sea, and it allocates certain quantities of spring water to Jordan near Lake Tiberias. It also stipulates that Israel and Jordan will cooperate to "find" an additional 50 MCM of water for Jordan. The agreement created the permanent Joint Water Committee (JWC) to oversee implementation. The JWC provides a setting in which the parties can work together to deal with unforeseen challenges.

This identification of the actionable space—flexibility in storing water in Lake Tiberias, cooperating to find additional water and creating the JWC to oversee governance—shows how the combination of strategic ambiguity, collaborative adaptive management and a strong desire to find feasible solutions can produce a relatively sustainable water agreement.

To manage complex water problems, we need to reframe the traditional *specify–design–implement* management approaches to become *try–select–adapt* approaches. Instead of searching for *optimal solutions* to address complex problems we need to look for *optimal space* in which certain solutions are actionable given the constraints imposed by context. While transferable solutions are often difficult to identify, this search for an actionable space in which to implement new arrangements, such as the one discovered for the Israeli–Jordanian water sharing, can be operationalized to manage complex water problems.

Acknowledgments

The work presented in this chapter was supported, in part, by two grants from the US National Science Foundation (RCN-SEES 1140163 and NSF-IGERT 0966093).

Note

1   This section has liberally borrowed material from Choudhury and Islam (2015).

References

Agnew, Clive and Phillip Woodhouse. 2011. *Water Resources and Development*. New York: Routledge.
Anderson, P. W. 1972. "More Is Different." *Science* 177(4047): 393–96.
Anderson, P. W., K. J. Arrow and D. Pines, eds. 1988. *The Economy as an Evolving Complex System*. Proceedings of the Santa Fe Institute, vol. 5. Redwood: Addison-Wesley.
Arrow, Kenneth. 1951. *Social Choice and Individual Values*. New York: Wiley.
Bar-Yam, Yaneer. 1997. *Dynamics of Complex Systems*. Reading, MA: Addison-Wesley.
Barabási, Albert-László. 2003. *Linked: How Everything Is Connected to Everything and What It Means for Business, Science and Everyday Life*. New York: Penguin.
Biswas, Asit K. 2004. "Integrated Water Resources Management: A Reassessment." *Water International* 29(2): 248–56.
Brinkel, Johanna, Mobarak H. Khan and Alexander Kraemer. 2009. "A Systematic Review of Arsenic Exposure and Its Social and Mental Health Effects with Special Reference to Bangladesh." *International Journal of Environmental Resources Public Health* 6(5): 1609–19.
Cilliers, Paul. 1998. *Complexity and Postmodernism*. London: Routledge.
Choudhury, Enamul and Shafiqul Islam. 2015. "Nature of Transboundary Water Conflicts: Issues of Complexity and the Enabling Conditions for Negotiated Cooperation." *Journal of Contemporary Water Research & Education* 155(1): 43–52.
Fleck, John. 2015. "Can We Retire 'Water Flows Uphill Toward Money'?" JFleck at Inkstain (blog), July 9. http://www.inkstain.net/fleck/2015/07/can-we-retire-water-flows-uphill-toward-money/.
Furtado, Bernardo Alves and Patrícia Alessandra Morita Sakowski. 2014. "Complexity: A Review of the Classics." Discussion Papers 2019a, Instituto de Pesquisa Econômica Aplicada—IPEA.
Harvey, Charles F., Christopher H. Swartz, A. B. M. Badruzzaman, Nicole Keon-Blute, Winston Yu, M. Ashraf Ali, Jenny Jay, Roger Beckie, Volker Niedan, Daniel Brabander,†, Peter M. Oates, Khandaker N. Ashfaque, Shafiqul Islam, Harold F. Hemond and M. Feroze Ahmed. 2002. "Arsenic Mobility and Groundwater Extraction in Bangladesh." *Science* 298(5598): 1602–06.
Hildago, Cesar. 2015. *Why Information Grows: The Evolution of Order, from Atoms to Economies*. New York: Perseus Books.
Islam, Shafiqul and Amanda C. Repella. 2015. "Water Diplomacy: A Negotiated Approach to Manage Complex Water Problems." *Journal of Contemporary Water Research & Education* 155(1): 1–10.
Itzkowitz, Gary. 1996. *Contingency Theory: Rethinking the Boundaries of Social Thought*. New York: University Press of America.
Kauffman, Stuart. A. 1993. *The Origins of Order: Self-Organization and Selection in Evolution*. New York: Oxford University Press.

Kumar, Praveen. 2015. "Hydrocomplexity: Addressing Water Security and Emergent Environmental Risks." *Water Resources Research* 51: 5827–38. doi:10.1002/2015WR017342.

Lewin, Roger. 1992. *Complexity: Life at the Edge of Chaos*. Chicago: University of Chicago Press.

Lorenz, Edward N. 1963. "Deterministic Nonperiodic Flow." *Journal of the Atmospheric Sciences* 20(2): 130–41.

Maguire, Steve, Bill McKelvey, Laurent Mirabeau and Nail Oztas 2006. "Organizational Complexity Science." In *The SAGE Handbook of Organizational Studies*, 2nd edn. S. Clegg, C. Hardy, W. Nord and T. Lawrence (eds.). Thousand Oaks, CA: Sage, 165–214.

Maguire, Steve. 2011. "Constructing and Appreciating Complexity." In *The SAGE Handbook of Complexity and Management*. Allen, P., Maguire, S. and McKelvey, B. (eds.). London: Sage, 79–92.

Maguire, Steve, Peter Allen, and Bill McKelvey. 2011. "Complexity and Management." In *The SAGE Handbook of Complexity and Management*. Allen, P., Maguire, S. and McKelvey, B. (eds). London: Sage, 1–26.

Mainzer, Klaus. 1994. *Thinking in Complexity: The Computational Dynamics of Mind, Matter and Mankind*. Berlin: Springer-Verlag.

Moore, Michele-Lee. 2013. "Perspectives of Complexity in Water Governance: Local Experiences of Global Trends." *Water Alternatives* 6(3): 487–505.

Nicolis, Gregoire and Ilya Prigogine. 1989. *Exploring Complexity: An Introduction*. New York: Freeman.

Priscoli, Jerome D. and Aaron T. Wolf. 2009. *Managing and Transforming Water Conflicts*. Cambridge: Cambridge University Press.

Ostrom, Elinor. 2005. *Understanding Institutional Diversity*. Princeton: Princeton University Press.

Narayanan, N. C. and J.-P. Venot. 2009. "Drivers of Change in Fragile Environments: Challenges to Governance in Indian Wetlands." *Natural Resources Forum* 33(4): 320–33.

Ramalingam, Ben and Harry Jones with Toussainte Reba and John Young. 2008. "Exploring the Science of Complexity: Ideas and Implications for Development and Humanitarian Efforts," 2nd edn. Working Paper 285a. London: Overseas Development Institute.

Rittel, Horst W. J. and M. M. Webber. "Dilemmas in a General Theory of Planning." *Policy Sciences* 4: 155–69.

Saravanan, V. S., G. T. McDonald and P. Mollinga. 2009. "Critical Review of Integrated Water Resources Management: Moving Beyond Polarised Discourse." *Natural Resources Forum* 33: 76–86.

Schnurr, Maria. 2006. "Global Water Governance: Managing complexity on a Global Scale." In *Water Politics and Development Cooperation: Local Power Plays and Global Governance*. Waltina Scheumann, Susanne Neubert and Martin Kipping (eds.). Berlin: Springer, 107–20.

Von Bertalanffy, Ludwig. 1968. *General System Theory: Foundation, Development, Applications*. New York: George Braziller.

Wallis, P. J., R. L. Ison and K. Samson. 2011. "Identifying the Conditions for Social Learning in Water Governance in Regional Australia." *Land Use Policy* 31: 412–21.

Chapter Two

LEVERAGING DIPLOMACY FOR RESOLVING TRANSBOUNDARY WATER PROBLEMS

Benjamin Pohl and Ashok Swain

Abstract

Water is a fundamental precondition for human life. Yet, much of it transcends state borders via shared river and lake basins or groundwater aquifers. The resulting political, economic, social and environmental interdependencies give water resources the crucial potential either to foster cooperation or exacerbate conflict. Transboundary waters thus constitute a promising entry point for diplomats aiming for high peace dividends, warranting strong interest from foreign policy makers. Yet there is a lack of agency at the international level. As a result, the international community faces huge challenges when it comes to systematically taking early action to both respond to emerging crises and reinforce cooperation. This chapter, therefore, argues that foreign policy makers can and should do more to realize these dividends by exerting political leadership in fostering intrabasin cooperation and integration, connecting and reinforcing appropriate institutional structures for coordinated and cross-sectoral comprehensive engagement and for strengthening the diplomatic track of transboundary cooperation on water. This chapter proposes a number of specific instruments of engagement. Yet, in the end, strengthening the governance of transboundary waters hinges on strengthening and connecting the institutions, national and international, that can channel political will into coherent action.

Introduction

The global water crisis is of such a magnitude that it is growing into an issue of common global concern.[1] This perspective puts the focus on water systems shared by two or more countries. Transboundary waters constitute a pivotal but underappreciated issue in global politics. The territories of 148 nations fall within international river basins that include the political boundaries of two or more countries (UN Water 2016). These international river basins cover 46 percent of the Earth's land surface (UN Water 2016), host about 40 percent of the world's population and account for approximately 60 percent of global river flow (Wolf et al. 2003). Shared waters are, therefore, often of critical importance for riparian states, that is, the states that comprise part of a transboundary basin. As a consequence, the use and apportionment of water in these basins is both a source of conflict risks and a chance for institutionalizing cooperative behavior. However, national politics often complicates improving the management of shared rivers. The widespread practice of hiding political issues behind a facade of integrated water resources management (IWRM) terminology (Jeffrey and Gearey 2006) has arguably contributed to the failure of achieving desirable governance in many transboundary basins (Conca 2006; Earle et al. 2015).

The management of international rivers in various parts of the world cannot follow a particular golden principle because the value of water and its demand and supply varies from one basin to another (Swain 2004). Thus, it can be safely argued that the "one-shot approach of management within the context of IWRM is far too simplistic to be useful, or applicable" for sustainable management of international rivers (Varis, Tortajada, and Biswas 2008). In spite of its huge significance for global peace and development, the available knowledge on how to manage trans-boundary waters is quite weak (Earle, Jägerskog, and Öjendal 2010). Moreover, the institutions we have on governance of international rivers are increasingly insufficient because of increased demand for, and a decreasing or volatile supply of fresh water. Adding to the problem, the threat of global climate change has begun to undermine the existing regimes and institutions of water sharing and management of international rivers (Gleick 2009; Drieschova et al. 2009; Swain 2012).

Given the political nature of water management in international river basins and the stakes that this entails in terms of conflict prevention, regional stability, environmental peacemaking and international security, transboundary water governance is a domain that should elicit great interest in the foreign policy community. Yet, so far the challenges and opportunities of transboundary waters for these pivotal foreign policy objectives remain insufficiently appreciated and utilized among foreign and security policy makers. Water governance remains largely a topic for technical and development cooperation. Yet, there are limits to what technical cooperation alone can achieve in conflictive political contexts. The management of transboundary waters is often eclipsed by basin politics which, in turn, is frequently complicated by power asymmetries (Zeitoun and Warner 2006). This poses difficult challenges for the water and development communities whose instruments are hardly prepared for enticing powerful states to play a constructive role. In this context, this chapter therefore stresses the need for more systematic engagement by foreign policy makers on transboundary waters. Focusing on politically contested waters, which are predominantly located in the Global South, we seek to underline the potential of diplomacy for building on technical collaboration so as to foster regional integration and facilitate stability and peace in the contested basins.

Transboundary Waters: Conflict and Peace

Conflicts over transboundary water threaten international peace and security (Swain 2013). Many international basins face increasing demand as a result of demographic pressures, industrialization and urbanization. Simultaneously, in many cases supply recedes because of both earlier mismanagement and the impact of climate change (Falkenmark and Jägerskog 2010). Climate change and natural climate variability observed over comparable time periods (Swain et al. 2011) might not only decrease supply through changes in precipitation and greater evaporation, but also increase the demand for water, for example for irrigation and cooling. Conflicting interests over the use of increasingly scarce water resources may also directly contribute to interstate tensions. Less than half of the world's transboundary basins are subject to any formal agreement as to how to divide and manage freshwater resources (Giordano et al. 2013). There are even fewer water bodies with institutionalized cooperation: only 116 out of 276 transboundary basins have ever had a river basin organization (RBO) (Schmeier 2013). The mere existence of treaties or RBOs is, moreover, insufficient for conflict prevention because they often lack robust conflict-management mechanisms.

So far, we have not witnessed clear-cut "water wars" (Wolf 2003; Swain 2004). However, there are strong links between water mismanagement and risks of social and political instability, and they may increase due to the impacts of climate change. Prominent examples for such risks include the events preceding the civil war in Syria; the consequences of floods and droughts in Pakistan; and the volatile situation in Central Asia (Gleick 2014; Redclift and Grasso 2013). Such instability has the potential to contribute to interstate conflicts as well as to intrastate conflicts that, in turn, might become internationalized. In this context, foreign policy engagement on transboundary water governance can help to contain the conflict potential of shared waters. Especially in cases in which water is highly politicized, third-party engagement may be a necessary, though rarely if ever a sufficient condition for defusing and transforming tensions.

Yet, water can also present a "bridging resource." The need to cooperate on water management might help to generate the political space necessary to address other contentious issues. It could thus serve as a tool for achieving positive spillovers in terms of regional peace and cooperation. This peacemaking can take place on the foundation of water-induced cooperation. Water cooperation has the potential to transform the mistrust and suspicion between and among countries to bring opportunities for shared gains and set up a model of reciprocity. It can also pave the way for greater interaction, interdependence and societal linkages. The diffusion to other areas of bilateral cooperation over water resources is not an uncommon phenomenon (Conca and Dabelko 2002).

However, water cooperation does not occur easily; nor will it automatically have peace-enhancing effects. Conca and Dabelko (2002) find a strong basis for the general proposition that collaboration over natural resources such as water can have positive spin-offs for peace. They see two pathways for peacemaking over water resources. The first path involves transforming the more immediate problems of mistrust, uncertainty, suspicion, divergent interests and short time horizons that typically accompany conflictual situations. A second pathway, consistent with the broader understanding of peace as "the unimaginability of violent conflict," focuses less on narrow, short-term interstate dynamics and more on the broader pattern of trans-societal relations. In other words, cooperation would be pursued as an objective in itself, diffusing from water across other areas of international interaction. Such "spill-over" advantages of water-induced cooperation have been witnessed in different parts of the world. Cooperation over the Rhine

River was a source of the idea of the European Union project. Similarly, cooperation on the Mekong River has contributed to regional cooperation in Southeast Asia. Finally, the relationship between India and Bangladesh has become more dynamic after both countries signed a long-term agreement on Ganges water sharing in 1996.

The changes and uncertainty that climate change entails for many basins might bring about new opportunities for cooperation. Climate change has repeatedly been described as a threat multiplier for unstable regions around the globe (Rüttinger et al. 2015). The impact of climate change is likely to be felt primarily through the water cycle, reinforcing many worrying trends regarding water scarcity, salinization and flood risks. Yet climate change adaptation efforts also imply opportunities. Financial means for adaptation and adaptation discourses that stress the need to address impending deterioration in the supply and quality of freshwater may simultaneously help to address current mismanagement of water resources (Tänzler, Carius and Maas 2013). Climate change adaptation pressures could thereby generate positive spillovers, serving as an entry point for building trust and engaging in politically difficult regions. In short, vulnerability to water and climate changes may constitute not only threat, but also opportunity multipliers.

Transboundary Water and Hydro-Diplomacy

Multilateral and bilateral donors have facilitated international treaties in many basins and have supported the establishment of river basin organizations (RBOs). The World Bank-supported Indus Waters Treaty, the Mekong River Commission and the Nile Basin Initiative are probably the best-known RBOs set up to promote cooperation between riparian states. Many RBOs continue to be financed primarily by international donor agencies (Trondalen 2013). The rationale that "preventing conflicts from arising in the first place is far more efficient than interventions at a later stage" (Pohl et al. 2014) constitutes a convincing call for engagement in transboundary freshwater basins. Yet, we argue that for technical cooperation to realize its full potential, it also needs political support to overcome inertia and vested interests, to ensure broad ownership and legitimacy and to convince political decision makers of the necessity and benefits of cooperation.

Foreign policy makers could play an instrumental role in nudging governments into cooperative behavior or in providing assistance to extend such behavior. Yet, the advantages of greater foreign policy engagement on transboundary waters should not be taken as a plea for foreign policy makers to always barge in. Many countries resist external engagement on water-sharing issues. For example, India has been reluctant to allow the involvement of external actors in its bilateral discussions with Nepal and Bangladesh on the Ganges, and Turkey has resisted US involvement regarding the Euphrates–Tigris basin. The United States, in turn, resisted the involvement of external actors in its water relations with Mexico and Canada (Jägerskog, Swain and Öjendal 2014).

There are also risks insofar as foreign policy makers might make things worse by inserting perceived outside agendas, whether it is balancing against rivals, putting environmental concerns above economic development or economic self-interest. Greater Western engagement in the Mekong basin has probably contributed to keeping China away from being a member of the Mekong River Commission (Earle et al. 2015). Apart from complicating local disputes by drawing them into a great-power balancing logic, foreign engagement may also fuel conflict by emboldening local actors to overplay their hands in the assumption that they might ultimately be able to achieve a better deal. Interventions designed to strengthen (cooperation between) weaker riparians might thus enable a standoff rather than a breakthrough. This arguably happened in the case of the Nile: donor support achieved the intended objective of greater cooperation, but primarily between upstream countries and possibly at the expense of hardening their collective attitude toward achieving a compromise between Egypt and Sudan. The ultimate outcome of the Nile water conflict is unclear at the point of this writing, but there is at least a possibility of a conflictive result, to which the international community's well-intentioned intervention would then have contributed.

Moreover, even successful foreign policy engagement as measured by an intergovernmental agreement to cooperate may fall short because it might paradoxically encourage conflict at the subnational level as societal groups that feel disadvantaged by, or excluded from, international deals may object (Conca 2012). The 1960 Indus agreement has thus created a number of internal conflicts in India and Pakistan. In electricity-starved Jammu and Kashmir, for example, the state legislature in 2002 called for review and annulment of the treaty in a near-unanimous resolution because the Indus Waters Treaty assigns the three rivers running through Jammu and Kashmir to Pakistan, hindering the exploitation of hydropower resources (Swain 2009). Recently, the Indian central government's attempts to cooperate with Bangladesh over the Teesta River water have been successfully opposed by its state of West of Bengal (Swain 2010). In Eastern Africa, the apparent understanding reached between the governments of Ethiopia and Kenya on the Gilgel Gibe III dam—whose operation might damage livelihoods and increase violence around Lake Turkana—has apparently ignored the concerns of local communities (Ackva et al. 2015). In short, neither foreign policy engagement nor international cooperation is a panacea. Yet there are many instances in which greater engagement on the part of foreign policy makers—and greater cooperation between the foreign, development and technical communities—could contribute to mitigating conflict over shared waters and to fostering greater regional peace and cooperation.

Current efforts to support international collaboration on transboundary waters are characterized by a relative underinvestment in the diplomatic pillar as compared with the techno-managerial pillar of cooperation. Yet, relying on technically focused management institutions is often not enough because basin management is affected by basin politics. It is insufficient to identify the solution with the greatest overall benefits and leave it to riparian governments to divide these among themselves because benefits will often be perceived as not accruing equally to countries or particular constituencies. To achieve real progress, therefore, any outside support in identifying possible benefits of cooperation needs to be carefully embedded in diplomatic and political engagement.

Political tensions between riparian governments present a key challenge to would-be mediators and demand that the latter also engage at a political level. Although disputes over water often represent real conflicting interests, they are also frequently used as an instrument to achieve other political ends. If there is political interest in conflict, technical solutions to water governance issues will obviously hardly help to overcome such conflicts. Greater diplomatic sensitivity regarding intra-basin politics can then help in negotiating water cooperation, whereas lack of political perspective might lead to missteps and failures on the part of the donor community.

Basin politics are usually complicated by power asymmetries, and hegemons frequently refuse to be drawn into multilateral basin institutions, although, in other cases they may promote them (see below). Therefore, attempts to institutionalize cooperation in transboundary river basins sometimes need external political support that foreign policy makers are well placed to provide.

International Community and Transboundary Water Management

The international community's objectives regarding transboundary water governance essentially comprise the need to facilitate the containment and resolution of conflicts in the short term, to manage resources such that conflicts are avoided in the longer term and to harness the water-cooperation mechanisms for the purpose of stronger regional cooperation and development. Although primary responsibility for cooperation rests with basin countries, external assistance can facilitate the negotiation of international river water agreements. The Indus River Agreement between India and Pakistan is one of the success stories of such external intervention. The World Bank played a significant role in bringing these two hostile countries to an agreement in 1960 to share and develop the Indus River. Recently, the World Bank has also facilitated river agreements in southern Africa. The Asian Development Bank also played a significant role in bringing the four lower riparian countries of the Mekong River Basin into a cooperative agreement. Where riparian countries have been able to establish cooperative arrangements, successful and sustainable development programs follow (Kirmani and Rangeley 1994). The establishment of cooperative arrangements on shared river systems also brings further international collaboration and external assistance.

Water scarcity and a lack of financial and technological resources are gradually encouraging riparian countries in the developing world to cooperate in order to enhance benefits. As constraints on the resource grow, the opportunity cost for not cooperating becomes clearer. The increasing scarcity of available fresh water per capita and the lack of financial strength in many developing countries may gradually encourage basin countries to cooperate in order to increase the benefit of shared water. Basin-based development of irrigation, hydropower or flood control projects can provide riparian countries with greater net benefits than what they could have achieved through purely state-centric development. This should be obvious from the fact that national borders provide artificial constraints—from the point of view of optimizing basin resource management—on where to locate hydropower dams or where to grow agricultural produce. The resulting net benefits from cooperative planning can then be shared and traded.

In this context, the international community might assume a facilitating role in encouraging early collaborative and participatory efforts among basin states, especially in terms of facilitating water resource sharing agreements. Riparian countries in the developing world are often unable to establish institutional cooperative arrangements because of their concerns regarding existing and future water rights. Mutual suspicion and uncertainties about reciprocal action tend to obstruct constructive engagement. To overcome these obstacles, the international community can provide impartial external assistance, including various types of guarantees against "misbehavior," to launch the process of cooperation (Subramanian et al. 2012). Gradually, this assistance can facilitate mutual trust and confidence among the basin riparian states in order to achieve collective action. The external support could be used as incentives for riparian states to come together (Priscoli 1994). The basin-based institutional arrangement on shared rivers can bring all riparian countries greater benefits than would be achieved through state-centric development. With the proper institutional support, international rivers could potentially become the cornerstone of regional cooperation instead of being a source of fierce conflict.

To be effective enough there is a need for a more coordinated and strategic approach among external actors. However, there is as yet no established institutional home for international engagement on transboundary water issues. The multitude of, and divergence between, international stakeholders results in a lack of agency when it comes to transboundary waters related crises. There are numerous international institutions that deal with various aspects of transboundary waters, but their lack of political clout and fragmentation (across UNEP, UNDP, UN Water, UNESCO including UNESCO IHP, the UNECE Water Convention plus other regional conventions, even when looking only at the UN system) prevent them from assuming such agency (Van Genderen and Rood 2011). As a result, the international community faces huge challenges when it comes to systematic early action.

Therefore, the international community should seek to strengthen existing institutions for coordination so as to better facilitate political processes in transboundary basins. These networks should go beyond donor coordination to allow for concerted foreign policy approaches. Moreover, before engaging in transboundary basins international actors need to beware of the inherent difficulties, which often necessitate long-term commitment. International actors should focus on achieving a basic political settlement between the key parties—if need be at the expense of a more comprehensive settlement.

There is a need for an international institutional platform dedicated to supporting transboundary water issues in promoting, coordinating and supporting relevant initiatives, including in the fields of standards for data collection and dissemination, monitoring of process developments, leveraging financing and perhaps for serving as a source of arbitration (ODI 2001). This would allow the institution to engage in proactive hydro-diplomacy at all stages, from conflict prevention via crisis management to conflict resolution. A similar institution, the Water Cooperation Facility, was proposed by UNESCO and the World Water Council in 2003 but stalled due to insufficient support (Carius 2005).

More recently, the Shared Waters Partnership (SWP) was created in 2010 under the UNDP's umbrella. It provides technical and financial support to diplomatic initiatives aimed at addressing transboundary water related conflicts and may thus become the nucleus of international agency. However, like many similar initiatives, it represents a supply-side solution whose demand-side counterparts (especially in terms of basin stakeholders) often hesitate to mandate it. To become a driver of agency on hydro-diplomacy, the SWP's challenge will consist of building up an institutional structure that maintains close contacts with all major donors such that the latter will proactively inform and use the SWP while avoiding a situation in which coordination requirements overwhelm its ability to engage quickly and proactively.

The challenge of strengthening agency on hydro-diplomacy has also been taken up by the Geneva Water Hub, established in 2015 to better understand and prevent water-related tensions, thereby to strengthen both multi-sectoral and transboundary cooperation. Its High-Level Panel has been tasked to develop a set of proposals for the international community to that effect, including strengthening the global architecture, and is slated to report back in 2017.

Finally, UN Water has emerged as an interagency mechanism that aims to coordinate other UN agencies working on water issues. Given its limited resources and standing as a mechanism, however, it is questionable whether UN Water can play a role in coordinating proactive hydro-diplomacy. In the absence of an established international organization willing and able to take on this task, an informal network of policy makers engaged in transboundary water issues could seek to fulfill a similar role and assign both emerging conflicts and conflict-resolution opportunities to an appropriate coalition of actors.

A call for greater foreign policy engagement also raises the question of the appropriate trigger for intervention. Should international negotiators only act on an invitation from all basin states, from one or a group of riparians or on their own initiative? There is no one golden rule on this. An invitation from all concerned parties is of course the ideal backdrop to external involvement, but it will likely be forthcoming primarily in cases in which basin countries seek international support to be able to develop their shared water resources for increased benefits. Clearly, the international community should support such ventures, but that leaves questions about the difficult and potentially dangerous cases unattended. If an invitation originates from only a subgroup of states in the basin, the respective external actor should evaluate whether its intervention would help foster agreement across the basin or might trigger a backlash from the remaining basin states. In the case of the latter, the appropriate response may be to turn down any formal role, redirect the requesting authorities to another mediator more widely perceived as impartial, or to prompt them to broaden the number of actors invited for mediation so as to balance perceived partiality. Intervening without the request from basin governments may sometimes be appropriate, but the respective external actor's appetite for risk would have to be substantially greater.

A crisis may be harder to defuse once it has gone public and positions have hardened, yet simultaneously it is difficult to muster sufficient political energy and local ownership for setting up anticipatory crisis-management mechanisms. Many of the challenges of transboundary water governance—quality, control and the distribution of varying water quantities—are long-term. At the same time, some other future challenges cannot be foreseen, in particular climate-change-induced uncertainties. Thus, it is crucial to establish cooperative institutions that are sufficiently resilient and ideally self-reinforcing (Wolf et al. 2003).

The lack of any effective external agency to address questions of transboundary water cooperation is particularly palpable when it comes to engaging riparian hegemons. These hegemons often do not participate in the multilateral basin-management institutions, but their shadow clearly looms large. In this context, one crucial task for foreign policy makers can be to somehow establish an effective dialogue with the basin hegemon as to how its global water policy objectives could be systematically reconciled with the need for a predictable framework for conflict management in transboundary basins.

Hegemony, not necessarily an obstacle to cooperation, can facilitate that cooperation (Zeitoun and Warner 2006). Brazil, for example, has facilitated the Amazon Cooperation Treaty and supported the respective river basin organization. Similar dynamics unfolded in the Danube Basin. The economically strongest basin countries, Germany and Austria, have from the beginning been active supporters of cooperation, especially in the context of the International Commission for the Protection of the Danube River (ICPDR).

The positions emanating from hegemony are not necessarily static. Both India and China have long embraced the position that building water infrastructure on international rivers is not a multilateral issue. India is now experiencing the consequences of such a position, as a downstream country to China in the Ganges-Brahmaputra Basin. This experience may prompt India to recalibrate its own discourse vis-à-vis its weaker neighbors—an opportunity for diplomatic engagement on the part of the international community, with the objective of nudging India toward a more multilateral approach.

Moreover, hegemony in a basin is itself not necessarily static but can be challenged. China's recent emergence as an alternative banker for large dams has begun to bring a strategic shift to intra-basin power dynamics in many parts of Africa (Swain 2012). One example is Ethiopia, which for a long time had been unable to use a greater share of the Blue Nile water because Western donors and multilateral lenders refused to finance such projects—no doubt in part because the West politically prioritized Egypt over upstream countries (Seide and Lind 2013). The advent of alternative sources of finance has contributed to changes in the balance of power and may yet force Egypt to adopt a more cooperative stance. China's emerging role is changing the landscape of international water sharing, and these changes imply both challenges and opportunities. Above all, they underline that shaping global hydro-politics is not only a technical, but also a diplomatic challenge.

Need for a Strategic Approach

As the stakes of hydro-diplomacy rise in many basins, there is a need for more coordinated and strategic approaches within and among donor countries to demonstrate credible and long-term commitment to cross-sector and transnational cooperation, not least by creating institutions tasked with enabling such collaboration. Thus far, transboundary water policy is not treated as sufficiently strategic to warrant greater ex ante coordination, both within and among governments. The lack of strategic coordination in supporting cooperation on transboundary waters shows, for example, in the fact that nine US programs were involved in supporting the Mekong River Commission, apparently in part without knowing about each other. As another example, Norwegian companies designed dams in Ethiopia with Norwegian state guarantees, just while Norwegian diplomats tried to prevent Ethiopia from taking such provocative actions. To achieve more effective coordination, governments will need to give their bureaucracies greater incentives to invest in strategic thinking about transboundary water issues—at least within, but preferably also across, the relevant line and foreign ministries. A strategic approach to transboundary waters should also see basin-wide thinking take root in embassies. Preventing future crises necessitates greater awareness of the importance of transboundary water governance among high-ranking foreign policy makers, so that they can seek to intervene before it is very costly or simply too late.

A more strategic approach regarding foreign policy objectives implies a number of challenges for development policy. Fundamentally, there might be a need to adjust overall objectives insofar as they encapsulate trade offs between different goals, such as peace building, economic growth and sustainable development. Different objectives often imply diverging priorities. Much development aid is still focused on supporting economic growth and energy production, which introduces a certain bias for large water infrastructure projects. Such large projects may however be detrimental to peace building, as in post-conflict Burma/Myanmar where big dams on the Salween are creating many new conflicts (Peel 2014).

Most people in fragile countries depend on agriculture and, thus, on the availability of water for irrigation. A strategy intended to bolster peace through water might thus redirect donor investments into a more sustainable use of water in agriculture (Troell and Weinthal 2014). Beyond these seminal goal trade-offs, there is also the perennial issue of earmarked funding which complicates any strategic approach to designing and adapting projects depending on political priorities. Development cooperation at its best is longer-term and focused on achieving measurable improvements for the local population. Its increasing reliance on metrics, however, introduces a bias for easily measurable outcomes that grates against a focus on political, and more ambiguous, objectives. Moreover, the focus on metrics contributes to a built-in self-interest in finishing projects, irrespective of whether these projects ultimately still represent the politically most appropriate outcomes.

These challenges for development policy are reflected similarly in the foreign policy arena. More *ad hoc* and driven in its priorities by exogenous events, the structural objectives of foreign policy are frequently eclipsed by crisis management. While development agencies can crucially support foreign policy objectives, diplomats' tolerance for the seesaw of politics can, in turn, facilitate and complement the structural foreign policy embedded in development work. Less driven by the need to finish a project, foreign policy makers can wait, attend to long-winding processes and prepare the ground with the relevant governmental and societal actors until the opportunity for an agreement on transboundary waters arises.

To realize such synergies, governments need to create institutionalized forms of cooperation that systematically and transnationally connect their officials from different fields. The challenges of a coherent and strategic approach apply not only to individual donors. They relate to the entire community of donors and investors, which comprises not only governments, but also international organizations, non-governmental organizations and private companies. When it comes to water infrastructure projects, coordination needs to be maintained over long periods of time, thereby ensured at an institutional level that will survive individual actors.

Building Capacity

Building capacity at the personal and institutional level is crucial to achieving sustained and sustainable cooperation in many transboundary basins. Hence, the international community should play a supportive role in building the necessary domestic and inter-national legal institutions to improve transboundary water governance, including institutional resilience to climate change. Support should also be provided for better domestic water-management practices so as to contribute to lesser pressure on limited and decreasing resources. Locally legitimate water management practices can directly contribute to preventing intergroup conflict and will likely have positive indirect effects as well in terms of enhancing governmental legitimacy, especially in the context of fragile states. In many cases, this could contribute to better transboundary cooperation by enabling national-level institutions in terms of technical capacity and confidence to devise cross-border deals.

Support for building institutions is also important at the international level. It should include advocacy for bilateral and multilateral consultations, especially in basins that lack a strong institutional structure, so as to achieve transparency on policy responses and a shift toward climate-adaptation approaches that are not exclusively domestic. Joint risk assessments offer particular opportunities in this respect. They serve not only to gain a shared understanding of the challenges, but also as a confidence-building measure that may in time prepare for joint management of shared waters rather than mere allocation or compensation mechanisms (although they admittedly also risk revealing inequalities that instead fuel the conflict, underlining the need for a careful diplomatic approach). In the absence of overarching political rapprochement or strong outside guarantees, trust can usually be built only through long-term transparency in terms of sharing data and intentions regarding future infrastructure, which in turn requires long-term engagement.

The first step in this engagement requires support for the principles embodied in the UN Convention on the Law of the Non-Navigational Uses of International Watercourses. The jury is still out on whether this convention's entry into force in August 2014 will change the politics of transboundary waters. Ensuring that the convention's provisions become the undisputed point of reference in conflict mediation and pertain to as large an area as possible remain worthy objectives. Yet, advocacy for the convention may not be an appropriate entry point for engaging in basin politics. For the international com-munity, the framework's principles may instead provide a good point of reference for initiating and extending basin-specific cooperative ventures without any reference to the convention, as they can also be linked to international customary law. A further way to strengthen multilateralism in hydro-diplomacy might finally consist in supporting the development and codification of international norms for transboundary groundwater resources.

Whereas legal norms and institutions can play a helpful role, they should not be seen as magic wands. The interpretation of general rules will often be contested, even where actors agree that their access to water should be subject to "equitable and reasonable utilization and participation." One area in which donors could make an impact in this respect is by training diplomats in the basics of water development and water experts in the basics of negotiation and mediation. Foreign policy officials usually lack technical knowledge to fully understand water-sharing dynamics, which has made them reluctant to enter into that territory. Water professionals, by contrast, often prefer a technocratic approach and do not want to securitize the water-sharing issue. Given the nature of the issue at stake, however, there is a need for an integrated approach that ensures the technical and political tracks move in sync. Moreover, the task of water professionals is no longer limited to basin-based regimes and institutions, but extends to achieving effective water management in the face of climate change. This entails a need to influence the supporting pathways from local, national and international policies and practices.

Climate change is rapidly emerging as a critical issue in the negotiation processes around the sharing of international water. In the past, a few negotiators trained specifically to deal with water issues, could effectively cover water-sharing matters. Today, however, water management knowledge is not only increasingly relevant to an increasing range of fields (such as energy generation, food production, human rights and health issues). Negotiators also need to have at least a basic grasp of the possible impacts of climate change on water availability and demand (such as precipitation patterns, glacier melting, temperature increase, rising sea water encroaching on fresh water systems); on the possible impacts of climate change on livelihoods, society, and on countries and regions; as well as on increasingly diversified climate change policy processes, including regional processes dealing with the impact of climate change on water resources and possible mitigation and adaption measures. In this context, they also need to be able to identify and classify important actors and groupings and their positions on climate change and water management issues. Many small developing riparian countries suffer from a lack of competent hydro-diplomats who can address climate change issues while carrying out negotiations over shared water resources in the context of existing power asymmetries.

Training by the international community should aim to give both water professionals and diplomats a better appreciation of the likely consequences of actions undertaken in the other policy domain so as to help them understand each other better and, ultimately, create shared expectations of what constitutes appropriate behavior. Specifically, the ability to design appropriate stakeholder and mediation processes should be systematically and widely diffused. Both donor and recipient country officials would profit from this type of training. This is particularly important when it comes to smaller and weaker riparian states: having parties negotiate on a level playing field should arguably create an enabling environment for sustainable cooperation (Troell and Weinthal 2014).

Apart from helping to prevent conflicts over water, efforts to enhance and systematize training might also provide an entry point for addressing ongoing water-related conflicts by taking officials out of their usual context to approach the issue from a fresh perspective. The international community might play an important role here by facilitating such talks, perhaps by inviting change agents from the respective governments abroad in order to allow for a substantial reflection on the larger issues at stake. The international community should also aim to move the debate from state rights to water to human rights to water. In some cases, that might help to de-securitize the water issue and encourage states to prioritize meeting basic human needs for water for their population. This is particularly important in countries recovering from a long period of violent conflict.

Building awareness and capacity requires financial support. Such funding is particularly necessary for the "soft," diplomatic aspects of hydro-politics. It would serve to initiate and support processes of capacity-building and confidence-building in transboundary basins, from collecting and agreeing on data to training of riparian officials and facilitating appropriate institutional settings (Trolldalen 1997). The effectiveness and sustainability of such engagement—legal-institutional, capacity-wise and financial—will clearly hinge on progress regarding stronger international agency and better coordination. Apart from strengthening transboundary water governance, improving knowledge and skills will also help to build the necessary societal awareness of the challenges and opportunities that shared waters imply.

Conclusion

Water is generally seen as a basic human need and public good. In 2010, the UN General Assembly explicitly recognized a human right to water through Resolution 64/292. Yet, in many regions, national water sectors are overwhelmed by the demands resulting from continuing population growth. Many developing countries suffer either a lack of availability (physical scarcity) or a lack of access (economic scarcity). This has become intertwined with the political debate over private versus public provision of water because the water sector needs significant reform and investment to provide sustainable solutions.

The combination of different pressures such as population growth, increasing per-capita consumption and the unfolding effects of climate change will further increase water scarcity. As climate change will contribute to long-lasting droughts and seasonal variation, people need a responsive state to respond to their water needs. When climate change makes it difficult for the state to meet water-demand challenges, conflicts over a narrowing resource base cannot be resolved, and instability and violent conflict within states may feed instability and conflict between states within the basin. Efficient and good water management in the face of climate change is therefore also a critical part of peace-building efforts—both to prevent countries from falling into armed conflict, and in helping them to not relapse after a period of violence.

There is no doubt that climate change poses extreme challenges to transboundary water resources management. Maarten De Wit and Jacek Stankiewicz (2006) demonstrated the dramatic potential effects of relatively small percentage changes in rainfall regimes due to climate change over the perennial drainage of rivers. In this context, transboundary water governance presents significant challenges and opportunities for foreign policy makers in terms of preventing conflict and harnessing opportunities for greater regional cooperation. However, awareness of the coming hydro-political challenges is currently not keeping up with the rising challenges, nor are the mediation and negotiation skills necessary to address them.

Despite the risk that climate-change-induced water scarcity poses to social wellbeing and economic growth, there has been alarmingly little progress toward managing freshwater sustainably in most countries. Significant economic and political resources are needed to develop technologies and infrastructure that provide better water management at the basin, national and transboundary levels. Global climate change and the resource deficit underlie many of the risks and uncertainties confronting the developing countries today.

Foreign policy makers have a range of useful tools to help alleviate the political consequences of transboundary water governance problems, and to foster greater cooperation. They can use their political mandate, connections and leverage as well as their ability to frame issues such that this context becomes enabling rather than disabling. This ability could and should become a greater asset in the international community's arsenal for addressing transboundary water governance issues. To make that happen requires express political interest and, given the substantial investment on the technical side of water infrastructure projects, there is a strong case for investing more on the diplomatic side.

Note

1 This chapter has borrowed liberally from the report: Pohl, Benjamin; Alexander Carius, Ken Conca, Geoffrey D. Dabelko, Annika Kramer, David Michel, Susanne Schmeier, Ashok Swain and Aaron Wolf. 2014. *The Rise of Hydro-Diplomacy. Strengthening Foreign Policy for Transboundary Waters.* Berlin: adelphi.

References

Ackva, Johannes, Benjamin Pohl, Adrien Detges, Camille Defard, Brooke Noorbergen and Jannis Rustige 2015: *The ECC Factbook*, https://factbook.ecc-platform.org/conflicts/security-implications-gilgel-gibe-iii-dam-ethiopia. Berlin: adelphi. Accessed March 21, 2016. https://factbook.ecc-platform.org/.

Carius, Alexander. 2005. *Brauchen Wir Eine International Water Cooperation Facility?* Accessed July 24, 2014. Www.Blog.Krium.De/Index.Php/Themen/Gender-231/Einfuehrung/141-Newsletter/Newsletter-Januar-2005/1796-Brauchen-Wir-Eine-Internationale-Water-Cooperation-Facility.

Conca, Ken. 2012. "Decoupling Water and Violent Conflict." *Issues in Science and Technology* 29: 39–49.

Conca, Ken and Geoffrey D. Dabelko, eds. 2002. *Environmental Peacemaking*. Baltimore: Johns Hopkins University Press.

De Wit, Maarten and Jacek Stankiewicz, "Changes in Surface Water Supply across Africa with Predicted Climate Change." *Science* 311: 1918–21.

Drieschova, Alena, Mark Giordano and Italy Fishhendler. 2009. "Climate Change, International Cooperation and Adaption in Transboundary Water Management." In *Adapting to Climate Change: Threshold, Values, Governance*, edited by W. Neil Adger, Irene Lorenzoni and Karen O'Brien, 384–98. Cambridge: Cambridge University Press.

Earle, Anton., Anders Jägerskog and Joakim Öjendal, eds. 2010. *Transboundary Water Management: Principles and Practice*. London: Earthscan.

Earle, Anton, Ana Cascao, Stina Hansson, Anders Jägerskog, Ashok Swain and Joakim Öjendal. 2015. *Transboundary Water Management and the Climate Change Debate*. London: Routledge.

Falkenmark, Malin and Anders Jägerskog. 2010. "Sustainability of Transnational Water Agreements in the Face of Socio-economic and Environmental Change." In *Transboundary Water Management: Principles and Practice*, edited by Anton Earle, Anders Jägerskog and Joakim Öjendal, 157–71. London: Earthscan.

Giordano, Mark; Alena Drieschova; James A. Duncan; Yoshiko Sayama; Lucia de Stefano and Aaron T. Wolf, 2013. "A Review of the Evolution and State of Transboundary Freshwater Treaties." *International Environmental Agreements: Politics, Law and Economics* 13: 1–22.

Gleick, Peter, ed. 2009. *The World's Water 2008–2009*. London: Island Press.

Gleick, Peter. 2014. "Water, Drought, Climate Change, and Conflict in Syria." *Weather, Climate & Society* 6: 331–40.

Jägerskog, Anders, Ashok Swain and Joakim Öjendal, eds. 2014. *Water Security* (4-Volume Set). London: Sage Publications.

Jeffrey, Paul and Mary Gearey. 2006. "Integrated Water Resources Management: Lost on the Road from Ambition to Realisation?" *Water Science & Technology* 53: 1–8.

Kirmani, Syed and Robert Rangeley. 1994. "International Inland Waters: Concepts for a More Active World Bank Role." World Bank Technical Paper No. 239. Washington, DC: The World Bank.

Peel, Michael. 2014: *Villagers Count Cost of Myanmar Dam Project.* March 21. Accessed July 24, 2014. Available Online at Www.Ft.Com/Cms/S/0/Edc7a794-Aff1-11e3-B0d0-00144feab7de. Html.

Pohl, Benjamin, Alexander Carius, Ken Conca, Geoffrey D. Dabelko, Annika Kramer, David Michel, Susanne Schmeier, Ashok Swain and Aaron Wolf. 2014. *The Rise of Hydro-Diplomacy. Strengthening Foreign Policy for Transboundary Waters.* Berlin: adelphi.

Priscoli, Jerome D. 1994. "Conflict Resolution, Collaboration and Management in International and Regional Water Resources Issues." Paper presented at the VIIIth Congress of the International Water Resources Association (IWRA), Cairo, November 1994.

Redclift, Michael R. and Marco Grasso, eds. 2013. *Handbook on Climate Change and Human Security.* Cheltenham: Edward Elgar.

Rüttinger, Lukas, Gerald Stang, Dan Smith, Dennis Tänzler, Janani Vivekananda, Alexander Carius, Oli Brown, Geoff Dabelko, Roger-Mark De Souza, Shreya Mitra, Katharina Nett and Benjamin Pohl. 2015. *A New Climate for Peace: Taking Action on Climate and Fragility Risks.* Berlin: adelphi, International Alert, The Wilson Center, and the European Union Institute for Security Studies.

Schmeier, Susanne. 2013. *Governing International Watercourses: River Basin Organizations and the Sustainable Governance of Internationally Shared Rivers and Lakes.* New York: Routledge.

Seide, Wondwosen Michago and Jeremy Lind. 2013. "Churning Waters: Strategic Shifts in the Nile Basin." *Rapid Response Briefing.* Brighton: Institute of Development Studies.

Subramanian, Ashok, Bridget Brown and Aaron Wolf. 2012. *Reaching Across the Waters. Facing the Risks of Cooperation in International Waters.* Washington, DC: World Bank.

Swain, Ashok. 2004. *Managing Water Conflict: Asia, Africa and the Middle East.* London: Routledge.

———. 2009. "The Indus II and Siachen Peace Park: Pushing India-Pakistan Peace Process Forward," *The Round Table: The Commonwealth Journal of International Affairs* 98: 569–82.

———. 2012. Global Climate Change and Challenges for International River Agreements." *International Journal on Sustainable Society* 4: 72–87.

———. 2010. "Environment and Conflict in South Asia: Water-Sharing between Bangladesh and India," *South Asian Journal* 28: 27–34.

———. 2013. *Understanding Emerging Security Challenges: Threats and Opportunities.* London: Routledge.

Swain, Ashok, Ranjula Bali Swain, Anders Themnér and Florian Krampe. 2011. *Climate Change and the Risk of Violent Conflicts in Southern Africa.* Pretoria: Global Crisis Solutions.

Taenzler, Dennis, Alexander Carius and Achim Maas. 2013. "The Need for Conflict-Sensitive Adaptation to Climate Change." In *Backdraft: The Conflict Potential of Climate Change,* 5–13. Washington DC: Woodrow Wilson Center.

Troell, Jessica and Erika Weinthal. 2014. "Harnessing Water Management for More Effective Peacebuilding." In: *Water and Post-Conflict Peacebuilding: Shoring Up Peace,* edited by Mikiyasu Nakayama, Jessica Troell and Erika Weinthal, 405–71. London: Earthscan.

Trolldalen, Jon Martin. 1997. "Troubled Waters in The Middle East: The Process Towards the First Regional Water Declaration Between Jordan, Palestinian Authority, and Israel." *Natural Resources Forum* 21: 101–8.

———. 2013. *Conflict Prevention and Peace Dividends Through Cooperation on Transboundary Water Management in SADC: Achieving Peace Dividends Through the Prevention of Water Conflicts,* 2nd edn. Geneva: Compass Foundation.

UN Water. 2016. "Statistics." Accessed March 18, 2016. http://www.unwater.org/statistics/en/.

Van Genderen, Ruben and Jan Rood. 2011. *Water Diplomacy: A Niche for The Netherlands.* The Hague: Netherlands Institute of International Relations "Clingendael."

Varis, Olli, Cecilia Tortajada and Asit K. Biswa. 2008. *Management of Transboundary Rivers and Lakes.* Berlin: Springer.

Wolf, Aaron T., Shira B. Yoffe and Mark Giordano. 2003. "International Waters: Indicators for Identifying Basins at Risk." *Water Policy* 5: 29–60.

Zeitoun, Mark and Jeroen Warner. 2006. "Hydro-hegemony—a Framework for Analysis of Trans-Boundary Water Conflicts." *Water Policy* 8: 435–61.

Part II

TOOLS, TECHNIQUES, MODELS AND ANALYSES TO RESOLVE COMPLEX WATER PROBLEMS

Chapter Three

TEN BANKRUPTCY METHODS FOR RESOLVING NATURAL RESOURCE ALLOCATION CONFLICTS

Mahboubeh Zarezadeh, Ali Mirchi,
Laura Read and Kaveh Madani

Abstract

Bankruptcy methods aim to fairly allocate a stock among a group of agents, for cases in which the given amount is insufficient to fully satisfy all the agents' demands. Natural resource allocation problems have certain characteristics that justify the use of bankruptcy methods for fair resource allocation. The approach is particularly appealing for situations in which natural resources are scarce or are exhausted beyond their capacity to meet the demands of various beneficiaries. To underline the utility of the bankruptcy theory in solving natural-resource allocation problems, this chapter reviews ten bankruptcy rules, namely: Proportional (P), Adjusted Proportional (AP), Constrained Equal Award (CEA), Constrained Equal Loss (CEL), Talmud (Tal), Piniles (Pin), Constrained Egalitarian (CE), Random Arrival (RA), Minimal Overlap (MO) and Generalized IBN Ezra (GIE). We apply these bankruptcy rules to determine the share of the five coastal states of the Caspian Sea—Azerbaijan, Iran, Kazakhstan, Russia and Turkmenistan—from its oil and natural gas reserves. Since the developed bankruptcy allocation solutions do not necessarily converge to the same resource share value for each negotiating party, developing bankruptcy-based allocation solutions requires evaluating their acceptability in order to ensure practical relevance. Herein, we use the Bankruptcy Allocation Stability Index (BASI) to identify bankruptcy rules that can facilitate the decades-old Caspian Sea negotiation.

Introduction

Continuous socioeconomic growth across the world is increasing the pressure on natural resources. The demand for limited natural resources exceeds availability, creating a situation that parallels the financial bankruptcy problem. The financial bankruptcy problem occurs when an entity does not have sufficient assets to fully satisfy its debts. In essence, bankruptcy methods seek to find fair allocation solutions for distributing the remaining assets despite a shortage. Similar to the case of a bankrupt financial entity's failure, a natural resource stock, defined as the total amount of resource (e.g., the amount of water in a river), may be insufficient to meet the stakeholders' claims or their desired share of the resource (e.g., water demand or right). Structuring natural-resource conflicts as bankruptcy problems can be appropriate in cases in which allocation rules are obscure or nonexistent.

A classic question arises when a system does not have the capacity to satisfy the demands or claims of the beneficiaries: That is, how should the available resource be allocated? Answering this question is more challenging when the utility functions are not known and utility information (i.e., total satisfaction received from consuming goods and services) is not readily available. Even if such information exists, some parties may not accept it because often parties have a zero-sum (i.e., one party's gain or loss of utility is exactly balanced by other parties' utility losses or gains) view of the problem and prefer to bargain over the quantity of their share as opposed to the amount of utility gained (Madani et al. 2012). When utilities are ignored, the incremental share of cooperation is zero, and conventional cooperative game theory approaches cannot provide a solution (Madani et al. 2014a). For these cases, the problem can be solved via bankruptcy methods, which represent a different set of cooperative game theory solutions. The main challenge is then in developing a scheme to fairly allocate the remaining resources among the beneficiaries, who have different demand (claim) levels. Bankruptcy methods (O'Neill 1982; Dagan and Volij 1993) can be considered as unconventional cooperative game theory methods that attempt to fairly allocate the total deficit, or the difference between the total claim and the available resource, among the parties.

Different notions of fairness have resulted in the development of various bankruptcy rules, including Proportional (P), Adjusted Proportional (AP), Constrained Equal Award (CEA), Constrained Equal Loss (CEL), Talmud (Tal), Piniles (Pin), Constrained Egalitarian (CE), Random Arrival (RA), Minimal Overlap (MO), and Generalized IBN Ezra (GIE) (O'Neill 1982; Dagan and Volij 1993; Aumann and Maschler 1985; Dagan et al. 1997; Chun and Thomson 2000; Alcalde et al. 2002; Bosmans and Lauwers 2007). Despite their relevance for solving allocation problems, applications of bankruptcy methods in this topic have been limited. Kampas and White (2003) applied four bankruptcy rules—namely CEA, CEL, AP and Pin—in selecting permit allocation rules to solve an agricultural pollution control problem in a watershed in England. Sheikhmohammady and Madani (2008) used P, CEA and AP methods to allocate the energy resources of the Caspian Sea to its five littoral states. Madani and Dinar (2013) examined the performance of the P, CEA and CEL rules as exogenous regulatory institutions for sustainable management of common pool resources. Ansink and Weikard (2012) proposed a sequential allocation approach to solve river sharing problems using the classical P, CEA, CEL and Tal methods. Madani et al. (2014a) developed bankruptcy optimization models to solve a transboundary river bankruptcy problem in Iran with respect to the temporal and spatial sensitivity of water deliveries, using the P, AP, CEA and CEL rules. Mianabadi et al. (2014a) developed an alternative bankruptcy method for solving river bankruptcy problems with a special attention to the contribution of each riparian creditor. Sechi and Zucca (2015) used the P, AP, CEA, CEL and Tal methods for water allocation in a multipurpose water system in southern Sardinia, Italy.

The objective of this chapter is to review bankruptcy rules and apply them to the decades-old Caspian Sea resource allocation problem, showing how the stability of the bankruptcy allocation solution might be different, even though, in theory, they should all be fair. Following Sheikhmohammady and Madani (2008a) the reviewed bankruptcy rules are applied to the Caspian Sea negotiations as the context to discuss the applicability of bankruptcy rules for natural resource conflict resolution.

Bankruptcy Solution Methods

In a natural resource allocation system shortage occurs when the resource stock is not large enough to meet all the demands. This problem has the essential characteristics of "capital distribution" in the economic bankruptcy literature. Bankruptcy rules provide a convenient method for dividing the capital stock to partially meet the stakeholders' demands in a fair manner. There are different methods for solving bankruptcy problems with implications for natural resource allocation; ten of these methods are explained below.

Proportional (P)

This bankruptcy rule is based on Aristotle's principle for fair allocation (Dagan and Volij 1993), which suggests that each stakeholder's demand be met proportionately with the demand size:

$$x_i = \lambda c_i \quad \forall i \tag{1}$$

$$\lambda = \frac{E}{C} \tag{2}$$

where x_i: resource share of stakeholder i; c_i: stakeholder i's claim; λ: a coefficient based on which the share of individual stakeholders is calculated; E: total available stock, and C: total claims.

Adjusted Proportional (AP)

This bankruptcy rule calculates the share of individual stakeholders v_i with respect to total available stock and the sum total of other stakeholders' claims (Bosmans and Lauwers 2007; Curiel et al. 1988). The initial share of individual stakeholders is given by equation (3):

$$v_i = \text{Max}\left\{0, E - \sum_{j \neq 1} c_j\right\} \quad \forall i \tag{3}$$

where v_i: what is left for stakeholder I after satisfaction of the claims of the other stakeholders (initial share).

The initial share provides a basis for final allocation. In the next step, the share of each beneficiary can be calculated based on equation (4):

$$
x_i = \begin{cases}
v_i + \left(c_i^E - v_i\right)\left(\displaystyle\sum_{j\in N}\left(c_j^E - v_j\right)\right)^{-1}\left(E - \displaystyle\sum_{j\in N} v_j\right) & \text{if } C > E > 0 \\
c_i & \text{if } E = C \qquad \forall i \\
0 & \text{if } E = 0
\end{cases} \tag{4}
$$

where $c_i^E = \text{Min}\{c_i, E\}$ and N is the set of stakeholders.

Constrained Equal Award (CEA)

In this method the stakeholders' least claim is used as the basis for allocation (Dagan and Volij 1993; Aumann and Maschler 1985; Dagan et al. 1997; Bosmans and Lauwers 2007), where this value is increased with respect to available stock by adding λ value incrementally until the value of allocation approaches the claim of each stakeholder. If the available stock only meets the smallest claim for all parties, the total available stock is equally divided between them as described in Equation (5):

$$
x_i = \text{Min}\left(\lambda, c_i\right) \qquad \forall i \tag{5}
$$

Constrained Equal Loss (CEL)

To calculate the value of allocation based on this method a constant value (λ) is removed from stakeholder claims $(c_i - \lambda)$ so that conditions and restrictions of problems are met (Dagan and Volij 1993; Dagan et al. 1997; Chun and Thomson 2000; Bosmans and Lauwers 2007). This method defines the value of allocation by Equation (6):

$$
x_i = \text{Max}\left(0, c_i - \lambda\right) \qquad \forall i \tag{6}
$$

Talmud (Tal)

The Tal method is a combination of the CEA and CEL methods (O'Neil 1982; Bosmans and Lauwers 2007). Allocation rules are defined according to the amount of available resources, which is either smaller or larger than half the total claims. In the former case, the TAL method essentially becomes the CEA method, allocating resources based on half the total stakeholder claims. In the latter case, the CEL method will be used for resource distribution, first accommodating half the total stakeholder claims and then allocating the remaining resources using the CEL method.

$$x_i = \begin{cases} \text{CEA}\left\{\frac{1}{2}c_i, E\right\} & \forall i \quad \text{if } E \leq \frac{1}{2}C \\ \frac{1}{2}c_i + \text{CEL}\left\{\frac{1}{2}c_i, E - \frac{1}{2}C\right\} & \forall i \quad \text{if } E \geq \frac{1}{2}C \end{cases} \quad (7)$$

The Pin and CE rules described below belong to the Tal Family.

Piniles' (Pin)

The difference between this rule and the Tal rule is that the Tal method selects the CEL rule as the basis for allocation, while the Pin rule uses the CEA allocation (Dagan et al. 1997; Bosmans and Lauwers 2007).

$$x_i = \begin{cases} \text{CEA}\left\{\frac{1}{2}c_i, E\right\} & \forall i \quad \text{if } E \leq \frac{1}{2}C \\ \frac{1}{2}c_i + \text{CEA}\left\{\frac{1}{2}c_i, E - \frac{1}{2}C\right\} & \forall i \quad \text{if } E \geq \frac{1}{2}C \end{cases} \quad (8)$$

Constrained Egalitarian (CE)

In this method, if $E \leq \frac{1}{2}C$, the CE method selects the CEA rule as the basis for allocation and if $E \geq \frac{1}{2}C$, the constant value λ is compared with the claims of stakeholders (Bosmans and Lauwers 2007). The stakeholder's share will be the smaller value of the two or half the initial claim, whichever is larger. Equation (9) shows the mathematical formulation of this method.

$$x_i = \begin{cases} \text{CEA}\left\{\frac{1}{2}c_i, E\right\} & \forall i \quad \text{if } E \leq \frac{1}{2}C \\ \text{Max}\left\{\frac{1}{2}c_i, \text{Min}\{c_i, \lambda\}\right\} & \forall i \quad \text{if } E \geq \frac{1}{2}C \end{cases} \quad (9)$$

The following constraint (Equation 10) should be met for determining λ:

$$\sum_{i=1}^{N} \text{Max}\left\{\frac{1}{2}c_i, \text{Min}\{c_i, \lambda\}\right\} = E \quad \forall i \quad (10)$$

Random Arrival (RA)

In the RA method, the amount of claims of some stakeholders is supplied randomly until the resources are exhausted completely (Chun and Thomson; 2000; Bosmans and Lauwers 2007). Therefore, this method requires that a stakeholder ranking be defined. For example, if three stakeholders {a, b, c} exist, they can be ranked in six different ways (3!) as follows:

$$
\begin{cases}
a, b, c \\
a, c, b \\
b, a, c \\
b, c, a \\
c, a, b \\
c, b, a
\end{cases}
$$

In the first case, the claim of the stakeholder "a" is supplied completely, followed by that of "b" and then "c." This is continued according to the stock value until the stock falls short of the claim of stakeholder i. In this situation, the remaining stock is allocated to the last stakeholder. This procedure is repeated for different states and the final value of allocation is obtained by taking an average of the values allocated to each stakeholder in all states. Equation (11) summarizes this description:

$$
x_i = \frac{1}{n!} \sum_{\pi \in \Pi^N}^{N} \mathrm{Min}\left\{ c_i, \mathrm{Max}\left\{ E - \sum_{j \in N, \pi(j) < \pi(i)} c_j, 0 \right\} \right\}
\tag{11}
$$

where n is the number of stakeholders.

Minimal Overlap (MO)

For calculating resource allocation using MO, first, the existing claims must be sorted in an ascending order (O'Neil 1982). Then, the value $\frac{c_i}{n}$ is allocated to all stakeholders, where c_1 is the smallest claim. In the next step, the difference between the second smallest claim (c_2) and the smallest claim (c_1) is divided by $n-1$ is added to the initially allocated values to $n-1$ stakeholders (all stakeholders except the stakeholder 1 with the smallest claim). This is continued to stakeholder k^*. For the stakeholders $k^* + 1$ to n, the remaining stock is divided by the number of remaining stakeholders in this group $(n - k^*)$ and is added to the values obtained from the previous steps. The value of k^* is determined using equation (12) by maximizing t.

$$
\left(c_{k^*+1} - t\right) + \left(c_{k^*+2} - t\right) + \cdots + \left(c_n - t\right) = E - t \quad \text{and} \quad c_{k^*} < t \leq c_{k^*+1}
\tag{12}
$$

Once the value of k^* is known, equations (13) or (14) can be used for solving the problem with respect to the A and B conditions:

Condition A) $c_{k^*} < E \le c_{k^*+1} \le c_n$

$$x_i = \frac{c_1}{n} + \frac{c_2 - c_1}{n-1} + \frac{c_3 - c_2}{n-2} + \cdots + \frac{c_i - c_{i-1}}{n-i+1} \qquad \text{for } i = 1, 2, \ldots, k^*$$

$$x_j = x_{k^*} + \frac{E - c_{k^*}}{n - k^*} \qquad \text{for } j = k^* + 1, k^* + 2, \ldots, n \tag{13}$$

Condition B) $c_n < E$ and $c_{k^*} < E \le C_{k^*+1}$

$$x_i = \frac{c_1}{n} + \frac{c_2 - c_1}{n-1} + \frac{c_3 - c_2}{n-2} + \cdots + \frac{c_i - c_{i-1}}{n-i+1} \qquad \text{for } i = 1, 2, \ldots, k^*$$

$$x_j = \left(c_j - t\right) + x_{k^*} + \frac{t - c_{k^*}}{n - k^*} \qquad \text{for } j = k^* + 1, k^* + 2, \ldots, n \tag{14}$$

Generalized IBN Ezra (GIE)

GIE is an iterative step-wise bankruptcy-solution method (Alcalde et al. 2002). The sum of values allocated to each stakeholder at each step shows the total allocated value. First, the claims are sorted in an ascending order. Stakeholder 1 is considered as the stakeholder with the smallest claim, followed by stakeholder 2, with the next smallest claim, and so forth. Next, the values allocated to stakeholders in the previous step are subtracted from their claims and the available resource at the beginning of the previous step. The previous step is then repeated for the new values (unsatisfied claim and remaining resource). These stages are repeated until the stock is exhausted. Finally, for calculating the values allocated to any stakeholder, the values obtained from the above steps must be added. The value of allocation to each beneficiary can be calculated based on the following equation:

$$x_i = \sum_{k=1}^{i} \frac{\text{Min}\left(c_k^t, E^t\right) - \text{Min}\left(c_{k-1}^t, E^t\right)}{n - k + 1}, c_0^t = 0 \tag{15}$$

where c_k^t is the claim of stakeholder k in step t and E^t is the remaining stock in step t.

Caspian Sea

The Caspian Sea is the world's largest enclosed inland body of water. With an approximate surface area of 376,000 km², the sea is bounded by Azerbaijan, Iran, Kazakhstan, Russia and Turkmenistan (Figure 3.1). The fair allocation of the natural oil and gas resources has been a source of concern for the five littoral states since the collapse of the Soviet Union in 1991. To date, five division rules have been considered for solving the

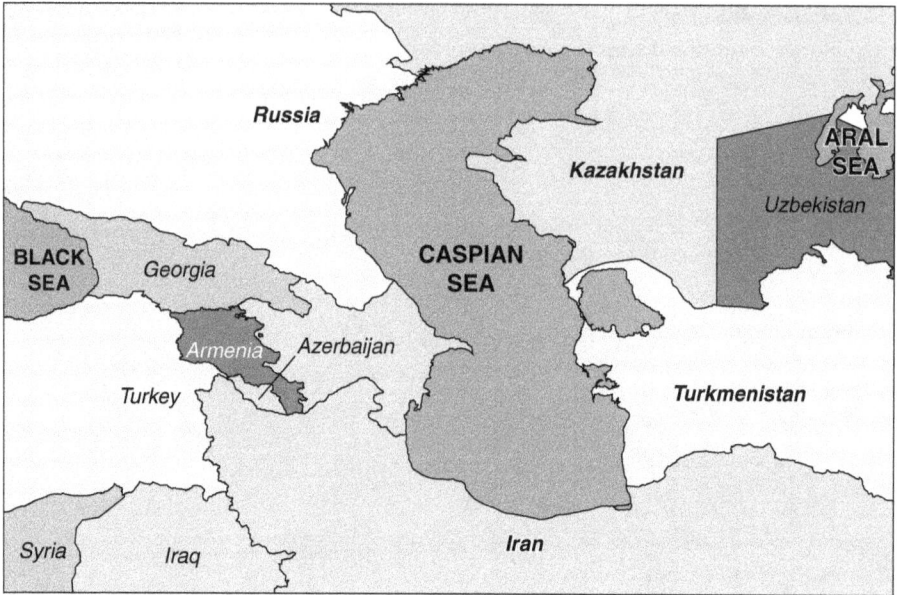

Figure 3.1 The Caspian Sea and its littoral countries.

problem (Sheikhmohammady and Madani 2008b). These include: common sovereignty between Iran and Russia; division based on the international Law of the Sea; dividing the sea to five equal parts (20 percent based) from the surface of the sea and sea bed; dividing the sea based on the Soviet Union divisions; and dividing the sea bed based on the international Law of the Sea and common division of the surface. These division methods result in different allocations of energy resources. Thus, the five littoral states have different orders of preference about the division methods, causing a deadlock in the Caspian Sea negotiations over the establishment of a governance system for managing and sharing the sea and its valuable resources.

Due to its significance as one of the most intractable international disputes over sharing natural resources, the Caspian Sea resource-sharing problem has been a popular subject of research in recent years. Sheikhhommady et al. (2006) used the Graph Model for Conflict Resolution (GMCR) to detect the likely resolution of the negotiations based on non-cooperative game theory concepts (Madani and Hipel 2011). Sheikhmohammady and Madani (2008a) used different social choice rules and fallback bargaining methods to predict the final resolution of the current negotiations if the parties want to select one of the five division methods, assuming that the negotiation parties are equally powerful. Since the claims of the littoral countries in the Caspian Sea negotiations exceed the sea's strategically important oil and gas reserves, Sheikhmohammady and Madani (2008b) suggested using bankruptcy methods for allocating the available resources, irrespective of the aforementioned five division methods that have been considered in the negotiations. This preliminary study used the P, AP and CEA rules for resource allocation without performing a strong stability analysis to single out the most stable allocation. Arguing that the negotiators' powers could affect the outcome of the negotiations, Sheikhmohammady and Madani (2008c) developed a descriptive framework for analyzing asymmetric multilateral negotiations, showing that the final results would be different from the previous studies, which assumed symmetry in powers. Madani and Gholizadeh (2011) used conventional cooperative game theory methods to determine the cooperative allocations of the energy resources, assuming that the parties' non-cooperative allocations in the status quo can be determined based on the old treaties between Iran and the Soviet Union. Due to the environmental significance of the Caspian Sea as the source of the 90 percent of the world's caviar, Imen et al. (2012) investigated the effects of including environmental resources on the allocation solutions using cooperative game theory. Their results, however, did not show a significant difference due to the low values considered for the ecosystem resources of the Caspian Sea in comparison with the value of its energy resources. Madani et al. (2014b) developed a negotiation support system (NSS) for the Caspian Sea problem with the ability to suggest optimal marine boundaries with respect to the locations and values of energy and environmental sources of the sea. Mianabadi et al. (2014b) examined the effects of the negotiators' risk attitudes on the outcome of the ongoing negotiations using risk-based, fuzzy and multi-attribute decision-making methods. As a synthesis study, Madani et al. (2014c) used multiple-criteria decision-analysis methods to find the optimal division method out of the methods studied or developed by previous studies, based on the social planner's perspective. Read et al. (2014) argued that conventional social planner types of solutions inherently assume perfect cooperation among the decision-makers and, therefore, their solutions might not be stable in practice. Their study highlighted the importance of examining the stability of allocation solutions in resource-allocation studies with multiple decision-makers.

This present study is an extension of the work done by Sheikhmohammady and Madani (2008b) with respect to the suggestions of Read et al. (2014). The ten bankruptcy rules are first applied to the Caspian Sea, an example of a natural resource bankruptcy problem. Then, the stability analysis method developed by Madani et al. (2014a) for bankruptcy problems is used to determine the most stable bankruptcy schemes for sharing the Caspian Sea energy resources.

Results

Table 3.1 shows the claimed shares of the five littoral states from the total value of the energy resources of the Caspian Sea. These values were previously calculated with the assumption that each country's claim is equivalent to the value of its energy (oil and gas) share under its most preferred division schemes out of the five schemes which have been seriously considered in the negotiations. As suggested by the table, the total claim of the five countries is about 33 percent higher than the total value of the Caspian Sea's accessible energy reserves, qualifying this problem to be solved using the bankruptcy methods.

Applying the ten bankruptcy methods (reviewed previously) to the Caspian Sea bankruptcy problem results in the allocations presented in Table 3.2.

Table 3.1 Claims of the five littoral countries from the Caspian Sea energy resources (Read et al. 2014).

Claim (%)	Country
20.00	Azerbaijan
20.00	Iran
44.17	Kazakhstan
20.00	Russia
28.43	Turkmenistan
132.6	Total

Table 3.2 Share percentile of the five littoral countries under different bankruptcy rules.

Country	P	AP	CEA	CEL	Pin	Tal	CE	RA	MO	GIE
Azerbaijan	15.08	14.61	20.00	13.48	16.74	13.48	19.48	14.53	13.48	13.58
Iran	15.08	14.61	20.00	13.48	16.74	13.48	19.48	14.53	13.48	13.58
Kazakhstan	33.31	35.39	20.00	37.65	28.83	37.65	22.08	35.56	37.65	37.49
Russia	15.08	14.61	20.00	13.48	16.74	13.48	19.48	14.53	13.48	13.58
Turkmenistan	21.44	20.77	20.00	21.91	20.95	21.91	19.48	20.85	21.91	21.75

As expected, different bankruptcy rules have yielded different share percentiles of the energy resources. This is due to the different notions of fairness resulting in the development of different bankruptcy rules. Logically, each party prefers the bankruptcy method that allocates the highest share to it. This would result in a potential conflict over selection of a bankruptcy solution that would be acceptable to all parties. Stability analysis can be helpful in these cases to identify the most stable (achievable) allocation and its potential acceptability in multi-participant decision-making problems (Read et al. 2014; Dinar and Howitt, 1997; Teasley and McKinney. 2011; Madani and Dinar 2012; Asgari et al. 2014; Madani and Hooshyar 2014).

Stability Analysis

The Caspian Sea beneficiaries have the following preference orders over the ten bankruptcy solutions, based on the value of their shares. The higher the value, the more popular the allocation solution:

Azerbaijan: CEA>CE>Pin>P>AP>RA>GIE>MO=Tal=CEL
Iran: CEA>CE>Pin>P>AP>RA>GIE>MO=Tal=CEL
Kazakhstan: Tal=MO=CEL>GIE>RA>AP>P>Pin>CE>CEA
Russia: CEA>CE>Pin>P>AP>RA>GIE>MO=Tal=CEL
Turkmenistan: Tal=MO=CEL>GIE>P>Pin>RA>AP>CEA>CE

The plurality rule suggests that the most popular solution is the most promising one in these situations. However, as discussed by Madani et al. (2014a), popularity of an alternative does not guarantee its stability. They suggested the Bankruptcy Allocation Stability Index (BASI) (Madani et al. 2014a) as a stability indicator in bankruptcy problems. This index is the modified version of the stability index suggested by Loehman et al. (1979), used to measure the acceptability of cooperative game theory solutions.

The first step in calculating BASI is calculation of the Bankruptcy Power Index of each stakeholder (BPI$_i$), based on Equation (16):

$$\text{BPI}_i = \frac{x_i - v_i}{\sum_{j \in N}\left(x_j - v_j\right)} \quad i \in N, \quad \sum_{i \in N}\text{BPI}_i = 1 \tag{16}$$

where v_i—the sum of the conceded asset (resource) to stakeholder i by all other players—is calculated based on Equation (3).

BASI is equal to the coefficient of variation of BPIs:

$$\text{BASI} = \frac{\sigma_{\text{BPI}}}{\overline{\text{BPI}}} \tag{17}$$

where σ_{BPI} is the standard deviation of stakeholders' BPIs and $\overline{BPI}$ is the mean BPI. The potential acceptability of a solution is expected to be higher if it has a lower BASI value.

Table 3.3 shows the stability analysis results based on the BASI method. According this method, allocation based on the Pin rule is the most stable allocation, followed by the P rule-based allocation. These rules distribute power more evenly across all parties. In this way, no single party is likely to leave the negotiation. On the other hand, the CEL, Tal and MO rules result in the least acceptable allocation in the Caspian Sea resource-allocation problem, mainly because of the imbalance of power caused by the high powers assigned to Kazakhstan and Turkmenistan relative to the other countries.

Table 3.3 BPIs and BASI under different bankruptcy methods.

	Method Country	P	AP	CEA	CEL	Pin	Tal	CE	RA	MO	GIE
BPI	Azerbaijan	0.17	0.17	0.23	0.15	0.19	0.15	0.22	0.16	0.15	0.15
	Iran	0.17	0.17	0.23	0.15	0.19	0.15	0.22	0.16	0.15	0.15
	Kazakhstan	0.25	0.27	0.10	0.29	0.20	0.29	0.12	0.27	0.29	0.29
	Russia	0.17	0.17	0.23	0.15	0.19	0.15	0.22	0.16	0.15	0.15
	Turkmenistan	0.24	0.23	0.23	0.25	0.24	0.25	0.22	0.24	0.25	0.25
	BASI	0.18	0.22	0.26	0.30	0.09	0.30	0.20	0.22	0.30	0.29
	Ranking	2	4	6	8	1	8	3	5	8	7

Conclusion

As human-centered development continues to increase demands and competition over stressed natural resources, bankruptcy theory will be an important tool for solving natural-resource allocation problems. This chapter presented the mathematical descriptions for ten different bankruptcy methods that are applicable for resolving natural-resource allocation conflicts. To show their applicability, these bankruptcy methods were used to solve the well-known Caspian Sea resource-allocation problem. This resource-allocation problem is of strategic geopolitical importance and, after the collapse of the Soviet Union in 1991, has complicated the negotiations over the fair allocation of the largest landlocked body of water on Earth. Since the various stakeholders prefer different allocation methods, stability analysis was performed to identify the most feasible solutions that negotiators may agree on. Results suggest that for the Caspian Sea, the Pin rule may be the most feasible and fair allocation according to bankruptcy and stability analyses. The shares of the countries under this rule are the closest to their shares under the condominium governance regime, which has been found by previous studies to be the best solution under cooperation.

Acknowledgments

The first author would like to thank the Hydro-Environmental and Energy Systems Analysis (HEESA) Research Group for their extensive support during her stay at the University of Central Florida (UCF) as a visiting scholar.

References

Alcalde, José, Maria del Carmen Marco and José A. Silva. 2002. "Bankruptcy Games and the Ibn Ezra's Proposal." Working Papers. Serie AD. 2002–28, Institu to Valenciano de Investigaciones Económicas, S.A.

Ansink, Erik and Hans-Peter Weikard, 2012. "Sequential Sharing Rules for River Sharing Problems." *Social Choice and Welfare* 38: 187–210.

Asgari, Sadegh, Abbas Afshar and Kaveh Madani. 2014. "Cooperative Game Theoretic Framework for Joint Resource Management in Construction." *Construction Engineering Management* 140(3): 04013066.

Aumann, Robert J. and Michael Maschler. 1985. "Game Theoretic Analysis of a Bankruptcy Problem from the Talmud." *Economic Theory* 36(2): 195–213.

Bosmans, Kristof and Luc Lauwers. 2007. "Lorenz Comparisons of Nine Rules for the Adjudication of Conflicting Claims." Center for Economic Studies: Discussion papers ces0705, Katholieke Universiteit Leuven, Centrum voor Economische Studiën.

Chun, Yeh and William Thomson. 2000. "Replication Properties of Bankruptcy Rules." *Mimeo.*

Curiel, Imma. J. and Michael Maschler, Stef. H. Tijs. 1988. "Bankruptcy Game." *Zeitschrift für Operations-Research* 31: A143–A159.

Dagan, Nir, Roberto Serrano and Oscar Volij. 1997. A Non-cooperative View of Consistent Bankruptcy Rules. *Games and Economic Behavior* 18: 55–72.

Dagan, Nir and Oscar Volij. 1993. "The Bankruptcy Problem: A Cooperative Approach." *Mathematical Social Sciences* 26: 287–97.

Dinar, Ariel and Richard E. Howitt. 1997. "Mechanisms for Allocation of Environmental Control Cost: Empirical Tests of Acceptability and Stability," *Journal of Environmental Management* 49(2): 183–203.

Imen, Sanaz, Kaveh Madani and Ni-Bin Chang. 2012. "Bringing Environmental Benefits into Caspian Sea Negotiations for Resources Allocation: Cooperative Game Theory Insights." *World Environmental and Water Resources Congress 2012.* 2264–71. doi: 10.1061/9780784412312.228.

Kampas, Athanasios and Ben White. 2003. Selecting Permit Allocation Rules for Agricultural Pollution Control: A Bargaining Solution. *Ecological Economics* 47(2–3): 135–47.

Loehman, Edna, Jacquelyn Orlando, John Tschirhart and Andrew Whinston. 1979. "Cost Allocation for a Regional Wastewater Treatment System." *Water Resources Research* 15: 193–202.

Madani, Kaveh and Ariel Dinar. 2012. "Cooperative Institutions for Sustainable Common Pool Resource Management: Application to Groundwater." *Water Resources Research* 48(9): W09553.

Madani, Kaveh and Keith W. Hipel. 2011. "Non-Cooperative Stability Definitions for Strategic Analysis of Generic Water Resources Conflicts." *Water Resources Management* 25(8):1949–77. doi: 10.1007/s11269-010-9743-4.

Madani, Kaveh, M. Zarezadeh and S. Morid. 2014a. Resolving Conflicts over Trans-boundary Rivers Using Bankruptcy Methods. *Hydrology and Earth System Sciences* 18: 3055–68. doi:10.5194/hess-18-3055-2014.

Madani, Kaveh, O. M. Rouhani, A. Mirchi and S. Gholizadeh. 2014b. "A Negotiation Support System for Resolving an International Trans-Boundary Natural Resource Conflict." *Environmental Modeling and Software* 51: 240–49. doi: 10.1016/j.envsoft.2013.09.029.

Madani, Kaveh, M. Sheikhmohammady, S. Mokhtari, M. Moradi and P. Xanthopoulos. 2014. "Social Planner's Solution for the Caspian Sea Conflict." *Group Decision and Negotiation* 23(3): 579–96. doi: 10.1007/s10726-0139345-7.

Madani, Kaveh and Ariel Dinar. 2013. "Exogenous Regulatory Institutions for Sustainable Common Pool Resource Management: Application to Groundwater." *Water Resources and Economics* 2–3: 57–76. doi:10.1016/j.wre.2013.08.001.

Madani, Kaveh and S. Gholizadeh. 2011. "Game Theory Insights for the Caspian Sea Conflict." In *Proceeding of the 2011 World Environmental and Water Resources Congress*, edited by R. E. Beighley II and M. W. Kilgore, Palm Springs, CA: ASCE, 2815–19.

Madani, Kaveh and M. A. Hooshyar. 2014. "Game Theory-Reinforcement Learning (GT-RL) Method to Develop Optimal Operations Policies for Multi-Operator Reservoir Systems," *Journal of Hydrology* 519: 732–42.

Madani, Kaveh and Jay R. Lund. 2012. "California's Sacramento-San Joaquin Delta conflict: From Cooperation to Chicken." *Water Resources Planning Management* 138: 90–99.

Mianabadi, Hojjat, Erik Mostert, Mahdi Zarghami and Nick van de Giesen. 2014. "A New Bankruptcy Method for Conflict Resolution in Water Resources Allocation." *Journal of Environmental Management* 144, 152–59.

Mianabadi, Hojjat, Majid Sheikhmohammady, Erik Mostert and Nick van de Giesen. 2014. "Application of the Ordered Weighted Averaging (OWA) Method to the Caspian Sea Conflict." *Stochastic Environmental Research and Risk Assessment* 28(6): 1359–72.

O'Neill, Barry. 1982. "A Problem of Rights Arbitration from the Talmud." *Mathematical Social Sciences* 2: 345–71.

Read, Laura, Kaveh Madani and Bahareh. Inanloo. 2014. "Optimality versus Stability in Water Resource Allocation Negotiations." *Environmental Management* 133: 343–54.

Sechi, Giovanni M. and Riccardo Zucca. 2015. "Water Resource Allocation in Critical Scarcity Conditions: A Bankruptcy Game Approach." *Water Resources Management* 29(2): 541–55. doi: 10.1007/s11269-014-0786-9.

Sheikhmohammady, Majid, D. Marc Kilgour and Keith W. Hipel. 2006. "Negotiations over the Caspian Sea: Preliminary Graph Model Analysis." In *Proceeding of Group Decision and Negotiation (GDN) 2006*, edited by C. Seifert and C. Weinhardt C. Karlsruhe, Germany, 100–102.

Sheikhmohammady, Majid and Kaveh Madani. 2008a. Sharing a Multi-national Resource through Bankruptcy Procedures. Proceeding of the 2008 World Environmental and Water Resources Congress, Honolulu, Hawaii. Edited by Babcock R. W. and Walton R., American Society of Civil Engineers; 1–9. doi: 10.1061/40976(316)556.

————. 2008b. Bargaining Over the Caspian Sea—the Largest Lake on the Earth. World Environmental and Water Resources Congress; 1–9. doi: 10.1061/40976(316)262

————. 2008c. A Descriptive Model to Analyze Asymmetric Multilateral Negotiations. Proceeding of 2008 UCOWR/NIWR Annual Conference, International Water Resources: Challenges for the 21st Century and Water Resources Education, Durham, North Carolina.

Teasley, Rebecca L., and Daene C. McKinney. 2011. "Calculating the Benefits of Trans-boundary River Basin Cooperation: The Syr Darya Basin," *Water Resources Planning and Management* 137(6): 481–90.

Chapter Four

FLEXIBLE DESIGN OF WATER INFRASTRUCTURE SYSTEMS

Melanie Wong Turlington, Richard de Neufville and Margaret Garcia

Abstract

Future water needs and sources of supply are often highly uncertain. This is especially the case in the many areas under stress from climate change and rapid, variable urban development. Innovations in water treatment technologies, and price volatility in energy and other resources combine to aggravate the uncertainties concerning the risks and opportunities associated with planning investments in water infrastructure systems. The problem is that conventional techniques for the planning and design of water supply systems do not adequately deal with these important uncertainties. This chapter presents "Flexible Design" as an approach to deal with the situation, and demonstrates its effectiveness in a case study, inspired by Singapore, of a long-term design of the water supply system for a major city. The results show that Flexible Design can significantly increase overall performance compared to conventional design methods—a 15 percent increase in expected present monetary value, while decreasing downside risk and increasing upside opportunities. Flexible Design achieves these improvements by carefully analyzing possible futures using Monte Carlo simulation and designing the system with modules that managers can implement as and when needed. Flexible Design is shown to be an effective tool and is increasingly used in infrastructure development.

Introduction

The future is uncertain. Technologies, needs, policies, economies, and environments change frequently. Some changes are quantifiable, and we attempt to make best estimates of what will happen, but we will always encounter game-changing events and unanticipated new technologies.

When decision-makers consider needs for future water infrastructure, they often use a conventional, deterministic approach. They assume that current conditions will not change and that we can safely rely on a forecast of long-term requirements. However, the water supply and demand system is complex. Natural, societal, and political forces interact to drive changes in both supply and demand over time. Our ability to predict future conditions is low when dealing with such complexity, leading to forecasts that are rarely accurate.

For projects in transboundary watersheds or other multi-stakeholder water systems, reaching consensus among stakeholders on highly uncertain future conditions is a key challenge; there is a tendency to downplay uncertainty to make the problem more tractable and to advocate for a vision of the future convenient to dominant interests (Lempert and Kalra 2011). There are many challenges to transboundary water planning and management including competing interests, the presence of multiple boundaries and levels, uncertain hydrological conditions and water demands. Water diplomacy addresses these challenges by drawing insights from theory and practice to develop methodologies to manage the system adaptively (Islam and Susskind 2013). Flexible Design cannot answer all of the challenges of multi-stakeholder water management; however, in the spirit of water diplomacy it offers an adaptive approach to planning and designing water infrastructure that recognizes the full range of uncertainties, both hydrological and human.

We need an approach to design that acknowledges uncertainty and attempts to learn from experiences by adjusting our decisions over time. This is what a Flexible Design approach aims to do. Flexible Design is a new approach to infrastructure planning and design that broadly has the potential to be an effective tool to address uncertainties without being paralyzed by them. It recognizes that the future contains many possibilities, risks, and opportunities. It aims to protect investments from downside losses, such as financial vulnerability to shrinking water demand. It also enables decision-makers to take advantage of upside opportunities, such as installing more efficient or cheaper technology (De Neufville and Scholtes 2011). Explicitly acknowledging and planning for uncertainty can encourage tolerance among stakeholders with competing views, enabling projects to move forward without stakeholder agreement on exact future conditions (Peterson et al. 2003). This offers an advantage for projects with multiple stakeholders, including those in transboundary watersheds. This chapter defines Flexible Design for water resource systems, and illustrates its use and value through a case analysis.

Recognizing Uncertainty in Planning

Planners for water infrastructure systems should consider all the factors that affect water supply and demand. Their objective is to align their decisions with good knowledge of these factors. However, there are multiple sources of uncertainty: lack of data; inability to predict future demand accurately; and uncertainty about natural and physical processes of the water cycle in the face of changing environmental, economic and technological settings (Coates et al. 2012). We now need to acknowledge the uncertain effects of climate change, which can affect water supply and demand through variability in the mean temperature and more extreme events, such as floods and droughts. Changes in land use, groundwater levels, and urbanization also drive variability in a water resources system. As the demand for water resources and the vulnerability of aging infrastructure increases, the effects of natural disasters may become more disruptive (Islam and Susskind 2013). We also do not know how individuals and institutions will react to market and technical innovations. Finally, competing stakeholders with conflicting water demands have different levels of political will, financial resources and institutional practices (Tropp and Joyce 2012).

A population and its water needs directly affect demand. Thus, unpredictable population growth rates can significantly impact forecasts. For example, consider the dramatic changes for Delhi. In 2007, the United Nations (UN) forecast its 2025 population as 22.5 million. Yet by 2011, Delhi already comprised 22.7 million people. The UN adjusted its 2025 forecast to 32.9 million, 46 percent higher than its forecast four years previous (UN, 2011). Conversely, it is possible to overestimate future population. After the fall of the Iron Curtain, for example, several cities in East Germany built large sewage systems to cope with anticipated significant population growth, which did not occur. The oversized facilities then led to very high wastewater treatment tariffs (Peters et al. 2011). Despite our best efforts, there is still great uncertainty in population forecasts.

Water equipment technology is becoming more energy efficient (Migliore 2010). Reverse osmosis technology for seawater desalination is a notable example. In 1995, it delivered potable water for about $0.80/m³ (Malek et al. 1996). Ten years later, Singapore's Tuas Seawater Desalination plant advertised a cost of $0.39/m³ (Black and Veatch 2006).

Conventional Approaches to Water Resources Planning

Conventional approaches are "predict-and-plan." They do not systematically consider uncertainty (Quay 2010). They are typically predicated upon a specific set of assumed possibilities. Examination of the textbooks in water resources documents the case (Table 4.1). The words "uncertainty" or "risk" rarely appear or are only referenced in the context of supply risk and rainfall data, structural failure risk, or environmental risk assessment. The engineering texts largely focus on standardized engineering practices

Table 4.1 Textbook literature in water resources engineering, planning and management.

Area of study	Author, date	Uncertainty in		Provides support for		
		Water supply	Water demand	Scenario planning	Decision analysis	Flexibility valuation
Water resources engineering	Mays, 1996	X	X		X	
	Prakash, 2004	X				
Water resources planning and management	Petersen, 1984	X				
	Stephenson, 2003	X	X			
	Loucks and van Beek, 2005	X	X		X	
	AWWA, 2007	X		X		
Flexible design	de Neufville and Scholtes, 2011	X	X	X	X	X

based on historical rainfall and flow data, assuming that the local hydrology will not change, and that its statistical characteristics will remain stationary (Milly et al. 2008).

The planning texts significantly focus on forecasting demand using historical data on consumption, production, weather, demographics, conservation and rate structure and pricing—yet due to great uncertainty about these values in a constantly changing environment, the forecasts are frequently wrong. Water supply planning methods acknowledge the uncertainty generated by variable streamflow and have traditionally dealt with this uncertainty by using a streamflow model fit to replicate the streamflow statistics of historic observations, to generate synthetic streamflow sequences (Thomas and Fiering 1962; Salas et al. 1980), or to sample from a distribution of streamflow to conduct a Monte Carlo simulation (Yeh 1985). For planning purposes, the outputs of such analyses are often summarized as the maximum yield that can be drawn a specified high percentage of the time (i.e., 95 percent)—this is often termed a safe or firm yield. These terms hide two complicating facts: that hydrologic systems change over time under natural and human influence, and that low-probability streamflow events do occur regularly, if infrequently. Master plans typically project water demand for 20 years but generally do not state the degree of uncertainty or the implications of imprecise predictions (Grayman 2005). In addition, practitioners typically use most-likely parameter values and ignore their uncertainty. Although researchers increasingly grapple with the propagation of parameter uncertainty through hydrological and water systems models (Steinschneider et al. 2012; Vano et al. 2014), these insights do not yet inform design practice. While conventional methods may once have been adequate, they are not sufficient for the current dynamic environment.

Decision Analysis and Scenario Planning

Over time, conventional planning and engineering practice have moved toward formally considering uncertainty through rigorous approaches such as decision analysis and scenario planning, as Table 4.1 suggests. The following review of these approaches to water system planning and design is not intended to be all-inclusive. It focuses rather on approaches employed in practice by utilities or water agencies. Many methods have been proposed to deal with uncertainty in water management but have not been widely used in practice for planning and design. For example, researchers use system dynamics to explore how feedback between technical, hydrological and social systems impact supply and demand in water systems; however, the expansion of model scope is matched by a decrease in model detail, resulting in system dynamics models being poorly suited for informing design or operational decisions (Winz et al. 2008; Mirchi et al. 2012). Multistage programming has also been widely used in water systems operations but has not been used in practice to inform design (Huang and Loucks 2000).

Decision analysis is a rigorous probabilistic method for evaluating decisions under uncertainty. Decision analysis has been used for many years in water resources and sewer planning (Barsugli et al. 2012). Analysts commonly use its decision trees and influence diagrams to visualize the alternatives, decision objectives, and uncertainties (WUCA 2010). Its great merit is that it explicitly models possible future developments, looks to define the probability that these might occur over multiple stages, and calculates the possible results. Even though this is complicated, modern computer software can rapidly complete most of the calculations involved. Unfortunately, however, decision analysis has a profound conceptual drawback that makes it inappropriate for most decisions concerning the design of water resources (and indeed of any form of public services). The difficulty with decision analysis in practice is that it leads to an optimal strategy on the assumption that neither the public nor the planners care particularly about risks, that everyone is "risk-neutral." Decision analysis "plays the averages" and leads to the plan or design with the highest average performance, even if has significant downside risks. This is generally unacceptable: planners and the public generally prefer to buy some insurance, or pay something extra, to avoid catastrophic consequences (Herman et al. 2014). Decision analysis does not meet this requirement.

Several American water utilities have used decision analysis to inform investment decisions. These include the Seattle Public Utilities, Tampa Bay Water and the Portland Water Bureau. Seattle Public Utilities, for example, used decision analysis to assess the impact of uncertainties in demand forecasts, source development, new regulations and climate change (WUCA 2010). The utility supplies water to 1.45 million people in its greater metropolitan area and draws the majority of its water supply from two rivers (Seattle Public Utilities 2007). While supplies are sufficient to meet near-term demands, projected decreases in snow pack and summer streamflows because of climate change point to potential water stress. The decision analysis examined a series of supply alternatives such as development of the Snoqualmie Aquifer, new pump stations to enable reservoir operation changes, and reclaimed water. The results informed both the selection and timing of alternatives and led to the decision to change reservoir operation policies and delay development of additional supplies (WUCA 2010).

Scenario planning identifies critical uncertainties and driving forces and then develops a range of potential futures. It helps coordinate decision-making with concrete actions. For example, Royal Dutch Shell used the approach in the 1970s to identify scenarios that would influence the price of oil. This preparation helped guide the company's decisions within the next decade and facilitated knowledgeable management responses during an oil price crisis (Means et al. 2005). A limitation of scenario planning is that planners must predetermine the scenarios or futures.

Two American water utilities, in Denver and Phoenix, have used scenario planning. In 2008, Denver Water created an Integrated Resource Plan that combined traditional scenario planning with prioritized factors that would contribute to future uncertainty (Quay 2010). Scenario planning allowed Denver Water to identify and rank critical uncertainties and facilitated "out-of-the-box" thinking in defining a spectrum of potential impacts associated with reductions in the amount of water available from the Colorado River (Denver Water 2009). Stakeholder participation included discussions of scenarios and priorities with the Denver Water board, customers and regional water providers. Denver Water grouped factors into five possible scenarios and then conducted future analyses that explored the range of potential climate-change impacts. Finally, they identified signposts that would alert planners when a certain scenario is likely to occur, such as reservoirs reaching certain levels, indicating that water demand will likely exceed available supplies.

Phoenix Water Services Department (PWS) used scenario planning to examine possible impacts of climate change on normal and drought conditions. Phoenix has historically experienced highly unpredictable wet and dry periods, ranging from 10 to 100 years long (Quay 2010). Tree-ring research shows that 20–30 year droughts repeatedly occurred in the region over the past thousand years (Woodhouse et al. 2010). In its 2005 plan, PWS defined ranges of future possibilities for three factors: delivery of surface water, regional growth and development and consumers' water-use behavior. Combined with a range of drought conditions, spatial growth patterns and levels of consumer use, PWS generated 144 scenarios of supply, demand, and the resulting water budgets. Stakeholder involvement included regional water suppliers, various interest groups and the city council. Through their analyses, they identified a portfolio of robust strategies that would work both in the short term and in worst-case water shortages over a 30-year drought. They estimated trigger (or tipping) points at which they would need to mandate demand reductions or provide more water supplies. Since PWS adopted the 2005 plan, it has monitored trends, growth and demand, and has adjusted trigger points based on new information (Quay 2010).

Scenario planning has the advantage of getting planners and other stakeholders to consider the range of possible futures, and of focusing attention on monitoring developments to identify trigger or tipping points that call for some change in plan or practice. In short, scenario planning gets planners thinking about what might happen and when to take notice. However, this process does not examine or evaluate in any detail what might be done. Thus, scenario planning is not sufficient for design purposes.

Researchers have developed analytical methods such as Robust Decision-Making, the Scenario Neutral Approach, and Decision Scaling to deal with the uncertainties of non-stationary hydrologic systems (Lempert and Groves 2010; Prudhomme et al. 2010; Weaver et al. 2013). These methods focus on understanding the vulnerability of alternatives to potential future conditions. They combine the concept of multiple possible futures of scenario planning with Monte Carlo simulation to sample across distributions. Robust Decision-Making, for example, assigns distributions for streamflow within a scenario but not to the scenarios. The analysis then leads to distributions of reliability, cost or other metrics for each alternative scenario combination. Decision-makers might use this information to judge which alternatives perform well enough over a range of future conditions. However, scenario-based design does not account for the way water management can adapt to changes. Although these methods can consider the sequential implementation of infrastructures and policies triggered by changing conditions, the results are not linked to the design of individual infrastructures. Hence, this extension of scenario planning is similarly insufficient to inform design.

Adaptive Management

Adaptive management aims to help managers reduce uncertainty and develop strategies that can respond to unanticipated events. It involves observing the effects of a project or policy with a view to improving future decisions. However, it goes beyond simply reacting to change by implementing a structured process. Adaptive management emphasizes "learning and adjusting" by developing alternative system models and taking note of observable triggers or tipping points (Medema et al. 2008). It can be either passive or active: passive approaches reduce uncertainty by using a single design or plan and adjusting hypotheses over time; active approaches use multiple designs to test competing hypotheses (Walters and Holling 1990; National Research Council 2004).

Adaptive management began in the United States in the 1970s, applied to the management of natural resources and ecological planning (Walters and Hilborn 1978). The US Army Corps of Engineers (USACE) started to use the approach in the 1990s, primarily to restore natural systems affected by development. The USACE applied adaptive management to the Kissimmee River and the Everglades Restoration Project in Florida, and the Assateague and Poplar Islands of Maryland. It proposed adaptive management for improving navigation on the Upper Mississippi by modifying the construction schedule for locks based on observable demand changes (Galloway 2006). It has prioritized further demonstration projects to test and validate adaptive management (USACE 2009). Overall, however, the USACE has not emphasized systematic exploration of uncertainty drivers and their implications for the planning process.

The Murray-Darling Basin Authority (MDBA), responsible for managing water resources for an agricultural region in southeastern Australia, has incorporated adaptive management into their decision-making and planning processes. It aims to balance the environmental health of the basin through sustainable diversion limits (SDLs) to conserve streamflow. It monitors and evaluates progress toward its goals and uses these results to feed into future improvements. Specifically, it reviews the SDLs over time and adjusts them as needed. The MDBA also uses adaptive management on a local level, which includes actions such as releasing floodwater to a forested area to relieve regional flooding (Murray-Darling Basin Authority 2011).

The great merit of adaptive planning is that it recognizes the value of modifying plans in the face of new understanding or knowledge. Given that this is how we live our day-to-day lives, this should not seem like much of an advance. However, it truly is a significant achievement in the context of conventional planning, which suggests a fixed plan! Yet adaptive planning cannot be the whole answer to designing in the context of uncertainty. It is deficient on two counts. First, it does not lead to specific designs. Most importantly, while adaptive planning encourages adaptation of existing designs, it does not guide planners to create original designs that facilitate adaptation. This is a crucial difference: it is one thing to tell planners to change if things are going wrong and quite another to provide them with the facility to make desirable changes easily.

Flexible Design

Flexible Design provides a complete approach to the development of water resources in the context of inevitable uncertainties (see Table 4.1). It builds upon and supplements existing approaches, incorporating features that together provide the range of desirable elements for effective planning and development (De Neufville and Scholtes 2011).

By building upon existing methods, Flexible Design can be readily applied in practice. Specifically, Flexible Design

- Recognizes multiple scenarios and the need to adapt to events, as does adaptive planning;
- Analyses uncertainties similarly to decision analysis, but without the need for complicated decision trees; and
- Presents the range of possible consequences of alternative design strategies, allowing planners and stakeholders to choose a proper balance of risk and rewards.

The Concept: Flexible Design directly addresses central development issues of deciding what, where and when to build. Recognizing the desirability of adapting the infrastructure to the inevitable uncertainties in water supply and demand, it designs capabilities into the system to facilitate easy reaction to new risks and opportunities. The capability to create insurance against uncertainties and to identify ways to adapt water infrastructure to changing conditions is a central feature of this approach to planning and design.

By way of illustration, consider an urban water system that we expect will eventually have to serve a much greater demand. A Flexible Design will provide initial expansion, less complete than what we expect to require, but with the capability to expand if, how, when and where needed. Such a design might call for procuring and preparing land to build additional facilities. If the demands grow rapidly, we are ready to expand in time. If growth is slower than expected, we avoid unnecessary expenses. If treatment technology becomes much cheaper, we can adopt this opportunity and avoid being stuck with obsolete production. In short, Flexible Design lowers the risk of overbuilding (one win) and enables us to take advantage of new opportunities (another win), while reducing the initial commitment of capital. Flexible Design thus leads to a great formula: "greater value for lower cost!"

A Flexible Design approach encourages learning from the information we gain over time and sets us up to be strategic with our next move. The concept is similar to that of a game of chess: we consider the range of possible futures and develop our responses to these situations, while moving one step at a time as we observe actual changes. This strategic approach is general. It can be applied at different spatial scales (national, state, local) and across decision-making capacities (planning, design and management). For a comparison to the methods reviewed, see Table 4.2.

The development of Flexible Design involves two conceptually different efforts:

- One aims to reduce possible risks and losses. In essence, it buys insurance by creating ways to avoid bad consequences. In financial terms, it creates "puts."
- The other seeks to increase possible gains from new opportunities. In financial terms, it establishes "calls."

Real Options Analysis

Creating flexibility in design is similar to developing "real options" in finance. An option provides "the right, but not the obligation" to do something, such as to buy or to sell an asset (a call or a put). It is "real" from a financial perspective because it refers to doing

Table 4.2 Comparison of methods for planning and design under uncertainty.

	Decision Analysis	Scenario Planning	Adaptive Management	Flexible Design
Representation of uncertainty	Probabilities	Scenarios	Not explicit, but anticipates change	Probabilities, anticipates further change
Strengths	Provides a structured method to consider uncertainty	Easy to implement, increases awareness of uncertainties	Encourages change in response to changing conditions	Plans for change in response to changing conditions, informs design
Weaknesses	Cumbersome with multiple dimensions of uncertainty, focuses on mean performance	Only a small number of scenarios explored	Does not inform design decisions	Not all infrastructure projects amenable to flexibility

something with a physical asset (such as expanding it or shutting it down) in contrast to plain options that deal only with financial transactions (see, for example, Trigeorgis 1996).

Real options analysis provides a way to value flexibility, and thus to judge what kinds of flexibility are justified or desirable. It rests upon sophisticated analyses derived from conditions that economists believe to prevail in fully functioning transparent markets. It is very powerful when those conditions apply (Jeuland and Whittington 2014). However, the practical use of real options for infrastructure planning is limited, because the products of water resource systems do not typically trade in markets.

Sources of Flexibility in Design

In terms of infrastructure design, there are three main sources of flexibility: the size of the facility; its functions or capabilities; and its location. Flexibility in infrastructure size facilitates adding capacity as demand increases. Designers may achieve this by simply having extra space for an addition. More sophisticatedly, they might pre-position some essentials for future additions. They might, for example, install a foundation to enable a plant extension to proceed without disturbing the operation of the existing plant. Or they could design a dam so that they could raise its crest to retain more water. Modular design is typical way to achieve flexibility in capacity. Many desalination plants are now modular—for example the Ramble Morales Desalination Plant in Almería, Spain, as Figure 4.1 shows. Similarly, configurations of smaller dams and reservoirs can serve as a modular alternative to large reservoirs (Jeuland and Whittington 2014). Phasing and modularity are not new strategies in infrastructure projects, but setting a plan for construction over time is not itself Flexible Design. In Flexible Design, simulation models are used to determine the size of the first phase and to test decision rules for implementation of future phases, but the timing of future phases is not decided upfront.

Figure 4.1 Ramble Morales Desalination Plant in Almería, Spain (David Martínez Vicente 2005).

Flexibility in function refers to the possibility of adding to or changing the functions of a facility. For example, we might consider changing the level of water purity from a desalination or wastewater treatment plant. We might, for example wish to upgrade production from agricultural to potable water.

Flexibility in location indicates the ability to add capacity or capability to a system at different sites throughout the service area. Insofar as the distribution of services requires effort—such as pumping if the system is not gravity-fed—placing infrastructure near the customers has the potential for reducing distribution costs that could contribute significantly to overall operational costs. For example, building smaller water-reclamation plants near customers as a city grows can decrease both the operational costs of pumping waste and reclaimed water and the capital cost of the collection and distribution systems. Beyond costs, a distribution of assets can increase the reliability of the system by decreasing its vulnerability to one-point failures.

Drivers of Value

Three main features drive the overall present cost of any system, and thus affect the desirability of flexibility from an economic point of view. These are: the time value of money; the degree of economies of scale; and the learning rate for production. Strong economies of scale favor big, inflexible infrastructure. The time value (or cost) of money and the learning rate of production favor the use of flexible, modular designs that we roll out incrementally as the need arises.

The time value of money means that money spent now is more costly than the same amount paid later on—think of the financing costs we can defer if we manage to spend money far into the future. This reality favors Flexible Design, which phases in additional capacity and capabilities later on, when we clearly need them. From an economic perspective, there is great value in delaying investments until required.

Economies of scale describe the phenomenon that, *when facilities are operating at full capacity*, it is cheaper per unit capacity to build and operate larger facilities than smaller ones. Economies of scale exist in all kinds of production facilities, in particular for water supply infrastructure. This fact drives the impulse to design and build large, inflexible infrastructure well in advance of demonstrated need. The problem is that the "build big and take advantage of economies of scale" approach is risky. Economies of scale assume the facilities are operating around capacity. When this is not the case, economies of scale can become diseconomies of scale! When we correctly prorate the cost of unused capacity over the sub-capacity amounts being produced, average costs can rise quickly. When demand is lower than expected, the cost per unit of production for a larger plant can become very high and easily become greater than the cost of smaller plants. This risk counteracts some of the value of economies of scale. In short, the risk associated with economies of scale favor Flexible Design.

Learning rates refer to the common observation that, as we produce more items, the cost to build subsequent items becomes cheaper. Thus, unit costs decrease with increasing production. While the literature labels the phenomenon as learning, it may occur because companies alter design elements, introduce new technologies, or find ways to eliminate waste in the production process. Thus, the literature also refers the phenomenon in terms of progress curves, experience curves and learning by doing. A common formulation is that the unit cost decreases by a fixed percentage (the learning rate) each time the amount produced doubles. Researchers have quantified learning rates from data on cost reduction with experience for many industries. Since energy can be such a large factor in the treatment and distribution water, the cost-reduction data in that field may be especially pertinent (McDonald and Schrattenholzer 2001).

Limitations

There are limitations to every methodology. One significant limitation of Flexible Design is that some types of infrastructure are better suited for Flexible Design than others because they can be easily designed to be built in modules or phases. Another potential limitation is that Flexible Design requires assigning distributions to each of the uncertain variables, which can be challenging in cases of deep uncertainty. The main value of Flexible Design is as a tool for incorporating uncertainty explicitly in an analysis rather than ignoring uncertainty. However, determining how much uncertainty to incorporate and what dimensions of uncertainty to include (population growth rates, energy prices, rainfall, etc.) is a significant challenge, and no definitive answer exists. Therefore,

Flexible Design aids designers and planners in considering uncertainty but cannot cover all aspects of uncertainty.

Flexibility Analysis

Monte Carlo simulation is the central tool for flexibility analysis. It is both very simple and very powerful. It has the great merit of producing not only an objective measure of the value of any design, but also a distribution of that value so that planners and stakeholders can appreciate the risks of any design—and especially the risk reduction achieved by Flexible Design. Monte Carlo simulation works with any model of the performance for the proposed design. A wide range of software and programming languages can be used to build a simulation model to compute the distribution of costs and benefits over time. The analysis joins the model of performance with the available estimates of uncertainty about key parameters, such as the assumed maximum, minimum, and most likely levels of water consumption. The Monte Carlo simulation then simultaneously samples all these distributions proportionately, thus creating one scenario that it applies to the model of performance to obtain its value. The simulation repeats this process as many times as desired to derive an estimate of overall performance of that design. This happens very quickly—a laptop can simulate thousands of scenarios in minutes. The case study illustrates the process in detail.

Example Applications

While Flexible Design is relatively new approach to water resources, it has been piloted in several cases. We summarize four examples below.

Sydney, Australia: An analysis investigated flexibility in design of water supply facilities for Sydney (Borison and Hamm 2008). It compared two fixed strategies and one flexible strategy. The fixed strategies committed to either building desalination plants or importing water. The flexible strategy reflected the possibility of recycling water. This analysis determined that the flexible strategy both increased expected value by $400 million (or 20 percent over the better fixed strategy) and reduced risk by $500 million, measured by standard deviation—more value with less risk!

Hamburg, Germany: Researchers at the University of South Florida examined flexible strategies for the design of urban drainage systems (Eckart 2012). As these facilities have an operational life up to 80 years, predictions for future drainage are highly speculative. Considering various future scenarios, the group generated inflexible and flexible designs. They then filtered these alternatives to identify the eight most promising designs with varying levels of flexibility (due to modular design, decentralized structure, real time control and scalability). Measuring value in terms of life-cycle costs, they found that the optimum solution was a flexible design that minimized regret with regard to performance, effort of change and range of change.

South Australia: Engineers in Singapore investigated the relative value of a conventional fixed design of production facilities compared to an innovative flexible design. The fixed design centered on the creation of a single large plant enjoying potential economies of scale if demand eventually grew sufficiently large. The flexible design featured modular plants that could be both implemented, as actual overall demand grew, and distributed across the region to where the demand was occurring. Although from the engineering perspective the smaller production modules appear inefficient compared to a large plant, the flexible design greatly outperformed the big fixed design on overall economic grounds. The flexible design deferred or avoided expenses and cut distribution costs by providing services where needed (Cardin et al. 2015).

Flexibility Analysis Method

The essence of the flexibility analysis is to investigate system performance under uncertainty, and over time, using Monte Carlo simulation to track the consequences for performance of strategies for adapting the design to actual conditions. The analysis carries the distribution of uncertainties throughout, thus providing distributions of value for each strategy considered. The process consists of the five following steps.

1. *Create a model of performance over time.* The model includes equations tracking changes in system performance in response to changes over time. The model calls upon formulas and distributions to obtain the appropriate values of performance, for any time and scenario, as a combination of uncertain parameters.
2. *Identify performance metrics desired for comparison of alternatives.* Given the importance of the time value of money, the net present value (NPV) is an obvious metric to be used. The model can, however, calculate a variety of other metrics as desired.
3. *Identify key inputs, uncertainty distributions and development strategies or paths.* Key inputs are variables influencing the performance metrics. Variables that are uncertain over the planning horizon are characterized by distributions. Uncertainty in the specification of fixed parameters can be evaluated using sensitivity analysis. Development strategies or paths are implementation approaches. They can be planned sequences of infrastructure development or, in the case of flexible paths, a set of rules specifying the conditions for implementation of infrastructure.
4. *Simulate performance using Monte Carlo simulation.* The simulation will sample each of the several uncertain parameters, thus creating a scenario for the model to analyze. By sampling proportionately to the distributions, the simulation generates a distribution of the performance. This leads both to the calculation of average value of performance and to an assessment of the downside risks (and of upside opportunities).
5. *Display performance using a multiple criteria table and target curves.* The flexibility analysis recognizes that decision-makers need to consider both overall expected performance *and* the risks and opportunities associated with alternative designs. The analyst's duty is then to make these tradeoffs evident so that stakeholder can make an informed judgment about the relative value of average and extreme performance.

Detailed Case Study

This chapter describes an example framework of a Flexible Design approach. It acknowledges that with limited knowledge about the future, engineers and planners must decide what technology to invest in, how to size facilities and when to build infrastructure. This framework intends to be applicable across a wide range of water resources systems.

The case is representative of a large number of metropolitan areas. Many rapidly growing cities have limited land and opportunities to create reservoirs. They must plan and develop their water supply in the face of great uncertainties concerning: size of the population; associated demand for water; energy and other costs of supply and distribution; and improvements in technology. This case is loosely based on Singapore, but is not a precise representation of that situation. It indicates how flexibility analysis proceeds, and what it can accomplish.

The model addresses future uncertainties, including the annual population growth rate, per-capita water use and the operating cost of desalination plants. It also considers financial benefits from economies of scale and learning effects. The methodology includes setting up the resource model, identifying system performance metrics, identifying key inputs and uncertainties and choosing development paths. Then the model uses a discounted cash flow Excel® model to define optimal development strategies that are flexible and responsive to future changes.

Background

Singapore is an island nation, densely populated with over 5 million people and highly constrained in land and natural rainfall (Tortajada 2006). The Public Utilities Board (PUB) is the national agency responsible for managing the planning, design, construction and management of water infrastructure. Singapore draws its water supply from four sources, known as "taps": imports from Malaysia; reservoirs; NEWater (high-grade reclaimed wastewater); and desalination plants (reverse osmosis of seawater). Table 4.3 presents the allocation of supply in 2010 and 2060 (aspirational) based on PUB's annual reports (PUB 2010).

The problem is to supply enough water at the least cost, to meet the objective of eliminating imports from Malaysia. The idea is to develop a design and implementation strategy for deciding what to build and when. Note the design tensions inherent in this situation:

- Economies of scale motivate the desire to build large treatment plants—but these may be long overbuilt if demand rises slowly, and they may lock in obsolete technology;
- Smaller additions to capacity may have higher unit costs—but these costs will diminish with learning, and later implementation can incorporate technological advances.
- The time cost of money impels delaying investments and building according to need.

The case study compares two approaches: the conventional approach, which designs around the most likely evolution of demand, supply and cost; and Flexible Design, which

Table 4.3 Sources of Singapore water (based on PUB 2010).

Tap	Share of Total Supply (%)	
	2010	2060
Import	40	0
Reservoir	20	20
NEWater	30	50
Desalination	10	30

Table 4.4 Assumed current fixed and operating costs for case analysis.

Tap	Cost, S$ per cubic meter	
	Fixed	Operating
Import	–	0.003
Reservoir	1,329	0.25
NEWater	1,351	0.30
Desalination	1,818	0.49

adjusts the pace and choice of development according to how the scenarios actually unfold. The flexible approach thus generates design sequences appropriate to each scenario. There could be as many designs as scenarios, but in practice several sequences of size and type of developments will suit several scenarios.

System Data

Table 4.4 shows the fixed (new construction) and operating costs for the four water sources used in the analysis. These derive from historical data, current literature and local press reports. The analysis assumed that the existing sources of water (the four "taps": import, reservoirs, NEWater, and desalination) would continue to fulfill the current daily demand of 1.7 million cubic meters (MCM). It also assumed that the quantity of imported water would decrease linearly to 0 percent by 2060.

Investment Costs

Two patterns characterize the cost of building capacity: economies of scale and learning (De Neufville and Scholtes 2011). Economies of scale describe the phenomenon that it is cheaper per unit to build larger facilities, according to Equation 1 (De Neufville and Scholtes 2011). The case study used historical data to develop the parameters A and K, shown in Table 4.5 (Wong 2013).

Average cost of capacity $= K \, (capacity)^{A-1}$ (1)

Table 4.5 Assumed economies of scale, learning rates and associated coefficients.

Tap	Learning rates		Economies of scale	
	L (%)	Slope, B	A factor	K (10^3)
Reservoir	5	−0.074	0.60	164
NEWater	10	−0.152	0.65	87
Desalination	20	−0.322	0.70	59

Learning rates describe how unit costs decrease with increasing experience: conventionally by reducing costs by a certain percentage as total capacity doubles according to Equation 2 (De Neufville and Scholtes 2011).

$$U_i = U_1 i^B 1 \tag{2}$$

where U_i is the production cost of the ith unit and U_1 is the initial cost. B is a function of the learning rate L determined by empirical observation, Equation 3 (De Neufville and Scholtes 2011).

$$B = \frac{\ln(100\% - L\%)}{\ln(2)} \tag{3}$$

The formula for the operating cost over time, O_T, evolves from the above in relation to the base operating cost O_B:

$$O_T = O_B i^{\frac{\ln(100\% - L\%)}{\ln(2)}} \tag{4}$$

The learning rates used in this analysis considered the relative development of each technology. As desalination has the highest potential to become more efficient through further research and development, it had the highest learning rate, as Table 4.5 shows. The model used the economies of scale factors and learning rates to calculate the fixed cost for new infrastructure over the planning period.

Step 1, create the model: The example analysis used a simplified screening model developed in Excel® that accounted for water demand, supply and costs from 2010 to 2060.

Step 2, identify performance metric: The analysis focused on net present value of the cost to build and operate the water infrastructure over the 50-year planning period of 2010–2060. The discount rate was 10 percent, which includes local inflation.

Step 3, identity inputs, uncertainties and development strategies: The demand parameters included the annual population growth rate, per-capita water usage, and a ratio of domestic to commercial water demand. The supply parameters included the land availability, fixed costs and operating costs of each water supply source. The model characterized uncertainty distributions (Table 4.6) for the

Table 4.6 Uncertainty distributions assumed.

Major Uncertainty	Distribution	Features
Population Growth, annual	Normal	Mean = 2% Standard deviation = 0.5%
Daily per capita use, cu. m.	Normal	Mean = 0.3 Standard deviation = 0.04
Desalination OPEX, 2060, S$	Weibull	Shape = 2.00 Scale = 2.00 Shift = 0.10

Table 4.7 Increments of capacity.

Tap	Daily Capacity, MCM	
	Small	Large
Reservoir	0.15	0.30
NEWater	0.11	0.22
Desalination	0.17	0.34

- annual population growth rate based on historical UN data (UN Data 2011);
- per-capita water usage, expected to double by 2060, to 3.4 MCM; and the
- operating cost for desalination plants in 2060, expected to be S$0.40 cu. m., based on a Weibull distribution skewed toward lower values, on the assumption that research and development efforts will lower this cost.

To examine the value of building increments of facilities in different sizes, the analysis considered two sizes, small and large. The capacity of the small modules was the same as recent investments in the technology. The capacity of large modules was double that of the small modules (Table 4.7).

The model automatically determined the build schedule between 2015 and 2060 in five-year periods. It could call for up to three facilities in each period, and spread out the timing and type of built facilities, making use of IF statements in the spreadsheet. The build schedule ensured that the total water supply met or exceeded water demand in each period. It used either small or large increments of capacity.

The analysis calculated the performance of six general development approaches. First, it created conventional designs based on deterministic values for all parameters. These established two benchmarks for establishing the value of flexible designs. The flexible development strategies also created build schedules based on the actual scenarios generated by the Monte Carlo simulation. Each scenario generated its own build schedule and measures of performance. One of the flexible strategies required the creation of one plant every ten years. This strategy provided insight into the long-term benefits of investing in desalination technology and is termed "forced desalination" in this analysis. Table 4.8 describes the six inflexible and flexible development paths.

Step 4, apply Monte Carlo simulation: Using the @RISK® software add-on to Excel®, the analysis ran five thousand iterations on each design approach and calculated NPV for each iteration.

Table 4.8 Description of inflexible and flexible development paths.

Development path	Path #	Increment
Inflexible	1	Small
	2	Large
Flexible	3	Small
	4	Large
Flexible with forced desalination	5	Small
	6	Large

Table 4.9 Multiple criteria analysis with mean, P5, P95, and mean value of flexibility.

Development path	Path #	Facility size	Value (billions of S$)			
			Mean	P5	P95	Flexibility
Inflexible	1	Small	6.03	3.96	9.16	–
	2	Large	5.54	3.68	8.33	–
Flexible	3	Small	5.52	3.6	8.41	0.51
	4	Large	4.69	3.07	7.14	0.85
Flexible + Desalination	5	Small	5.7	3.76	8.7	0.33
	6	Large	4.84	3.2	6.54	0.7
Best path			**4**	**4**	**4**	**4**

Step 5, display performance using table and target curves: The simulation generated distributions of performance for each design approach. These led to the calculation of both average and extreme values, such as the P5 and P95. The comparison between the greater average values of the flexible designs to that of the fixed designs determines the value of flexibility (Table 4.9).

As Table 4.9 indicates, the flexible large development path had the lowest cost when measured by mean value, P5 value, and P95 value. It also had the highest mean value of flexibility. This dominant design approach implies an average cost savings of 15 percent over the corresponding inflexible path that used deterministic planning. The flexible scenarios add S$330 to S$850 million in present value terms to the system when compared to their inflexible counterparts.

Target curves present the cumulative distribution of possible values associated with each development path. The x-axis reflects the measure of performance (NPV in this case) and the y-axis contains the probability that the realized performance will be lower than the target NPV. The six target curves in Figure 4.2 show the range of the inflexible and flexible paths' NPV performance value. Since the aim is to minimize the costs, the analysis prefers strategies that are consistently to the left of the other curves.

From the simulated results, the flexible "build large" development path consistently produced the lowest cost and showed stochastic dominance. The second most stochastically dominant development path was the flexible "build large" path with forced desalination.

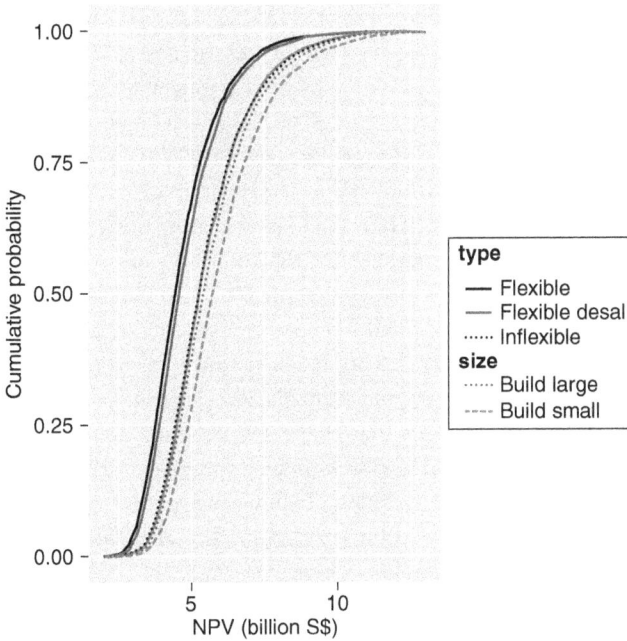

Figure 4.2 Cost target curves for inflexible and flexible development paths.

Discussion

This stylized study of Singapore's water supply system illustrates the flexible analysis method by examining a portfolio of infrastructure options, including reservoirs, NEWater (reclaimed water) plants and desalination plants to reduce the country's dependence on imported water. It analyzed four implementation approaches, including inflexible, large and small flexible and large and small flexible with forced desalination. The study showed that the flexible "build large" development path was the most cost-effective, demonstrating that Flexible Design can offer a cost savings over fixed design. The analysis employed Monte Carlo simulation to sample across distributions of population growth rates, changes in per-capita demand and desalination operating costs. This enabled examination of the full distribution of NPV costs for each implementation approach. From the distribution, it is evident that the flexible design outperforms the inflexible design for both mean NPV and extreme values such as the P5 and P95.

Conclusion

The Singapore case study demonstrates the flexible analysis method and the potential of Flexible Design to generate efficient, cost-saving designs for water resources infrastructure under uncertain conditions. While water resources engineers have long planned for the uncertainty of supply and have developed powerful tools to guide design, other uncertainties have received less attention. However, with growing and mobile populations, rapid technological and economic change and the potential for climate change to drive significant shifts in the hydrologic cycle, it is no longer sufficient to focus solely on supply-side uncertainties. The interaction of natural, societal and political forces drives changes in supply and demand. This complex system challenges our ability to predict future conditions, and history has shown that our forecasts are rarely accurate.

Therefore, the answer is not better predictions but to design for change. Flexible Design uses strategies such as: modular construction and inclusion of extra space or structural capacity to facilitate future change in function; localized expansion, or distributed expansion in order to respond to changing patterns of demand and supply; as well leveraging new technologies. This approach encourages learning from the information we gain over time, delays costly investments and sets us up to be strategic with our next move. The same uncertainties that challenge prediction can also challenge negotiations among stakeholders. When transboundary watershed negotiations fail to reach consensus over projected future conditions, planning can paralyze infrastructure projects. Flexible Design offers a way to move forward with a broad understanding of the range of potential conditions by planning for change.

References

AWWA. 2007. *Manual of Water Supply Practices: M50 Water Resources Planning* (2nd edn.). Denver: American Water Works Association (AWWA).

Barsugli, J. Joseph, Jason M. Vogel, Laurna Kaatz, Joel B. Smith, Marc Waage and Christopher J. Anderson. 2012. "Two Faces of Uncertainty: Climate Science and Water Utility Planning Methods." *J. Water Resources Planning and Management* 138(5): 389–95.

Black and Veatch. 2006. "Black and Veatch-Designed Desalination Plant Wins Global Water Distinction." Accessed February 22, 2016. http://www.edie.net/news/3/Black–Veatch-Designed-Desalination-Plant-Wins-Global-Water-Distinction/11402

Borison, Adam and Gregory Hamm. 2008. "Real Options and Urban Water Resources Planning in Australia." *Water Services Association of Australia*. Occasional Paper No. 20. April.

Cardin, Michel-Alexandre, Mehdi Bourani and Richard de Neufville. 2015. "Improving the Lifecycle Performance of Engineering Projects with Flexible Strategies: Example of On-Shore LNG Production Design." *Systems Engineering* 18: 253–68. doi:10.1002/sys.21301.

Coates, David, Daniel P. Loucks, Jeroen Aerts and Susan van't Klooster. 2012. "Working under Uncertainty and Managing Risk." In *World Water Development Report 4: Managing Water under Uncertainty and Risk*. Paris: UNESCO. http://www.unesco.org/new/fileadmin/MULTIMEDIA/HQ/SC/pdf/WWDR4%20Volume%201-Managing%20Water%20under%20Uncertainty%20and%20Risk.pdf

de Neufville, Richard and Stefan Scholtes. 2011. *Flexibility in Engineering Design*. Cambridge, MA: MIT Press.

Denver Water. 2009. *Solutions Magazine: Saving Water for the Future*. Accessed February 22, 2016. http://www.denverwater.org/docs/assets/DD81F7B9-BCDF-1B42-DBDA3139A0A3D32D/solutions1.pdf

Eckart, Jochen. 2012. "Flexible Urban Drainage Systems in New Land-Use Areas." PhD dissertation. University of South Florida. http://scholarcommons.usf.edu/etd/4033

Galloway, Gerald E. 2009. "Making the Transition: Moving Water Resources Planning and Management into the Twenty-First Century." In *The Evolution of Water Resource Planning and Decision Making*, edited by Clifford S. Russell and Duane D. Baumann, 259–84. Northampton, MA: Edward Elgar Publishing.

Grayman, Walter M. 2005. "Incorporating Uncertainty and Variability in Engineering Analysis." *J. Water Resources Planning and Management* 131(3): 158–60.

Herman, Jonathan D., Harrison B. Zeff, Patrick M. Reed, and Gregory W. Characklis. 2014. "Beyond Optimality: Multistakeholder Robustness Tradeoffs for Regional Water Portfolio Planning under Deep Uncertainty." *Water Resources Research* 50: 7692–7713. doi:10.1002/2014WR015338

Huang, Gui-Hai and Daniel P. Loucks. 2000. "An Inexact Two-Stage Stochastic Programming Model for Water Resources Management Under Uncertainty." *Civil Engineering and Environmental Systems* 17(2): 95–118. doi:10.1080/02630250008970277

Islam, Shafiqul and Lawrence Susskind. 2013. *Water Diplomacy: A Negotiated Approach to Managing Complex Water Networks.* New York: RFF Press.

Jeuland, Marc and Dale Whittington. 2014. "Water Resources Planning under Climate Change: Assessing the Robustness of Real Options for the Blue Nile." *Water Resources Research* 50: 2086–2107. doi:10.1002/2013WR01370.

Lee, Deborah H. 1999. "Institutional and Technical Barriers to Risk-Based Water Resources Management: A Case Study." *J. Water Resources Planning and Management* 125(4):186–93.

Lempert, Robert. J. and David G. Groves. 2010. "Identifying and Evaluating Robust Adaptive Policy Responses to Climate Change for Water Management Agencies in the American West." *Technological Forecasting and Social Change* 77(6): 960–74. doi:10.1016/j.techfore.2010.04.007.

Lempert, Robert and Nidhi Kalra. 2011. *Managing Climate Risks in Developing Countries with Robust Decision Making World Resources Report Uncertainty Series.* Washington, DC: World Resources Report.

Loucks, Daniel P., Eelco van Beek, Jery R. Stedinger, Jozef P. M. Dijkman and Monique T. Villars. 2005. *Water Resources Systems Planning and Management: An Introduction to Methods, Models and Applications.* Paris: UNESCO.

Malek, A., Mohammed. N. A. Hawlader and J. C. Ho.1996. "Design and Economics of RO Seawater Desalination." *Desalination* 105(3): 242–61.

Mays, Larry W. 1996. *Water Resources Handbook.* New York: McGraw-Hill.

McDonald, Alan and Leo Schrattenholzer. 2001. "Learning Rates for Energy Technologies." *Energy Policy* 29(4): 255–61.

Means, Ed, Roger Patrick, Lorena Ospina and Nicole West. 2005. "Scenario Planning: A Tool to Manage Future Water Utility Uncertainty." *J. American Water Works Association* 97(10): 68–75.

Medema, Wieske, Brian S. McIntosh and Paul J. Jeffrey. 2008. "From Premise to Practice: A Critical Assessment of Integrated Water Resources Management and Adaptive Management Approaches in the Water Sector." *Ecology and Society* 13(2): 29.

Migliore, Matt. 2010. "Key Trends in Water and Wastewater Treatment." *Flow Control: Solutions for Fluid Movement, Measurement and Containment.* http://www.flowcontrolnetwork.com/articles/key-trends-in-water-wastewater-treatment

Milly, Paul C. D., Julio Betancourt, Malin Falkenmark, Robert M. Hirsch, Zbigniew W. Kundzewicz, Dennis P. Lettenmaier and Ronald J. Stouffer. 2008. "Stationarity is Dead: Whither Water Management?" *Science* 319: 573–74.

Murray-Darling Basin Authority. 2011. Delivering a Healthy Working Basin: About the Draft Basin Plan. Accessed February 22, 2016. http://www.mdba.gov.au/sites/default/files/archived/proposed/delivering-a-healthy-working-basin.pdf.

National Research Council. 2004. *Adaptive Management for Water Resources Project Planning.* Washington, D.C.: The National Academies Press.

Peters, Christian, Heiko Seiker and Jochen Eckart. 2011. "Handling Future Uncertainties in Urban Drainage Planning Using Flexible Systems: The COFAS Method." SWITCH Conference. Retrieved from http://www.switchurbanwater.eu/outputs/pdfs/W2-1_GEN_PAP_Handling_future_uncertainties_COFAS_Method.pdf.

Petersen, Margaret S. 1984. *Water Resources Planning and Development.* Englewood Cliffs, NJ: Prentice-Hall.

Peterson, Garry D., Graeme S. Cumming and Stephen R. Carpenter. 2003. "Scenario Planning: A Tool for Conservation in an Uncertain World." *Conservation Biology* 17(2): 358–66.

Prakash, Anand. 2004. *Water Resources Engineering: Handbook of Essential Methods and Design*. Reston, VA: American Society of Civil Engineers Press.

PUB. 2010. *Water Works: PUB Annual Report 2009/2010*. Accessed January 15, 2013. http://www.pub.gov.sg/annualreport2010.

Quay, R., 2010. "Anticipatory Governance: A Tool for Climate Change Adaptation." *J. American Planning Association* 76(4): 496–511.

Salas, Jose D., J. W. Delleur, Vujica Yevjevich, and William L. Lane. 1980. *Applied Modeling of Hydrologic Time Series*. Chelsea, MI: Water Resources Publications

Seattle Public Utilities. 2007. *2007 Water System Plan: Our Water, Our Future*. Accessed February 22, 2016. http://www.seattle.gov/util/Documents/Plans/Water/WaterSystemPlan/Plan-2007/index.htm.

Sewilam, Hani and Guy Alaerts. 2012. "Developing Knowledge and Capacity." *World Water Development Report 4*. UN World Water Assessment Programme. Accessed February 22, 2016. http://www.unesco.org/new/en/natural-sciences/environment/water/wwap/wwdr/wwdr4-2012.

Steinschneider, Scott, Austin Polebitski, Casey Brown and Benjamin H. Letcher. 2012. "Toward a Statistical Framework to Quantify the Uncertainties of Hydrologic Response under Climate Change." *Water Resources Research* 48(11): W11525. November 20.doi:10.1029/2011WR011318. http://doi.wiley.com/10.1029/2011WR011318.

Stephenson, David. 2003. *Water Resources Management*. Meppel, The Netherlands: Krips the Print Force.

Thomas, Harold A. and Myron B. Fiering. 1962. "Mathematical Synthesis of Streamflow Sequences for the Analysis of River Basins by Simulation." 459–93. In: *Design of Water Resource Systems*, edited by Arthur Maass, Maynard M. Hufschmidt, Robert Dorfman, Harold A. Thomas, Stephen A. Marglin and Gordon M. Fair. Cambridge, MA: Harvard University Press.

Trigeorgis, Lenos. 1996. *Real Options: Managerial Flexibility and Strategy in Resource Allocation*. Cambridge, MA: MIT Press.

Tortajada, Cecilia. 2006. "Water Management in Singapore." *International Journal of Water Resources Development* 22(2): 227–40.

Tropp, Hakan and John Joyce. 2012. "Water and Institutional Change: Responding to Present and Future Uncertainty." In *World Water Development Report 4*, UN World Water Assessment Programme. Accessed February 22, 2016. http://www.unesco.org/new/en/natural- sciences/environment/water/wwap/wwdr/wwdr4-2012.

United Nations. 2007. World Urbanization Prospects, Highlights. Accessed February 22, 2016. http://www.un.org/esa/population/publications/wup2007/2007WUP_Highlights_web.pdf.

United Nations. 2011. *Population and Vital Statistics Report*. http://unstats.un.org/unsd/demographic/products/vitstats/Sets/Series_A_2011.pdf.

US Army Corps of Engineers. 2009. *Louisiana Coastal Protection and Restoration: Final Technical Report, Adaptive Management Appendix*. Accessed February 22, 2016. http://biotech.law.lsu.edu/la/coast/lacpr/default4467.html

Vano, Julie A., Bradley Udall, Daniel R. Cayan, Jonathan T. Overpeck, Levi D. Brekke, Martin Hoerling Tapash Das, Holly C. Hartmann, Hugo G. Hidalgo, Gregory J. McCabe, Kiyomi Morino, Robert S. Webb, Kevin Werner and Denis P. Lettenmaier. 2014. "Understanding Uncertainties in Future Colorado River Streamflow." *Bulletin of the American Meteorological Society* 95(1) (January): 59–78. doi:10.1175/BAMS-D-12-00228.1.

Vicente, David M. (2005). Rambla Morales Desalination Plant (Almería, Spain). Flickr. Accessed February 25, 2016. https://www.flickr.com/photos/somachigun/6816041887

Walters, Carl J. and Ray Hilborn. 1978. "Ecological Optimization and Adaptive Management." *Annual Review of Ecology and Systematics* 9: 157–88.

Walters, Carl J. and C. S. Holling. 1990. "Large-Scale Management Experiments and Learning by Doing." *Ecology* 71(6): 2060–68.

Weaver, Christopher P., Robert J. Lempert, Casey Brown, John Hall, David Revell and Daniel Sarewitz. 2013. "Improving the Contribution of Climate Model Information to Decision Making: The Value and Demands of Robust Decision Frameworks." *Climate Change* 4(1): 39–60. doi:10.1002/wcc.202.

Winz, Inez, Gary Brierley and Sam Trowsdale. 2008. "The Use of System Dynamics Simulation in Water Resources Management." *Water Resources Management* 23(7): 1301–23.

Wong, Melanie. 2013. "Flexible Design: An Innovative Approach for Planning Water Infrastructure Systems Under Uncertainty." S.M. Thesis, MIT Technology and Policy Program, Cambridge, MA. http://ardent.mit.edu/real_options/Real_opts_papers/MKWong%20Thesis%202013%20Final%20copy.pdf

Woodhouse, Connie A., David M. Meko, Glen M. MacDonald, Dave W. Stahle and Edward R. Cook. 2010. "A 1,200-Year Perspective of 21st Century Drought in Southwestern North America." *Proceedings of the National Academy of Sciences* 107(50): 21283–88. doi:10.1073/pnas.0911197107.

WUCA, Water Utility Climate Alliance. 2010. Decision Support Planning Methods: Incorporating Climate Change Uncertainties into Water Planning. Accessed February 22, 2016. http://www.wucaonline.org/assets/pdf/pubs_whitepaper_012110.pdf.

Yeh, William W-G. 1985. "Reservoir Management and Operations Models: A State-of-the-Art Review." *Water Resources Research* 21(12): 1797–1818.

Chapter Five

EXTREME VALUE ANALYSIS FOR MODELING NONSTATIONARY HYDROLOGIC CHANGE

Arpita Mondal and P. P. Mujumdar

Abstract

Robust communication of hydrologic risk is one of the most important factors for ensuring a sustainable water future by virtue of directly influencing infrastructure designs. Under transient conditions, the risk of hydrologic extremes needs to be properly assessed, taking into consideration the associated uncertainties. This chapter describes the methods for characterizing nonstationary behavior in hydrologic extremes within the framework of statistical extreme value theory (EVT), with particular emphasis on estimation of design flood quantiles and their uncertainties under transient conditions. Applications of such methods are demonstrated through a real-world case study for extreme rainfall at a location in India. Toward the end, we also include a discussion on the associated caveats and existing alternate scientific views on the notion of nonstationarity.

Introduction

Hydrologic extremes such as floods constitute complex water problems that affect societies through direct and indirect interactions with natural and man-made systems. While such extremes are rare by their very definition, they need to be all the more properly investigated for robust estimation of associated risks. Rapidly changing conditions such as those due to land use or land cover changes, human-interventions and climate change can aggravate the risk of such extremes (Sivapalan and Samuel 2009), thereby emphasizing the necessity of a comprehensive analysis and modeling for an uncertain water future. For example, a warmer world under a greenhouse effect can hold more moisture leading to an intensification of the hydrologic cycle that can possibly lead to more severe floods (Milly et al. 2008). On the other hand, land-use effects such as urban development may lead to increases in peak runoff, even without changes in the total or mean flows. This issue is all the more consequential for a developing region such as India, where economic growth and rapid development are undeniable realities. Traditionally, hydrologic designs are based on the assumption of stationarity, which can be questioned under transient conditions. It is common for hydrologists to estimate the 100-year flood based on the statistical theory of extreme values, extrapolating the model derived from observed values for possibly less than a hundred years. Changes in climate, land-use or land-cover or other causal physical conditions require the model to account for the nonstationarity observed historically in the floods for such an extrapolation. Nonstationarity in observed extremes affects not only the design quantiles but also the uncertainties associated with their estimation. This chapter focuses on developing statistical extreme value models for studying the hydrologic extremes of floods and estimation of flood return levels and their uncertainties under change.

Although the idea of considering observations of real-world physical processes as realizations of random variables has been criticized (Klemes 1986, 2000), hydrologic frequency analysis is fairly widespread in engineering practices and the return periods and corresponding return levels have served as useful design tools. For example, the ten thousand year return flood level that is used for dike design in the Netherlands (Botzen et al. 2009) represents the flood that is expected to be exceeded once in every ten thousand years, with an annual chance of occurrence of 1/10,000. Alternately, it can also be understood that the time it takes, on an average, between two floods of this magnitude is ten thousand years. In a nonstationary setting, however, the annual chances of exceedance may vary with time. For example, the 100-year flood in a 2020 climate may be exceeded once every 80 years in a 2060 climate if annual floods are increasing with time. Thus, the risks associated with hydrologic designs vary from one time period to another and need to be explicitly accounted for.

Extremes require different treatment as compared to the bulk majority of the data, as they exhibit typical tail behaviors that control the probability of exceedance of rare events. In hydrology, while non-parametric tests for detecting change (such as trend tests or change-point tests) have been commonly used, they do not account for tail behavior of extremes per se. Although the theory of extremes suffers from assumptions, it provides a more efficient and physically meaningful approach (Katz et al. 2002). As an example, it can present information about change in a form useful for hydrologic designs such as shifts in return levels. Alternately, some studies apply standard regression analysis to time series of hydrologic extremes after transformation to an approximate normal distribution (Vogel et al. 2011). This approach fundamentally fails to allow for a heavy tail, which is reported to be markedly present in hydrologic variables (Papalexiou and Koutsoyiannis 2013; Cavanaugh et al. 2015). Nonstationary extensions of the extreme value theory have only recently been implemented to analyze transient behavior of hydroclimatic processes (Katz et al. 2002; 2013; Kharin and Zwiers 2005; Sillmann et al. 2011; Towler et al. 2010; Mondal and Mujumdar 2015a; 2015b).

Statistical extreme value models can be broadly classified into two approaches—the block maxima approach, which considers maximum of each, sufficiently large, block of values, and the peak-over-threshold (POT) approach, which considers excesses above a sufficiently high threshold. In this chapter, all theoretical formulations and examples are shown using the former approach, although extension of similar ideas to the latter approach is straightforward. An example of characterization of extreme daily flows can be observed in Figure 5.1 for two consecutive years. The annual maxima approach, as is obvious, leads to a single value per year, whereas the POT approach gives the excesses above a sufficiently high threshold, as shown. It may be noted that in the POT approach, dependence between the daily flows above threshold need to be accounted for, or the dependent values above the threshold need be declustered (Coles 2001). Threshold selection is also an important criterion in the POT approach, and specific methods need to be adopted for an appropriate choice of threshold.

As explained in detail in the next section, the Generalized Extreme Value (GEV) distribution family forms the appropriate model for block maxima. Nonstationarity is introduced in this model by allowing the parameters of the GEV distribution to vary as a function of a covariate such as time. Methods of estimation of parameters of such a time-varying model, based on Maximum Likelihood Estimation (MLE) and Bayesian approaches, are described.

The notions of return period and return level, on the other hand, become ambiguous in the nonstationary setting as the annual probability of exceedance is no longer constant and the one-to-one correspondence between a return level and a return period cease to exist. Only a handful of studies attempt to present a framework for hydrologic risk communication under nonstationarity (Cooley 2013; Salas and Obeysekera 2013; Serinaldi 2014; Rootzén and Katz 2013; Cheng and AghaKouchak 2014), using methods within the extreme value theory framework to define design levels and to obtain the uncertainties in their estimates. In this chapter, we include a description of methods for defining return periods and return levels under nonstationarity and for obtaining uncertainties associated with them, along with a discussion on alternate means for risk communication. An example application is shown for annual maximum daily rainfall at a location on the west coast of India.

It may also be noted that conflicting views on the notion of nonstationarity can be observed in hydrologic literature, with some studies emphasizing that the property of stationarity applies only to the modeling world of stochastic processes, and not to real-world observations (Koutsoyiannis 2006, 2013). Very recently, it has been reported that a nonstationary model for hydrologic extremes might increase the overall uncertainty (Serinaldi and Kilsby 2015) and, hence, it may be advisable to retain stationary models as benchmarks for hydrologic practice (Montanari and Koutsoyiannis 2014). A detailed discussion highlighting these issues and possible directions that can be explored to address them are also included in this chapter.

Background on Extreme Value Theory for Hydrologic Frequency Analysis

For a collection of random variables $\{X_1, X_2, \ldots, X_n\}$ that are independent and identically distributed (iid) with a common cumulative distribution function (cdf) F, such as daily flows or daily rainfall values over n days, we are interested in the statistical behavior of the maximum $M_n = \max\{X_1, X_2, \ldots, X_n\}$. If n is the number of observations in a

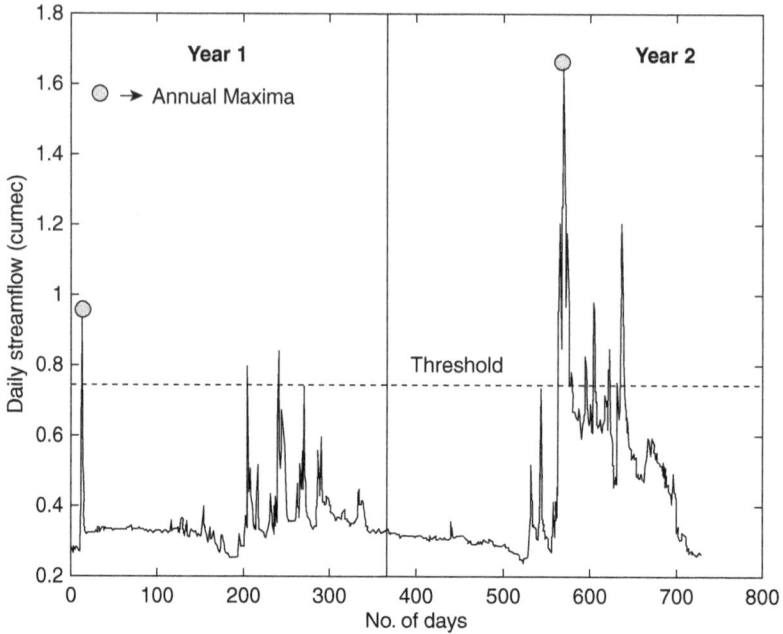

Figure 5.1 An example of characterization of extreme daily flows over two years. If a sufficiently high threshold is chosen, one may have more than one value a year; however, dependence between these values need to be specifically accounted for. The annual maxima for these two years are also shown.

year, then M_n represents the annual maximum daily flow or rainfall values. In theory, the distribution of M_n can be precisely known for all values of n:

$$Pr\{M_n \leq z\} = Pr\{X_1 \leq z, X_2 \leq z, ..., X_n \leq z\} = \{F(z)\}^n. \tag{1}$$

However, since the distribution function F is unknown, even a small error in its estimation might lead to large errors in the distribution of M_n. The alternative approach is to find approximate family of models for F^n, which can be estimated using data of only the maxima. This is analogous to approximating the distribution of sample means by the normal distribution, following the Central Limit Theorem. It is theoretically possible to *stabilize* the distribution of many random variables so that asymptotically their extreme values follow specific distributions. However, $F^n(z) \to 0$ as $n \to \infty$; unless F has a finite upper limit and z is greater than that. This upper limit may make the distribution of M_n degenerate to a point mass at that point, which can be avoided by choosing suitable normalizing constants $a_n > 0$ and b_n and looking at the normalized variable $M_n^* = \dfrac{M_n - b_n}{a_n}$. Appropriate choices of these constants stabilize the location and scale of the transformed extremes as n increases. The entire range of possible limit distributions for the transformed extremes is given by the extremal types theorem (Coles 2001) which states that the rescaled sample maxima M_n^* converge in distribution to a variable having a distribution within one of the three families termed as the *extreme value distributions* with the three types widely known as Gumbel, Frechet and Weibull families, respectively. These three families can be combined into a single family of models that give the distribution function of the GEV distribution, given by

$$
G(z; \mu, \sigma, \xi) = \begin{cases} \exp\left[-\exp\left(-\dfrac{z-\mu}{\sigma}\right)\right], & \xi = 0 \\[4mm] \exp\left[-\left(1+\xi\dfrac{z-\mu}{\sigma}\right)^{-\xi^{-1}}\right], & \xi \neq 0, \quad 1+\xi\dfrac{z-\mu}{\sigma} > 0. \end{cases} \tag{2}
$$

where μ, $\sigma > 0$ and ξ are the location, scale and shape parameters respectively. The shape parameter ξ influences the tail behavior, and positive, zero and negative values of the shape parameter result in the heavy-tailed (Frechet), unbounded light-tailed (Gumbel) and bounded finite tail (Weibull) extreme value distributions, respectively. This formulation holds regardless of the parent distribution F. Also, through inferences on ξ, the appropriate type of tail behavior can be estimated from the data. The probabilistic quantiles of the GEV distribution or the return levels, $z_{(1-p)}$, can be obtained by inversion of equation (2):

$$
z_{(1-p)} = \begin{cases} \mu - \dfrac{\sigma}{\xi}\left[1 - \left(-\log(1-p)\right)^{-\xi}\right] & \text{for } \xi \neq 0 \\[4mm] \mu - \sigma\log\left(-\log(1-p)\right) & \text{for } \xi = 0 \end{cases} \tag{3}
$$

where $0 < p < 1$, and $z_{(1-p)}$ is the return level corresponding to the return period $(m = 1/p)$, which is the flood magnitude expected to be exceeded once in every $1/p$ years on an average (Coles, 2001). Standard methods of estimation of parameters of the GEV distribution include the MLE and Bayesian approaches, which will be described in detail in the next section for the nonstationary case. The uncertainty associated with the estimation of can also be estimated by the Delta method and from the posterior predictive distribution for the MLE and Bayesian approaches, respectively, also shown in greater detail for the nonstationary case in the next section.

It may be noted that the iid assumption for the parent data, from which the block maxima series is derived, may not be valid for the daily flows or daily rainfall values. There are several studies in hydroclimatology in which such an assumption is made for the extreme value theory to be valid, not necessarily testing for the independence of the parent data (Katz et al. 2002; Katz 2013; Kharin et al. 2007; Towler et al. 2010). It is argued that the asymptotic extreme value distributions do not change even if the parent data are dependent, as long as certain relatively weak mixing conditions hold good (Leadbetter et al. 1983). Additionally, whether the asymptotic GEV distribution provides a reasonable description of the tail behavior of annual maximum flows, or rainfall values can be further tested by diagnostic plots such as the quantile or probability plots. These diagnostic plots can also be extended to the nonstationary setting, as described subsequently.

Extreme Value Models under Nonstationarity

One or more parameters of the GEV distribution can vary as a function of time under transient conditions, such as $\mu(t)$, $\sigma(t)$ and $\xi(t)$, for $t = 1, 2, 3, \ldots$, where t typically denotes the year if the extreme variable of interest is annual maxima. As an example, let us consider a nonstationary GEV model of the form

$$\mu(t) = \mu_0 + \mu_1 t, \quad \sigma(t) = \exp(\sigma_0 + \sigma_1 t), \quad \xi(t) = \xi \tag{4}$$

Typically, the shape parameter is kept constant as it is difficult to estimate and less likely to vary as a smooth function of time. The exponential in the expression for scale parameter is to ensure positive values always. The location and scale parameters in this model have linear trends. For example, the increase in location parameter may be caused by increases in mean river levels, while the increase in scale parameter could result from increasing variability of rainfall. At any point in time t, the GEV distribution conforms to the family given in Equation (2) derived from the extreme value theory.

It may be noted that an observed hydrologic extreme series may have both stationary and nonstationary periods with a regime shift dividing them. The starting point of the nonstationary period can significantly influence the trend so, therefore, detection of change point is also important. One possible way to address this issue is to apply change point detection tests and subsequently separate stationary and nonstationary models for the different regimes. For simplicity, we consider here a case in which a continuously increasing trend is observed throughout the period of record. This is indeed true for the case study application we illustrate later.

An exploratory analysis is typically carried out on the observed data for peak flows or extreme rainfall values for detection of the presence of trends. The non-parametric Mann-Kendall trend test provides one example of such a test (e.g., used by Cheng et al. 2014, and Westra et al. 2013). This test is based on the difference in magnitudes of paired values. The test statistic can be assumed to follow the normal distribution if the data are serially independent. Annual maxima series is likely to have insignificant autocorrelation values.

The suitability of the nonstationary GEV distribution is formally tested by a likelihood ratio test (Coles 2001). Let M_0 be the stationary sub-model with constant μ, σ and ξ, and M_1 be the nonstationary model with parameters $\mu(t), \sigma(t)$ and ξ. If $l_0(M_0)$ and $l_1(M_1)$ are the maximized log-likelihoods for M_0 and M_1, respectively, then the null hypothesis of M_0 can be rejected in favor of the validity of M_1 at the α level of significance if $2\{l_1(M_1) - l_0(M_0)\} > c_\alpha$ where c_α is the $(1-\alpha)$ quantile of the Chi-square distribution, whose degree of freedom equals the number of additional parameters in M_1, which in this example is 2.

If one has to choose between a number of candidate models (all nested), model selection criteria such as Akaike's information criterion (AIC) or the Bayesian information criterion (BIC) can be used. These criteria penalize the likelihoods for the number of parameters estimated. If any model M_i involves k parameters and the maximized log-likelihood for M_i is $l_i(M_i)$, then,

$$\text{AIC}(M_i) = 2l_i(M_i) + 2k, \quad \text{BIC}(M_i) = 2l_i(M_i) + k\ln N \tag{5}$$

where N is the sample size (total number of years for this example on block maxima).

As discussed earlier, diagnostic plots such as the quantile and the probability plots offer a means for model checking. In the nonstationary setting, to overcome the lack of homogeneity in the distributional assumptions, diagnostic checks are applied to a standardized version of the data, conditional on the estimated parameter values (Coles 2001). Such a standardization transforms the data to stationarity. As an example, for an estimated model $Z_t \sim \text{GEV}\left(\hat{\mu}(t), \hat{\sigma}(t), \hat{\xi}\right)$, the transformed variable

$$\tilde{Z}_t = \frac{1}{\hat{\xi}}\ln\left\{1 + \hat{\xi}\left(\frac{Z_t - \hat{\mu}(t)}{\hat{\sigma}(t)}\right)\right\} \tag{6}$$

follows a standard Gumbel distribution with $\mu = 0$, $\sigma = 1$. One can then substitute the estimates of the parameters in Equation (6) and generate a conventional quantile–quantile plot in terms of the transformed variable $\tilde{Z}_t$ as compared to a standard Gumbel distribution. It should be noted that the estimates substituted in Equation (6) are based on the sample, thereby making such a quantile plot only a crude method for model checking. Choice of the Gumbel distribution as a reference is appropriate, as it belongs to the GEV family (Coles 2001).

Parameter Estimation

The most common approaches for estimation of parameters of the extreme value distributions are MLE and Bayesian, since other methods (such as the Method of Moments, or L-Moment based methods) cannot be easily extended to account for time-varying parameters (Katz 2013).

The MLE method finds estimates of the parameters that maximize the probability of observing the data that is observed. If g represents the derivative of G with respect to z, and β denotes the vector of the GEV parameters, $\beta = (\mu_0, \mu_1, \sigma_0, \sigma_1, \xi)$, the likelihood function is given by (Coles, 2001):

$$L(\beta) = \prod_{t=1}^{N} g(z_t; \beta) \tag{7}$$

where z_t denotes the annual maximum one-day flow in the year t, and N is the number of years of observed record. Assuming independence of block maxima, it is convenient to work with the log-likelihood function and minimize the negative of that function to arrive at the estimates.

In the Bayesian approach, the knowledge brought by a prior distribution $p(\beta)$ is combined with the observation vector $\vec{z} = \{z_t\}_{t=1:N}$ to obtain the posterior distribution of parameters β. The Bayes theorem for estimation of GEV parameters under non-stationarity can be written as (Renard et al. 2013)

$$p(\beta | \vec{z}) \propto p(\vec{z} | \beta)\, p(\beta) \tag{8}$$

The resulting posterior distribution $p(\beta | \vec{z})$ summarizes information about parameters β whose inference is sought. Assuming independence of block maxima again, for the nonstationary model given by Equation (4), the likelihood function is given by

$$p(\vec{z} | \beta) = \prod_{t=1}^{N} p(z_t | \beta) = \prod_{t=1}^{N} p(z_t | \mu(t), \sigma(t), \xi) \tag{9}$$

where $p(z_t | \beta)$ is the probability density function of the GEV distribution evaluated at z_t.

Commonly used priors for the location and scale parameters are non-informative normal distributions (Cheng et al. 2014), while for the shape parameter a normal distribution with a standard deviation of 0.3 is suggested (Renard et al. 2013). Computations are easier if the priors are also assumed to be independent. For obtaining the posterior distribution of parameters, it is not possible to solve Equation (8) analytically. For generating a large number of realizations from the joint posterior distribution of the parameters, a Markov Chain Monte Carlo (MCMC) sampler is typically used. Values from the first few iterations are discarded to negate the effect of poorly chosen starting values at the beginning of the chain, and several methods are in use for monitoring MCMC convergence. For greater detail on principles of Bayesian estimation of parameters, the reader is referred to Gelman et al. (2014). The suitability of the nonstationary model M_1 over the stationary sub-model M_0 can be judged by computing the Bayes factor, based on the posterior distributions of sampled parameters, as follows:

$$B = \frac{p(M_0 | \vec{z})}{p(M_1 | \vec{z})} \bigg/ \frac{p(M_0)}{p(M_1)} = \frac{p(\vec{z} | M_0)}{p(\vec{z} | M_1)} \tag{10}$$

where both the numerator and the denominator can be expressed as Monte Carlo integration estimation for a given set of model parameters for each model. A value of $B < 1$ indicates that the nonstationary model M_1 is a better fit as compared to the stationary model M_0.

Estimation of Design Quantiles

The trends for a nonstationary model such as the one given by Equation (4), can be easily physically interpreted in terms of the corresponding transient or "effective" (Katz et al.

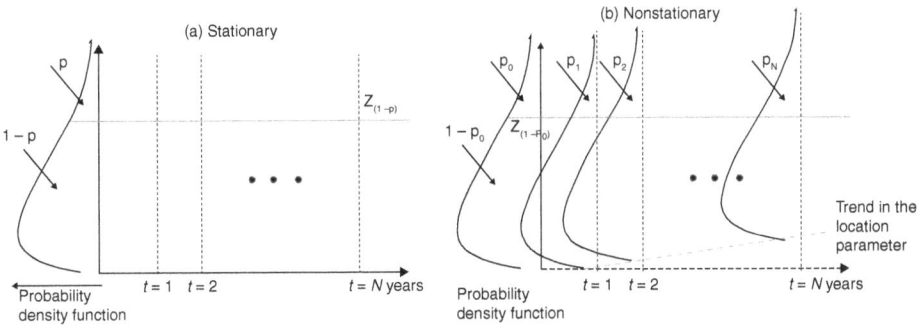

Figure 5.2 Schematic of design flood quantiles under (a) stationary and (b) nonstationary (with trend in location parameter) conditions. In the nonstationary case, the probability of exceedance corresponding to a fixed design quantile changes with time.

2002) return levels. These effective return levels are conditional quantiles such that the return level is allowed to vary from one time period to the next holding the probability of exceedance constant. The transient $(1-p)^{th}$ quantile of the nonstationary GEV distribution can thus be expressed as a function of the covariate by inverting equation the nonstationary GEV cdf, as given by

$$z_{(1-p)}(t) = \begin{cases} \mu(t) - \dfrac{\sigma(t)}{\xi}\left[1-\left(-\log\left(1-p\right)\right)^{-\xi}\right] & \text{for } \xi \neq 0 \\ \mu(t) - \sigma(t)\left[\log\left(-\log\left(1-p\right)\right)\right] & \text{for } \xi = 0 \end{cases} \tag{11}$$

Figure 5.2 shows the probability density functions for stationary and nonstationary conditions. For the stationary case, the probability of exceedance corresponding to a flood level is constant over time. On the other hand, trend in the location parameter in the nonstationary case leads to a shift in the probability density function, resulting in exceedance probabilities that vary with time.

For a typical example of annual maximum daily flows with trends, the median, 0.05th quantile and 0.95th quantiles obtained from a fitted nonstationary GEV distribution are shown in Figure 5.3. The 0.95th quantile from the fitted nonstationary GEV model represents the 20-year effective return levels ($p = 0.95 = 1 - 1/20$).

Effective return levels constitute a mode of risk communication that is akin to redefining the flood plain from one year to another, which may not be feasible for hydrologic designs. Therefore, precise definitions of return period (and the corresponding return level) are necessary in the nonstationary setting. Under stationarity, there are two interpretations of an "m-year event," which can also be extended to the nonstationary situations (Cooley 2013). The first interpretation is that the expected waiting time until the next exceedance of the event is m years, while the second is that the expected number of events in any m years is 1. In the stationary case, it is straightforward to show that both the two interpretations are correct and convey the same meaning.

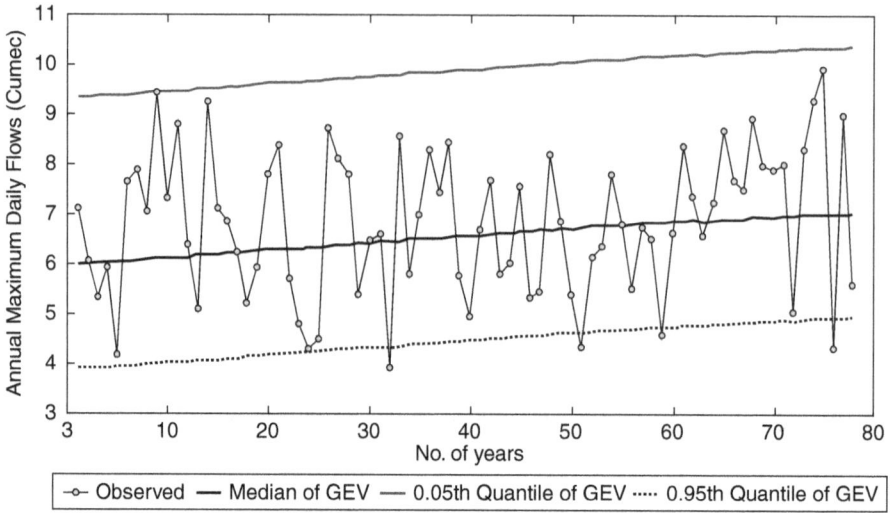

Figure 5.3 A typical example of transient observed annual maximum flows. The median, 0.05th quantile and 0.95th quantile obtained from the nonstationary GEV distribution are also shown.

For defining the m-year return level in terms of the expected waiting time (Cooley 2013; Salas and Obeysekera 2013), in the nonstationary case, let the random variable T represent the time at which a flood exceeding the level r will occur for the first time. Therefore, the probability that the first flood that exceeds r occurs at time t is given by

$$\Pr(T = t) = \Pr(\mathcal{Z}_1 \le r)\Pr(\mathcal{Z}_2 \le r)\cdots\Pr(\mathcal{Z}_{t-1} \le r)\Pr(\mathcal{Z}_t > r)$$
$$= \prod_{y=1}^{t-1} G_y(r)(1 - G_t(r)) \tag{12}$$

where $\mathcal{Z}_i$ represents the flood level at year i. Thus, the probability distribution of the waiting time follows a generalized geometric distribution whose parameter varies in time. The m-year return level r_m can be obtained by equating the expected value of this waiting time T with m years.

$$E(T) = \sum_{t=1}^{\infty} t \prod_{y=1}^{t-1} G_y(r_m)(1 - G_t(r_m)) = 1 + \sum_{t=1}^{\infty} \prod_{y=1}^{t} G_y(r_m) = m \tag{13}$$

where the right-hand side comes from a small algebraic exercise (Cooley 2013). Thus, unlike the stationary case, the m-year return level depends on the time varying nonexceedance probabilities.

Equation (13) is computationally challenging as it involves an infinite summation. However, since $G_y(z)$ is a monotonic decreasing function as $y \to \infty$, it is possible to bind the right-hand side by choosing an arbitrary, large number (Cooley 2013). The first hydrologic application of this definition of return level under nonstationarity is reported by Salas and Obeysekara (2013).

On the other hand, the expected number of events based definition aims to find the design quantile r_m such that the expected number of events in m years is one (Cooley 2013). Let the random variable Q denote the number of exceedances that occur in m years, beginning with the year $y = 1$ and ending with $y = m$. Unlike the stationary case, the distribution of Q is no longer binomial as the probability of exceedance changes with time. Let I be an indicator variable which takes a value of one if the flood level r is exceeded, and zero otherwise.

Hence,

$$Q = \sum_{y=1}^{m} I\left(Z_y > r\right)$$
$$\Rightarrow E(Q) = \sum_{y=1}^{m} E\left[I\left(Z_y > r_m\right)\right] = \sum_{y=1}^{m} Pr(Z_y > r_m) = \sum_{y=1}^{m} (1 - G_y(r_m)) = 1 \tag{14}$$

Under stationary conditions, in the Bayesian framework uncertainties in the return levels are obtained from the predictive distribution (Gelman et al. 2014). Application of the definitions of return periods and return levels for the nonstationary case in the Bayesian framework is not common, although it can be achieved by carrying out the above calculations with specific quantiles from the posterior distribution of the parameters.

Uncertainty in Estimation of Design Quantiles

The variance of return levels can be obtained by the Delta method (Oehlert 1992). In the nonstationary setting, the variance of the effective return level $z_{(1-p)}(t)$ given in Equation (11) varies as a function of time and is given by (Coles 2001)

$$\text{Var}\left(z_{(1-p)}(t)\right) \approx \nabla z_{(1-p)}{}^{T} V \nabla z_{(1-p)} \tag{15}$$

where V is the variance-covariance matrix of $(\mu_0, \mu_1, \sigma_0, \sigma_1, \xi)$ and

$$\nabla z_{(1-p)}{}^{T} = \left[\frac{\partial z_{(1-p)}}{\partial \mu_0}, \frac{\partial z_{(1-p)}}{\partial \mu_1}, \frac{\partial z_{(1-p)}}{\partial \sigma_0}, \frac{\partial z_{(1-p)}}{\partial \sigma_1}, \frac{\partial z_{(1-p)}}{\partial \xi}\right] \tag{16}$$

evaluated at the estimated values of $(\mu_0, \mu_1, \sigma_0, \sigma_1, \xi)$. The variance-covariance matrix V is obtained by inversion of the observed information matrix evaluated at the MLE estimates. The derivatives in Equation (16) can be computed numerically when their analytical forms become cumbersome. The variance of return level is used to compute the confidence intervals for $z_{(1-p)}$ assuming asymptotic normality of the MLE estimates that can be theoretically shown. There are other approaches to compute the confidence intervals of return levels such as the profile likelihood approach (Coles 2001); however, such approaches are not common in hydrologic applications (Obeysekera and Salas 2014). Other non-parametric methods often used for obtaining uncertainties associated with the return levels are the resampling based approaches (Kuczera 1999).

The Delta method can be extended to compute the uncertainties associated with return levels defined based on expected waiting time and expected number of events in the nonstationary setting, and the modified method is termed the Implicit Delta method (Cooley 2013). If $\hat{\beta}$ denotes the MLE estimates of the parameter vector β, and $V(\hat{\beta})$ is the approximated variance-covariance matrix associated with $\hat{\beta}$, then the m-year return level r_m is a function of the parameter vector, $r_m = f(\beta)$ whose gradient is given by ∇r_m. According to the Delta method, the variance of r_m is approximately given by

$$\mathrm{Var}\left(r_m\right) \approx \nabla r_m{}^T \, V\left(\hat{\beta}\right) \nabla r_m. \tag{17}$$

For either of the two interpretations in the nonstationary case, analytical expressions cannot be obtained for f, rather a generalized expression that can be obtained from Equations (14) and (13) of the form $m = s\left(\beta, r_m\right)$. If m_0 is the desired return period and r_{m0} is the corresponding return level, numerical derivatives of gradient of r_m can be obtained if s is known. Applying chain rule of differentiation (Cooley 2013),

$$\frac{\partial r}{\partial \beta_i}\Big|_{\beta=\hat{\beta};m=m_0} = \frac{\partial r}{\partial m}\Big|_{\beta=\hat{\beta};m=m_0} \, \frac{\partial m}{\partial \beta_i}\Big|_{\beta=\hat{\beta};m=m_0}$$

$$= \left(\frac{\partial m}{\partial r}\Big|_{\beta=\hat{\beta};r=r_{m_0}}\right)^{-1} \frac{\partial m}{\partial \beta_i}\Big|_{\beta=\hat{\beta};m=m_0} \tag{18}$$

allows computation of the necessary gradient.

Obeysekara and Salas (2014) also compute the bootstrap resampling-based confidence intervals for the return levels using the expected waiting-time based definition. Incidentally, this is the only study in hydrology that explicitly accounts for uncertainties in quantiles of time-varying hydrologic extremes. For the bootstrap method, a transformation of the observed peak flows to standard Gumbel distribution, as described by Equation (6), is first carried out. The sample of standardized residuals thus obtained may be used in a bootstrap scheme to obtain a sampling distribution of r_m. From the empirical frequency distribution of r_m, confidence intervals can be obtained corresponding to a desired confidence level.

Alternate Approaches for Risk Communication

It has been argued recently (Serinaldi 2014) that the notion of return period may lead to misinterpretations of hydrologic risk, even in the stationary and/or univariate cases, by virtue of it being a derived quantity and inadvertently hiding that there is a probability p to observe the hazardous event every year. The iid assumption necessary for defining return periods and corresponding return levels is also argued to be one of the most serious limitations of this approach. Instead, the probability of failure or risk represents the probability of observing at least one critical event in the lifetime of M years and allows one to compute a design value r_d with the appropriate probability. It can be shown that when design life M years = return period m years = 100 years, under iid assumption, the probability of at least one critical event occurring during the lifetime of the structure is 63 percent (Serinaldi 2014). Therefore, the probability of failure may not always be intuitively understood based only on return periods.

In the nonstationary setting, for both interpretations of return period based on expected waiting time and number of events, it is argued that the return period simply summarizes information on the average probability of exceedance, and hence becomes a redundant design criteria. Moreover, Rootzén and Katz (2013) argue that the definition of return period based on expected number of events fixes both the design life and the risk, while the expected waiting-time based definition is overly sensitive to the presence of trends or the lack of it, even after the design life period.

The probability of failure, or risk, on the other hand can apply to several situations (iid; independent, but not identically distributed; neither independent, nor identically distributed) and can be written as

$$p_M = 1 - \Pr\left(Z_1 \leq r_d \cap Z_2 \leq r_d \cdots \cap Z_M \leq r_d\right)$$
$$= 1 - H_M\left(Z_1 \leq r_d, Z_2 \leq r_d, \ldots, Z_M \leq r_d\right) = 1 - C_M\{G_1(r_d), G_2(r_d), \ldots, G_M(r_d)\} \qquad (19)$$

where H_M denotes the joint distribution of the set of random variables Z_i for $i = 1, \ldots, M$, G_i denotes the marginal univariate distribution of Z_i, and C_M is the copula describing the margin-free dependence structure (Serinaldi 2014). Under iid conditions, $G_i = G \, \forall \, i = 1, 2, \ldots, M$, and $C_M = \prod$. For the independent, but not identically distributed data, $C_M = \prod$, but G_i changes at each year.

According to Rootzén and Katz (2013), the basic information needed for hydrologic design must include i) design life of the structure (e.g., say, next 50 years) and ii) the probability of a hazardous event occurring within the design life of the structure. They propose a new quantity for hydrologic risk communication, which they term 'design life level', and can be thought of as a special case of Equation (19), for independent but not identically distributed data. For example, for winter season largest daily rainfall at a location in Australia, the 2011–2060 5% design life level is 121 mm (Rootzén and Katz 2013). This means that the maximum probability of exceedance of any storm having daily winter season rainfall in excess of 121 mm, in any given year in the design life period 2011–2060, is at most 5%. Note here that this is different from a design value based on 5% annual exceedance probability which actually corresponds to a 20-year storm, whose magnitude will be much lower than 121 mm.

While these alternate approaches overcome limitations, it is to be noted that they force computation of very high probabilities based on limited observation records. In the stationary case, for example, a 5% risk of failure over the design life of 20 years corresponds to a return period of about 390 years. For the Indian region, in particular, obtaining long records of observed hydrometeorologic variables might be difficult. Estimating such high tail quantiles from short observed records is likely to increase the uncertainty associated with these quantiles.

An Example Application

For illustration of application of the above-mentioned concepts, annual maximum daily rainfall values are considered at the point 15.5N, 73.5E near Panjim, Goa, India, as shown by the small box in Figure 5.4(a). A high-resolution gridded (1° latitude × 1° longitude) daily rainfall data for India for 1951–2003 (Rajeevan et al. 2008) is used for this purpose. Quality-controlled observed rainfall data from more than 1800 gauges is used to prepare this dataset that is increasingly being used in studies on the Indian rainfall (Mani et al. 2009; Neena and Goswami 2010; Ghosh et al. 2011; Mondal and Mujumdar 2015b; Vittal et al. 2013). Although a gridded dataset involves interpolation and smoothing, it can have fewer missing values as compared to station data (Jones et al. 2013), and are generally spatially and temporally continuous unlike measured point data. Thus, the dataset is suitable for analyzing long-term trends in extremes. The observed record of annual maxima is for the period 1951–2003. The grid box location centered at 15.5N, 73.5E lies in the western coast of India that receives prominent rainfall from the South West Indian Summer Monsoon. Coastal regions such as this may be susceptible to changes in large-scale circulation patterns due to natural or anthropogenic causes, thereby necessitating a thorough study of changes in hydrometeorological variables in such regions. Changes in extreme rainfall at this location may influence the risk of pluvial flooding affecting the urban regions that witness significant tourist footfall throughout the year. Although all calculations are shown in this example for annual maximum daily rainfall values, similar steps can be carried out for peak flows or other hydrologic variables of interest is such data is available.

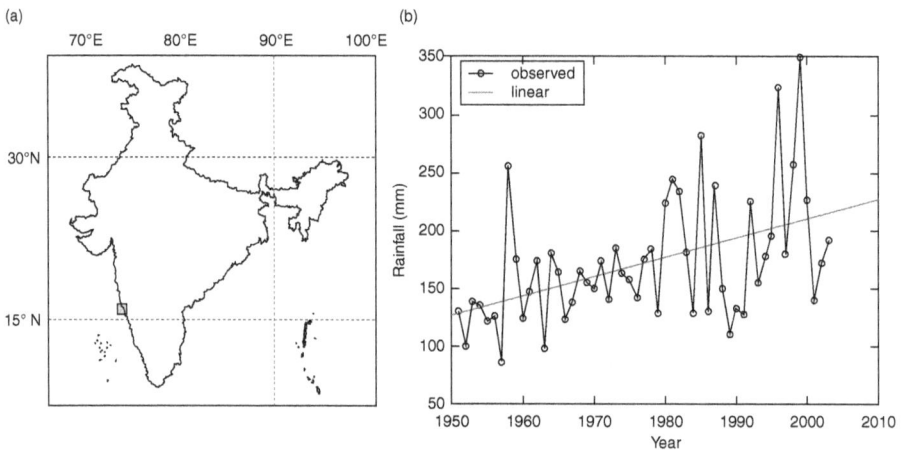

Figure 5.4 (a) Location of the point 15.5N, 73.5E. (b) Annual maximum daily rainfall (mm) at the point 15.5N, 73.5E. The linear best fit is also shown to have an increasing trend.

The annual maximum daily rainfall values at the given location is shown in Figure 5.4(b). A statistically significant increasing trend is found in the annual maximum daily rainfall values in this location using the non-parametric Mann-Kendall trend test (p-value = 0.0002).

Since this location receives maximum rainfall mostly in the monsoon season, assuming independence in the extremes is justified for this trend test. It now remains to be seen whether a nonstationary GEV distribution with trends in the parameters represents the observed rainfall maxima well.

Two candidate models are considered for fitting the observed annual maximum daily rainfall values—the stationary model $M_0 \sim$ GEV(μ, σ, ξ) and the nonstationary model $M_1 \sim$ GEV($\mu(t)$, σ, ξ), where $\mu(t) = \mu_0 + \mu_1\, t$, and t = no. of years since 1950. The open-source R package extRemes version 2.0–6 (available at http://cran.r-project.org/web/packages/extRemes/index.html) (Gilleland and Katz 2011) is used for the extreme value analysis. For this example application, all calculations using only the MLE method are shown, though a Bayesian analysis can also be conducted using the same package (albeit with some limitations). The details of the fitted models are provided in Table 5.1. The parameter μ_1 is found to be statistically significant as the likelihood ratio test shows significant evidence to reject the stationary model (p-value 0.0013). The standard errors of estimation are also shown. It is worth noting that the confidence interval around the shape parameter ξ includes zero in both the models suggesting a light tail. For the sake of simplicity, subsequent calculations are performed assuming the three parameter GEV distribution only. This does not change the computational results and the final conclusions.

The diagnostic plots for the nonstationary model are presented in Figure 5.5. The model simulated quantiles are found to reasonably match well with the observed quantiles when transformed to the standard Gumbel scale, except for very high quantiles. Figure 5.5(b) also shows the transient effective return levels corresponding to 2-year, 20-year and 50-year periods. The trends in the nonstationary GEV distribution are well reflected in the trends in these return levels, and they are found to capture the observed trends in annual maximum daily rainfall values. The design rainfall quantiles corresponding to 50-year and 100-year return periods, using various definitions, are presented in Table 5.2. The effective return levels are computed considering only the observed

Table 5.1 Details of the fitted GEV models for annual maximum daily rainfall at 15.5N, 73.5E.

	Parameters	Estimate value	Negative log-likelihood	Likelihood ratio test
Stationary model M_0	μ (SE) σ (SE) ξ (SE)	145.71 (5.95) 38.50 (4.42) 0.07 (0.10)	279.47	p-value = 0.0013, Likelihood ratio = 10.409
Nonstationary model M_1	μ_0 (SE) μ_1 (SE) σ (SE) ξ (SE)	121.3 (9.22) 0.99 (0.31) 34.76 (4.14) 0.08 (0.12)	274.27	Chi-square critical value = 3.84

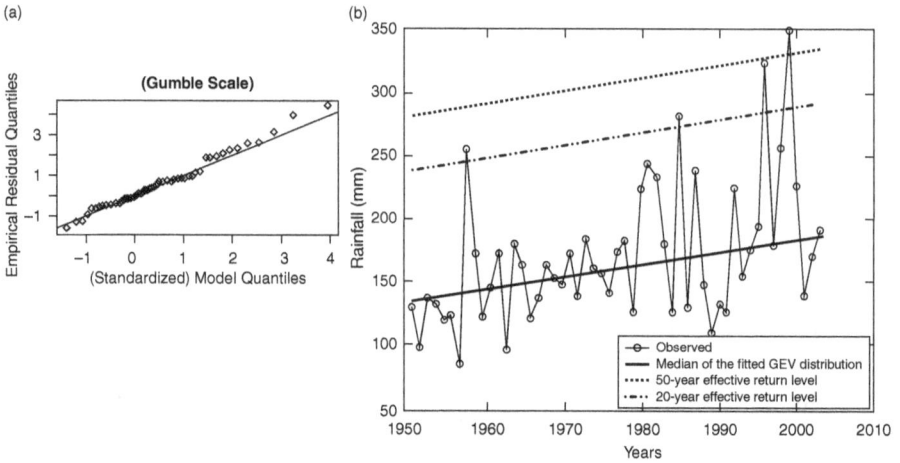

Figure 5.5 Diagnostic plots for the nonstationary GEV model for annual maximum daily rainfall at 15.5N, 73.5E. (a) Quantiles of standardized extremes versus those from standard Gumbel distribution. (b) Quantiles from model-simulated data as compared to observed annual maximum daily rainfall values (mm).

period of record (1951–2003). The second column of Table 5.2 presents the highest of the transient effective return levels (highest of all the return levels given by Equation (11)) for the period 1951–2003. This index is termed as the 'minimax design life level' by Rootzen and Katz (2013) who propose it as a measure of design level. The third column of Table 5.2 presents the design values obtained by equating the right-hand side of Equation (13) to 50 (100) for the first (second) row and solving for r_m. Similarly, the fourth column of Table 5.2 presents the design values obtained by equating the right-hand side of Equation (14) to 1, setting $m = 50$ ($m = 100$) for the first (second) row and solving for r_m.

It is also to be noted that the expected waiting time based return levels is sensitive to the choice of the starting period of the design life, and to the choice of time for which the trends are assumed to continue in the future. For the expected number of events-based definition of return levels, too, the starting year of the design life period needs to be specified. For both the expected waiting time and expected number of events based interpretations, in this example we assume the design life to start at 2000.

This example highlights the non-uniqueness of estimation of design extreme quantiles and it is observed that different definitions lead to different magnitudes of return levels, subject also to the assumptions about the design life and projected trends. The uncertainties in these design quantiles can be estimated using the methods described earlier in the section, Uncertainty in Estimation of Design Quantiles. The alternate measures of risk such as the design life level described earlier in Alternate Approaches for Risk Communication can also be obtained for this example and used for designs.

Table 5.2 Design quantiles of extreme rainfall (mm) at 15.5N, 73.5E from the fitted nonstationary GEV model.

Return period (also used as design life here)	Traditional stationary return level	Highest effective return level (1951–2003)	Expected waiting time based return level (assuming design life to start at 2000, trend to stop at end of design life)	Expected number of events based return level (assuming design life to start at 2000)
50 years	320	333.33	371.75	357.73
100 years	356.87	367.67	452.95	422.83

Discussion

This chapter describes and discusses statistical tools for developing extreme value models for nonstationary hydrologic extremes such as floods, focusing on the more common block-maxima approach. Of course, this does not do justice to the POT or the associated point process characterizations of extremes that are also extremely useful. Model development and selection procedures discussed here are not meant to be exhaustive, and the consideration of spatial and temporal dependencies in extremes is still an active area of scientific research. Several extensions are possible to the methods discussed, such as increasing model complexity, consideration of physically based covariates, detection of change in return levels of hydrologic extremes and so on. The principal aim of using nonstationary extreme value analysis is to explicitly account for transient behavior of extremes for hydrologic risk communication that influences infrastructure design or management decisions.

It is worth noting that hydrologic literature presents disagreements on the notion of nonstationarity in the ways it is detected and modelled. For example, in hydrology, it is common to characterize a given time series as stationary or nonstationary and, for modelling convenience, to break it down into deterministic and stochastic components (Salas 1993). Typical deterministic components may include trends, as we have observed in this chapter for extremes. Use of statistical tests, such as the Mann-Kendall trend test, has been criticized since the hypothesis under test is not really independent of the data that is already observed. Moreover, such tests tend to ignore dependence in the data and treat the observed physical processes as purely random (Von Storch 1995), thereby inflating the chances of rejecting the null hypothesis of no trend. Typically, large-scale variations (such as trends) are treated as deterministic components, while small-scale variations are modelled as random components; however, in hydrology, there seems to be no clear criteria for such considerations. Clearly, the small-scale variations observed in hydrologic time series are not measurement errors and do not come from repeatable experiments. Observed changes at both large and small temporal scales might be a part of larger natural variability. Some authors (e.g., Koutsoyiannis 2006) argue that treating the trend as deterministic reduces the uncertainty, which may inappropriately increase risks. Going a step further, recent studies (e.g., Montanari and Koutsoyiannis 2014) also point out that the notion of nonstationarity is applicable only to the world of stochastic models, and not to real-world observations.

Deterministic models also suffer from the danger of extrapolating beyond the range of data, and the choice of such a projection may greatly influence the design quantiles and their uncertainties. However, estimation of future risk for hydrologic designs inevitably involves extrapolation and the best practice perhaps consists of explicitly mentioning the assumptions and conveying the uncertainty to the best extent possible. In some situations, mechanistic approaches can be adopted in working with trends—for example, in rapidly urbanizing watersheds, the underlying causes of changes in annual maximum flows are sometimes identifiable and can inform the decision on how far to project the trend in the future. This may not, however, be universally practicable. For example, underlying causes for the increasing extreme rainfall at the example grid location on the west coast of India are hard to identify. The approaches discussed in this chapter essentially are data-driven and limitations of such approaches are duly admitted.

It is also worth mentioning that nonstationary models essentially consist of deterministic relationships between hydrologic variables and covariates. For example, hydrologic extremes are assumed to be linear functions of time all throughout this chapter. We not only assume that such a deterministic relationship remains unchanged throughout the design life period, but the future values of the covariates are also assumed to be deterministically known always. Also, the nonstationary statistical properties of the stochastic process that generates the hydrologic extremes are estimated from time properties of one realization of the random process, which is the observed time series, implicitly assuming an ergodic process, and thus contradicting the very basic notion of nonstationary processes. Some of these issues are discussed in detail by Serinaldi and Kilsby (2015) for hydrologic extremes. The issue of scaling is of growing interest in the hydrologic community; however, to adequately study scaling processes, one requires a long series of observed records that are hard to obtain for developing regions such as India. In general, it is recommended (Montanari and Koutsoyiannis 2014; Serinaldi and Kilsby 2015) that the use of a nonstationary model over its stationary counterpart be posed as a standard model-selection problem whereby one selects the best model that optimizes the tradeoff between bias in the design variables and their uncertainties. As far as hydrologic extremes are concerned, it can thus be concluded that in order to solve complex water problems for an uncertain future, one has to make the best use of the state of knowledge, data available as well as experience-based engineering wisdom.

Acknowledgments

The authors thank two anonymous reviewers for their suggestions and comments, which helped improve the chapter significantly. The authors also thank Dan Cooley, Rick Katz, Holger Rootzen and Francesco Serinaldi for helpful clarifications.

References

Botzen, Wouter J. W., Jeroen C. J. H. Aerts and Jeroen C. J. M. van den Bergh. 2009. Dependence of Flood Risk Perceptions on Socioeconomic and Objective Risk Factors. *Water Resources Research* 45(10): W10440.

Cavanaugh, N. R., Alexander Gershunov, Anna K. Panorska and Tomaz J. Kozubowski. 2015. The Probability Distribution of Intense Daily Precipitation. *Geophysical Research Letters* 42(5): 1560–67.

Cheng, Linyin and Amir AghaKouchak. 2014. Nonstationary Precipitation Intensity-Duration-Frequency Curves for Infrastructure Design in a Changing Climate. *Scientific Reports* 4: doi 10.1038/srep07093.

Cheng, Linyin, Amir AghaKouchak, Eric Gilleland and Richard W. Katz. 2014. Non-stationary Extreme Value Analysis in a Changing Climate. *Climatic Change* 127(2): 353–69.

Coles, Stuart. 2001. *An Introduction to Statistical Modeling of Extreme Values*. London: Springer-Verlag.

Cooley, Daniel. 2013. Return Periods and Return Levels Under Climate Change. In: *Extremes in a Changing Climate: Detection, Analysis, and Uncertainty*, edited by Amir AghaKouchak, David Easterling and Kuolin Hsu. 97–114. New York: Springer.

Gelman, Andrew, John B. Carlin, Hal S. Stern, David B. Dunson, Aki Vehtari and Donald B. Rubin. 2014. *Bayesian Data Analysis*. London: Chapman and Hall/CRC Press.

Ghosh, Subimal, Debasish Das, Shih-Chieh Kao and Auroop R. Ganguly. 2011. Lack of Uniform Trends but Increasing Spatial Variability in Observed Indian Rainfall Extremes. *Nature Climate Change* 2(2): 86–91.

Gilleland, Eric and Richard. W. Katz. 2011. New Software to Analyze How Extremes Change Over Time. *Eos, Transactions American Geophysical Union* 92(2): 13–14.

Jones, Mari R., Stephen Blenkinsop, Hayley J. Fowler, David B. Stephenson and Christopher G. Kilsby. 2013. *Generalized Additive Modelling of Daily Precipitation Extremes and Their Climatic Drivers*. Boulder: National Center for Atmospheric Research. http://nldr.library.ucar.edu/collections/technotes/TECH-NOTE-000-000-000-868.pdf, NCAR.

Katz, Richard W. 2013. Statistical Methods for Nonstationary Extremes. In: *Extremes in a Changing Climate: Detection, Analysis and Uncertainty*, edited by A. AghaKouchak, D. Easterling and K. Hsu, 15–37. New York: Springer.

Katz, Richard W., Marc B. Parlange and Philippe Naveau. 2002. Statistics of Extremes in Hydrology. *Advances in Water Resources* 25(8): 1287–1304.

Kharin, Viatcheslav V. and Zwiers, Francis W. 2005. Estimating Extremes in Transient Climate Change Simulations. *Journal of Climate* 18(8): 1156–73.

Kharin, Viatcheslav V., Francis W. Zwiers, Xuebin Zhang and Gabriele C. Hegerl. 2007. Changes in Temperature and Precipitation Extremes in the IPCC Ensemble of Global Coupled Model Simulations. *Journal of Climate* 20: 1419–44.

Klemeš, Vít. 1986. Dilettantism in Hydrology: Transition or Destiny? *Water Resources Research* 22(9S): 177S–188S.

Klemeš, Vít. 2000. Tall Tales About Tails of Hydrological Distributions. I. *Journal of Hydrologic Engineering* 5(3): 227–31.

Koutsoyiannis, Demetris. 2006. Nonstationarity Versus Scaling in Hydrology. *Journal of Hydrology* 324(1): 239–54.

———. 2013. Hydrology and Change. *Hydrological Sciences Journal* 58(6): 1177–97.

Kuczera, George. 1999. Comprehensive At-site Flood Frequency Analysis Using Monte Carlo Bayesian Inference. *Water Resources Research* 35(5): 1551–57.

Leadbetter, M. Ross, Georg Lindgren and Holger Rootzén. 1983. *Extremes and Related Properties of Random Sequences and Processes*. New York: Springer.

Mani, Neena J., Ettamal. Suhas, and B. N. Goswami. 2009. Can Global Warming Make Indian Monsoon Weather Less Predictable? *Geophysical Research Letters* 36(8): L08811.

Mani, Neena J. and B. N. Goswami. 2010. Extension of Potential Predictability of Indian Summer Monsoon Dry and Wet Spells in Recent Decades. *Quarterly Journal of the Royal Meteorological Society* 136(648): 583–92.

Milly, Paul C. D., B. Julio Betancourt, Malin Falkenmark, Robert Hirsch, Zbigniew Kunzewica, Dennis Lettenmaier, Ronald Stouffer, Michael Dettinger and Valentina Krysanova. 2008. "Stationarity Is Dead: Whither Water Management?" *Science*, 319: 573–74.

Mondal, Arpita and Pradeep P. Mujumdar. 2015a. Return Levels of Hydrologic Droughts Under Climate Change. *Advances in Water Resources* 75: 67–79.

———. 2015b. Modeling Non-stationarity in Intensity, Duration and Frequency of Extreme Rainfall over India. *Journal of Hydrology* 521: 217–31.

Montanari, Alberto and Demetris Koutsoyiannis. 2014. Modeling and Mitigating Natural Hazards: Stationarity Is Immortal! *Water Resources Research* 50: 9748–56.

Obeysekera, Jayantha and José D. Salas. 2014. Quantifying the Uncertainty of Design Floods Under Nonstationary Conditions. *Journal of Hydrologic Engineering* 19(7): 1438–46.

Oehlert, Gary W. 1992. A Note on the Delta Method. *The American Statistician* 46(1): 27–29.

Papalexiou, Simon M. and Demetris Koutsoyiannis. 2013. Battle of Extreme Value Distributions: A Global Survey on Extreme Daily Rainfall. *Water Resources Research* 49(1): 187–201.

Rajeevan, Madhaven, Jyoti Bhate and Ashok K. Jaswal. 2008. Analysis of Variability and Trends of Extreme Rainfall Events over India Using 104 Years of Gridded Daily Rainfall Data. *Geophysical Research Letters* 35: L18707.

Renard, Benjamin, Xun Sun and Michel Lang. 2013. Bayesian Methods for Non-stationary Extreme Value Analysis. In: *Extremes in a Changing Climate: Detection, Analysis, and Uncertainty*, edited by A. AghaKouchak, D. Easterling and K. Hsu, 39–95. New York: Springer.

Rootzén, Holger and Richard W. Katz. 2013. Design Life Level: Quantifying Risk in a Changing Climate. *Water Resources Research* 49(9): 5964–5972.

Salas, José D. 1993. Analysis and Modeling of Hydrologic Time Series. In: *Handbook of Hydrology*, edited by D. Maidment, 19.1–19.72. New York: McGraw-Hill.

Salas, José D. and Jayantha Obeysekera. 2013. Revisiting the Concepts of Return Period and Risk for Nonstationary Hydrologic Extreme Events. *Journal of Hydrologic Engineering* 19(3): 554–68.

Serinaldi, Francesco. 2014. Dismissing Return Periods! *Stochastic Environmental Research and Risk Assessment* 29(4) 1179–89.

Serinaldi, Francesco and Chris G. Kilsby. 2015. Stationarity Is Undead: Uncertainty Dominates the Distribution of Extremes. *Advances in Water Resources* 77: 17–36.

Sillmann, Jana, Mischa Croci-Maspoli, Maalak Kallache and Richard W. Katz. 2011. Extreme Cold Winter Temperatures in Europe Under the Influence of North Atlantic Atmospheric Blocking. *Journal of Climate* 24(22): 5899–5913.

Sivapalan, Murugesu and Jos M. Samuel. 2009. Transcending Limitations of Stationarity and the Return Period: Process-based Approach to Flood Estimation and Risk Assessment. *Hydrological Processes* 23: 1671–75.

Towler, Erin, Balaji Rajagopalan, Eric Gilleland, R. Scott Summers, David Yates and Richard W. Katz. 2010. Modeling Hydrologic and Water Quality Extremes in a Changing Climate: A Statistical Approach Based on Extreme Value Theory. *Water Resources Research* 46(11).

Vittal, Hersha, Subhankar Karmakar and Subimal Ghosh. 2013. Diametric Changes in Trends and Patterns of Extreme Rainfall over India from Pre-1950 to Post-1950. *Geophysical Research Letters* 40: 3253–58.

Vogel, R. M., Chad Yaindl and Meghan Walter. 2011. Nonstationarity: Flood Magnification and Recurrence Reduction Factors in the United States. *Journal of American Water Resources Association* 47: 464–74.

Von Storch, Hans. 1995. Misuses of Statistical Analysis in Climate. In: *Analysis of Climate Variability: Applications of Statistical Techniques*, edited by Hans von Storch and Antonio Navarra, 11–26, Berlin: Springer.

Westra, S., L. V. Alexander and F. W. Zwiers. 2013. Global Increasing Trends in Annual Maximum Daily Precipitation. *Journal of Climate* 26(11): 3904–18.

Chapter Six

THE WATER–FOOD NEXUS
AND VIRTUAL WATER TRADE

Joel A. Carr and Paolo D'Odorico

Abstract

Most of the human appropriation of freshwater resources on Earth is used for agriculture, and water availability is a major factor constraining mankind's ability to meet the future needs for food by the growing and increasingly burgeoning human population. Many countries are in conditions of chronic food deficit and rely on food imported from other regions of the world. What is the impact of the virtual transfer of water associated with international food trade? Here we analyze the global patterns of virtual water trade and evaluate how they affect the ethical and physical links between societies and the water resources that sustain them. In the last 25 years, societal reliance on trade has doubled. Today, about one fourth of the water used for food production worldwide is accessed through virtual water trade. The globalization of water prevents the emergence of famine in water-scarce regions, increases equality in access to water and allows for more water-efficient food production. It also entails, however, loss of environmental stewardship, increase in trade dependency and reduced societal resilience to drought.

Introduction

Human societies rely on freshwater resources for a number of activities, including drinking, household usage and industrial and agricultural production (e.g., Gleick, 1993; Rosegrant et al. 2009). Water consumption for food production exceeds by far all the other societal appropriations of freshwater resources (Falkenmark and Rockstrom 2006). In fact, about 85 percent of the human consumption of water is used for crop and livestock production. Locally, however, household and industrial uses can be predominant, particularly in major urban areas. Securing water resources for agriculture, while reconciling the competing water needs of growing cities and surrounding rural areas, is a major challenge of our time.

In the last few decades the unprecedented increases in global crop production, fueled by the recent availability of nitrogen fertilizers (Erisman et al. 2008), has led to the misconception that mankind will never again face a food crisis. Water remains, however, an important limiting factor controlling food production. With an ever-increasing global population, the ability to maintain adequate food supplies with limited water resources has become a pressing concern (Falkenmark and Rockstrom 2004). In fact, despite the development of new technology (e.g., new cultivars, irrigation techniques and water reuse methods), the human pressure on global freshwater resources has been increasing at alarming rates in response to population growth, changes in diet and new bioenergy policies (Rockstrom et al. 2012; US Congress 2007; EU Parliament 2009). At the regional scale water resources may be insufficient to meet the food demand of the local population. Already, some regions of the world (e.g., the Middle East) are in conditions of chronic water scarcity: the population already exceeds the limits imposed by the availability of freshwater resources for food production. When a society has a limited access to a vital resource such as water, it can either decide to use it more effectively by reducing waste or other losses, and/or try to import water from somewhere else. However, the amount of water required for food production is enormous and not easily transported. Therefore, instead of transferring water, societies trade food commodities (Hoekstra and Chapagain 2008). Thus, food trade is associated with a virtual transport of the water used for its production (e.g., Allan, 1998). Global food security strongly depends on virtual water trade (Figure 6.1). Through the intensification of trade, some regions have become strongly dependent on food produced with water resources available elsewhere.

If we use virtual water trade as a currency for food trade, countries can be self-sufficient (Figure 6.1) or rely on trade, depending on whether their virtual water exports are greater or smaller than their imports, respectively (Suweis et al. 2013). Trade dependency makes some countries particularly vulnerable to the uncertainties of the trade market such as during the recent food crises of 2007–2008 and 2010–2011 when the governments of some major food-producing countries decided to limit or ban their exports to contain the effects of the escalating food prices (e.g., Fader 2013). Due to the uncertainties of food trade, self-sufficiency and trade-dependency may be used as indicators of sustainable access to the water resources needed to ensure food availability to a given country. Therefore, in this chapter we investigate the global patterns of virtual water trade and human appropriation of freshwater resources and discuss their impacts on the resilience of the global food system, demographic growth and equality in the access to water for food production. It is important to note that we investigate only the dimension of food security related to water availability (through both domestic production and trade), without considering other socioeconomic factors determining access to food and water, or evaluating the nutritional aspects of food security.

Water and Food Crisis

About a decade ago research on water and food security started to alert us about the important constraints imposed by water availability on global food security (Gleick 2003; Falkenmark and Rockstrom 2004). At that time about 6.5 billion people lived on the planet (with 800 million undernourished) and roughly 4,500 km^3 y^{-1} of water was used for food production (Falkenmark and Rockstrom 2006). It was estimated that, to eradicate malnourishment and meet the needs of the additional 3 billion people expected to "appear" on the planet by 2050 the human appropriation of freshwater resources

1986

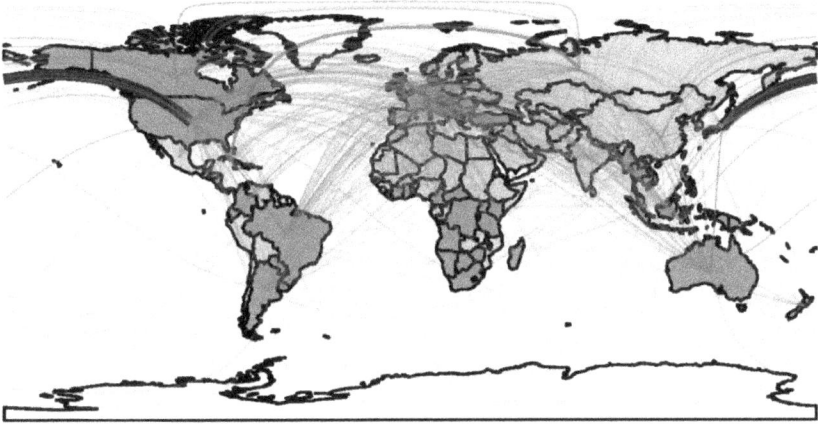

2011

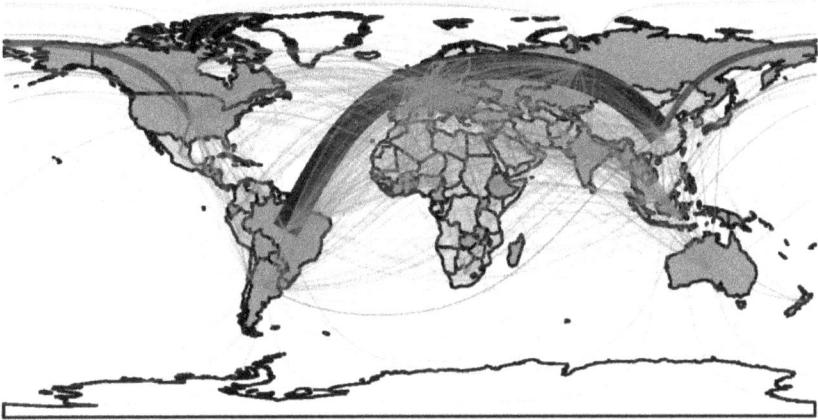

Figure 6.1 The global patterns of virtual water trade in 1986 (top) and 2011 (bottom). Net exporters are shown in dark gray; net importers in light gray and trade flows as lines with width proportional to the traded volumes of virtual water. Only trade flows that carry volumes of water greater than 95 percent of the total trade in 1986 are shown.

for food production should have increased by another 4,200 km^3 y^{-1} (Falkenmark and Rockstrom 2006). It is inconceivable that all of this water can come by increasing irrigation at the expense of surface water bodies and groundwater reservoirs. Rivers flowing through major agricultural regions are already overexploited and, due to agricultural withdrawals, many of them no longer reach the ocean (Richter 2014). Likewise, unsustainable use of aquifers (*overpumping*) is depleting groundwater resources worldwide (Konikow 2011). Thus, most of the water required to enhance agricultural production should come in the form of water savings by adopting *"more crop per drop"* technology, including the use of more efficient irrigation methods (e.g., drip irrigation), the choice of more suitable crop types and the development of new cultivars (Falkenmark and Rockstrom 2006). Moreover, new cultivation methods enhancing soil water storage and reducing evaporation can dramatically reduce unproductive water losses (Rockstrom 2002; Biazin et al. 2012).

This type of analysis stresses how water is becoming a major limiting factor for mankind's ability to produce enough food for everybody. In the meantime the human pressure on agro-ecosystems and their water resources has increased also because of the escalating demand for biofuels and the increase in food consumption per capita due to a shift to more water-demanding diets. In fact, as societies become more affluent, their populations tend to switch to diets richer in meat and other animal products (e.g., Office of Science 2011). Known as "Bennett's law" this phenomenon is associated with a dramatic increase in water consumption for food production because one calorie of meat-based food requires on average 4–10 times more water than one calorie of vegetable food (e.g., Falkenmark and Rockstrom 2004).

Altogether, population growth, changes in diets and increasing demand for agricultural products have led to an increase in water used for food production from 6.2×10^{12} m^3 y^{-1} in 1986 to 10.5×10^{12} m^3 y^{-1} in 2011 (Figure 6.2). Table 6.1 compares these volumes to major forms of human appropriation of freshwater resources and virtual water trade. Water for food includes both green water (i.e., rainwater transpired by crops) and blue water (i.e., irrigation water) and is therefore one order of magnitude greater than the rates of groundwater withdrawal and irrigation water consumption. We stress that the water footprint of food production is an important component of the global water cycle (Figure 6.3): in 2011 it accounted for roughly 14 percent of evapotranspiration from global land masses (estimated at 73.2×10^{12} m^3 y^{-1}, see Figure 6.3).

About 80 percent of the total water use for food production is accounted by 20 primary food commodities, including wheat ($\approx$12 percent, in 2011), rice ($\approx$11 percent), maize ($\approx$10 percent), cow milk ($\approx$9 percent), and cattle meat ($\approx$7 percent) (see Table 6.2).

Virtual Water Trade

The intensification of international trade (D'Odorico et al. 2014) has dramatically altered the way societies have access to water for food production. Trade allows us to consume food commodities produced in other regions of the world and virtually to rely on water resources available in those regions. As a result, there is the potential for an unbalance between the population inhabiting a certain area and the water resources available therein (Allan, 1998). Virtual water trade can help avoid mass migrations and water wars but creates a disconnect between consumers and the natural environment that supports them.

In recent years, research in hydrology has quantified the global patterns of virtual water trade (Hoekstra and Chapagain 2008; Suweis et al. 2011; D'Odorico et al. 2012) and investigated their temporal changes (Carr et al. 2012; 2013). This research has highlighted an intensification of virtual water trade over the last three decades. The global

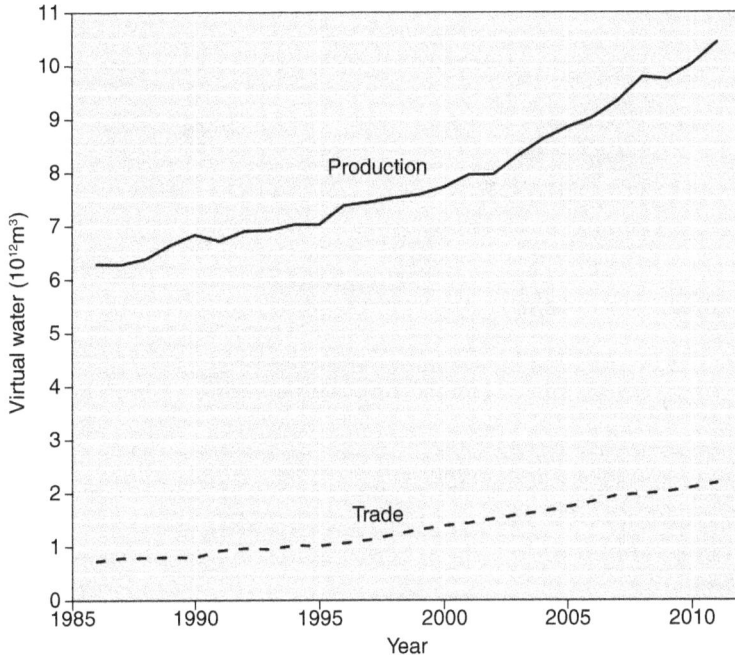

Figure 6.2 Recent changes in virtual water associated with global food production and trade (1986–2011).

Table 6.1 Comparison between the global rate of virtual water trade and other major virtual and real water fluxes in the Earth system.

	Annual flow (m³ y⁻¹)	Based on year	Source
Groundwater Depletion	0.14×10^{12}	2001–2008	*Konikow, 2011*
	0.28×10^{12}		*Wada et al., 2010*
Groundwater withdrawals	0.60–0.70×10^{12}		*Margat and Custodio, 2004*
	0.95–1.00×10^{12}		*Shah et al., 2007*
Groundwater used* for Irrigation	0.54×10^{12}	2000–2010	*Siebert et al., 2010*
Water grabbing**	0.45×10^{12}	2012	*Rulli et al., 2013*
Freshwater used*·** for food production	6.75×10^{12}	1996–2005	*Mekonnen and Hoekstra, 2011*
	10.5×10^{12}	2011	*This study*
Virtual water trade** (food only)	0.7×10^{12}	1986	*This study*
	2.2×10^{12}	2011	*This study*

The superscript * indicates a consumptive use of water, while ** denotes that both green and blue water are accounted for.

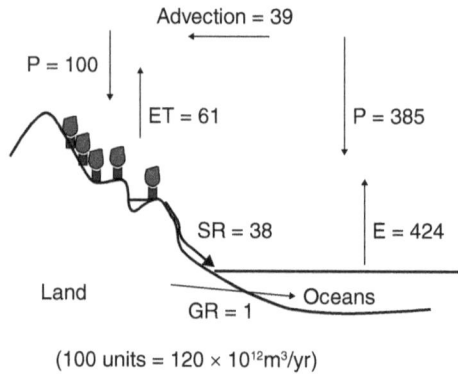

Advection = 39

P = 100

ET = 61

P = 385

SR = 38

E = 424

Land

GR = 1

Oceans

(100 units = 120 × 10¹²m³/yr)

Figure 6.3 A schematic representation of the global water cycle, where P is precipitation, ET evapotranspiration, E evaporation from the ocean, SR and GR surface and groundwater runoff, respectively (based on values from Chow et al., 1988).

Table 6.2 Production volumes (m^3) of virtual water in respective crop commodities for 1986 and 2011 along with the percentage (based on FAO (2015) and Mekonnen and Hoekstra (2011)).

Item	1986 Production	%	Item	2011 Production	%
Wheat	8.90×10^{11}	14.2	Wheat	1.21×10^{11} E+12	11.6
Cow milk, whole, fresh	6.70×10^{11}	10.6	Rice, paddy	1.12×10^{11} E+12	10.7
Rice, paddy	6.64×10^{11}	10.6	Maize	9.94×10^{11}	9.5
Cattle meat	5.07×10^{11}	8.1	Cow milk, whole, fresh	9.39×10^{11}	9.0
Maize	4.94×10^{11}	7.9	Cattle meat	6.84×10^{11}	6.6
Pig meat	2.89×10^{11}	4.6	Soybeans	6.20×10^{11}	5.9
Barley	2.51×10^{11}	4.0	Pig meat	5.19×10^{11}	5.0
Soybeans	2.08×10^{11}	3.3	Chicken meat	3.60×10^{11}	3.5
Sugar cane	1.89×10^{11}	3.0	Sugar cane	3.32×10^{11}	3.2
Sorghum	1.74×10^{11}	2.8	Hen eggs, in shell	1.93×10^{11}	1.9
Sugar beet	1.30×10^{11}	2.1	Sorghum	1.93×10^{11}	1.8
Chicken meat	1.13×10^{11}	1.8	Barley	1.93×10^{11}	1.8
Millet	1.09×10^{11}	1.7	Coconuts	1.52×10^{11}	1.5
Coconuts	1.03×10^{11}	1.6	Cassava	1.52×10^{11}	1.5
Oats	1.03×10^{11}	1.6	Rapeseed	1.43×10^{11}	1.4
Coffee, green	9.29×10^{10}	1.5	Sunflower seed	1.36×10^{11}	1.3
Hen eggs, in shell	9.00×10^{10}	1.4	Cottonseed	1.36×10^{11}	1.3
Cassava	7.17×10^{10}	1.1	Millet	1.34×10^{11}	1.3
Cottonseed	6.43×10^{10}	1.0	Coffee, green	1.25×10^{11}	1.2

network of food trade has become much more interconnected (Figure 6.1), while the water volumes that are virtually traded have more than trebled between 1986 and 2011 (Figure 6.3). In 1986 trade accounted for about 12 percent of the total human appropriation of freshwater resources for food production, and this percentage increased to 24 percent by 2011 (Table 6.1). Through trade humans virtually redistribute an amount of water comparable to roughly 3 percent of evapotranspiration from continental land masses. Twenty products account for 75 percent of the global virtual water trade in food commodities, including wheat ($\approx$12 percent, in 2011), soy beans ($\approx$11 percent), palm oil ($\approx$9 percent), cake of soy beans ($\approx$9 percent) and maize ($\approx$8 percent) (see Table 6.3). Moreover, the analysis of net exports shows that more than 50 percent of the volumes of virtual water traded in the global market are contributed (and controlled) by four countries (Carr et al. 2012): Brazil (16 percent), the United States (14 percent), Argentina (14 percent) and Indonesia (7 percent) (Figure 6.4).

Drivers of Virtual Water Trade

Empirical analyses of the virtual water trade network have shown how its structure can be explained in terms of the gross domestic product (GDP) of countries and annual rainfall on agricultural areas (Suweis et al. 2011). This research has highlighted some organizing principles underlying the structure of the virtual water network which is encoded in its topology. Virtual water trade establishes links among countries based on GDP: the

Table 6.3 Trade volumes (m³) of virtual water in respective crop commodities for 1986 and 2011 along with the percentage (based on FAO (2015) and Mekonnen and Hoekstra (2011)).

Item	1986 Trade	%	Item	2011 Trade	%
Wheat	1.31×10^{11}	17.9	Wheat	2.58×10^{11}	11.8
Coffee, green	6.64×10^{10}	9.1	Soybeans	2.08×10^{11}	10.7
Soybeans	5.12×10^{10}	7.0	Palm oil	1.60×10^{11}	9.3
Maize	4.71×10^{10}	6.4	Cake of soybeans	1.34×10^{11}	8.6
Cake of soybeans	3.55×10^{10}	4.9	Maize	1.17×10^{11}	8.2
Sugar raw centrifugal	3.28×10^{10}	4.5	Coffee, green	9.43×10^{10}	7.2
Cocoa beans	2.75×10^{10}	3.8	Chocolate Prsnes	8.81×10^{10}	7.2
Palm oil	2.42×10^{10}	3.3	Cattle meat	7.35×10^{10}	6.5
Barley	1.99×10^{10}	2.7	Cocoa beans	7.30×10^{10}	6.9
Cattle meat	1.98×10^{10}	2.7	Rapeseed	4.93×10^{10}	5.0
Sunflower oil	1.22×10^{10}	1.7	Sugar raw centrifugal	4.82×10^{10}	5.2
Cattle meat	1.21×10^{10}	1.7	Soybean oil	4.58×10^{10}	5.2
Sugar refined	1.18×10^{10}	1.6	Sunflower oil	4.39×10^{10}	5.2
Sorghum	1.03×10^{10}	1.4	Sugar refined	4.03×10^{10}	5.1
Soybean oil	1.02×10^{10}	1.4	Chicken meat	3.80×10^{10}	5.0
Chocolate Prsnes	9.69×10^{9}	1.3	Barley	3.12×10^{10}	4.3
Cassava dried	9.03×10^{9}	1.2	Cocoa butter	2.96×10^{10}	4.3
Tea	8.80×10^{9}	1.2	Flour of wheat	2.78×10^{10}	4.2
Coconut (copra) oil	8.64×10^{9}	1.2	Olive oil, virgin	2.64×10^{10}	4.2

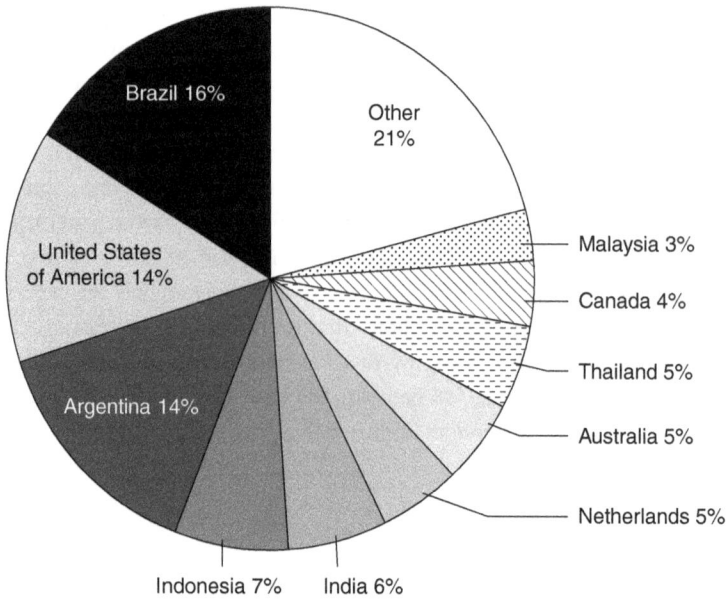

Figure 6.4 Contributions to net virtual water exports from major exporters.

number of trade partners of each country appears to depend on its social development status (Suweis et al. 2011), suggesting that the globalization of water is driven mainly by economics. However, some geographical factors such as distance, location, rainfall and agricultural area play a role in determining the magnitude of fluxes among countries (Suweis et al. 2011).

Even though there is no easy explanation for drivers of virtual water trade, population, GDP and geographical distance have been found to be major factors explaining virtual water flows, whereas the agricultural production of the exporting country seems to play only a minor role (Tamea et al. 2013). A temporal analysis of virtual water trade has highlighted how the dependence on GDP has decreased over time while, not surprisingly, distance has remained a major factor determining global trade patterns (Tamea et al. 2013). The analysis of the virtual water network shows, however, that its community structure can only partly be explained by geographic proximity. Many of the communities of trade (defined as groups of network nodes with dense internal connections and sparser connections to other communities) include countries that are not geographically contiguous but span around the world.

The global distribution of wealth (GDP) and population does not explain the strength of the community structure of the virtual water trade networks observed in the past few decades, but a statistical ("gravity") relationship accounting for the demand for and availability of virtual water was found to have higher explanatory power (D'Odorico et al. 2012).

Indeed, water availability for food production was found to play a role in shaping the patterns of virtual water trade. Water is a source of comparative advantage in the global trade of agricultural products in the sense that countries that are more water abundant tend to export products that are more water intensive (Debaere 2014). The traditional production factors (i.e., capital and labor), however, play a more important role than water in shaping the patterns of trade (Debaere 2014).

Methods for Network Reconstruction

The reconstruction of the trade of virtual water requires (a) the detailed description of trade of goods and commodities, and (b) the virtual water content of those commodities. Considering only water use for food commodities, production and detailed trade data for food commodities are freely available through the Food and Agricultural Organization of the United Nations (FAO) for numerous plant and animal commodities—beginning in 1961 for production and 1986 for detailed trade (FAO 2015). These data sets currently incorporate trade and production of roughly 574 primary and secondary commodities for plants and animals. Based on this trade data, for each commodity, m, and year t, a trade matrix, $\boldsymbol{T}_m(t)$ can be constructed, where the (i,j) entry of matrix $\boldsymbol{T}_m(t)$ is the export of commodity m from country i, to country j. A corresponding production vector, $p_m(t)$ can be created with entry i, corresponding to production of commodity m in country i. In order to convert to virtual water equivalents and develop an estimate of the blue and green virtual water footprint for each commodity m, at time t, in country i, is required.

The water footprint of a crop is typically estimated based on field scale evapotranspiration (integrated across a growing season) and crop yield (Mekonnen and Hoekstra 2010). In this manner the water footprint is comprised of both blue (surface and groundwater) and green (rainfall) components. A tertiary part of the water footprint—the *gray water* component—considers the volume of water required for dealing with the pollutant load due to nitrogen fertilization (Mekonnen and Hoekstra 2010; O'Bannon et al. 2014). These footprints can be estimated leveraging various FAO data sets, but temporal averages are readily available from Mekonnen and Hoekstra (2010).

Using these country-specific, time-averaged blue and green virtual water estimates for each commodity WC_m and converting the individual commodity trade matrices to water equivalents by $\boldsymbol{W}_m(t) = WC_m \bigcirc \boldsymbol{T}_m(t)$ (where the symbol "$\bigcirc$" represents the component-wise multiplication), the total water trade matrix is then simply $\boldsymbol{W}(t) = \Sigma_m \boldsymbol{W}_m(t)$. Production values of virtual water for each country can similarly be calculated with care taken to omit double accounting of water use by only using primary products rather than both primary and secondary commodities (Carr et al. 2015). The virtual water trade balance along each link (i,j) is calculated as the difference between imports and exports and expressed by the matrix $\boldsymbol{W}_{net}(t) = \boldsymbol{W}(t) - \boldsymbol{W}(t)^T$ that defines the net virtual water flow for each connection (i,j). Limiting water use to edible commodities, a shortened commodity list of approximately 253 commodities, 149 of which are primary, remain (Carr et al. 2015). Typically, the virtual water of production is calculated with reference to primary commodities only, whereas trade encompasses both primary and secondary commodities.

Limitations and Issues

One of the major concerns in any analysis of trade and production is the concept of double accounting. More specifically, the water used for production of primary goods (e.g., wheat) should either be removed from the virtual water content estimates of secondary goods (bread, pasta) or, alternately, the virtual water of production should be limited to only primary commodities. Typically, the latter case is leveraged as virtual water estimates of goods incorporate the entire lifecycle of the commodity in question (Mekonnen and Hokstra 2010). The addition of live animal trade and production is also non-trivial, as life cycles and age are not reported in the trade and production data, and carcass weights and yields vary, not only by country, but by animal breed as well. It is also important to note that water use for animals incorporates both the production of feed crops and byproducts of food processing. Similarly, some portion of production is retained as seed for subsequent crops. As such, when linking food production and trade to food security, feed and seed is typically removed. Another limitation in linking food security to food trade is the lack of detailed information on fish trade and production. While this does not affect volumes of virtual water trade and production, it does affect the evaluation of food security, especially when the virtual water of global diets and nutritional sufficiency are investigated (Gephart et al. 2014).

Trade, Inequality and the Right to Water

The global distribution of total renewable freshwater resources among countries is strongly unequal with a Gini coefficient (G), a common metric of inequality, of 0.65 (Seekell et al. 2011). A smaller inequality exists in the water resources used for food production (G≈0.32). Thus, inequality in the access to water originates from the natural distribution of resources (Seekell et al. 2011). Is such an inequality iniquitous? Natural distributions of resources exist, however; it is the way institutions deal with natural inequality that may be just or unjust (Rawls, 1999). Trade is one of the most effective mechanisms that can modify the (virtual) distribution of natural resources, including those that are crucial to food production. How does trade affect inequality in access to water for agriculture? Comparing patterns of virtual water distribution with and without international food trade we find that on average trade enhances equality because the Gini coefficient drops to about 0.25 once the effect of trade is accounted for (Carr et al. 2015). In fact, while trade has overall reduced inequality in the global distribution of (virtual) water, it has the potential to almost fully redress this inequality (Figure 6.5). However, the majority of trade does not *directly* redress inequality, implying that trade decisions could further improve the global distribution of (virtual) water resources.

The analysis of water justice should be based on a comprehensive evaluation of the causes of inequality (i.e., natural, or human-induced?), whether countries with more limited access to freshwater resources have enough to meet their needs, whether those needs constitute human rights and to what extent those rights entail obligations by the state or international community.

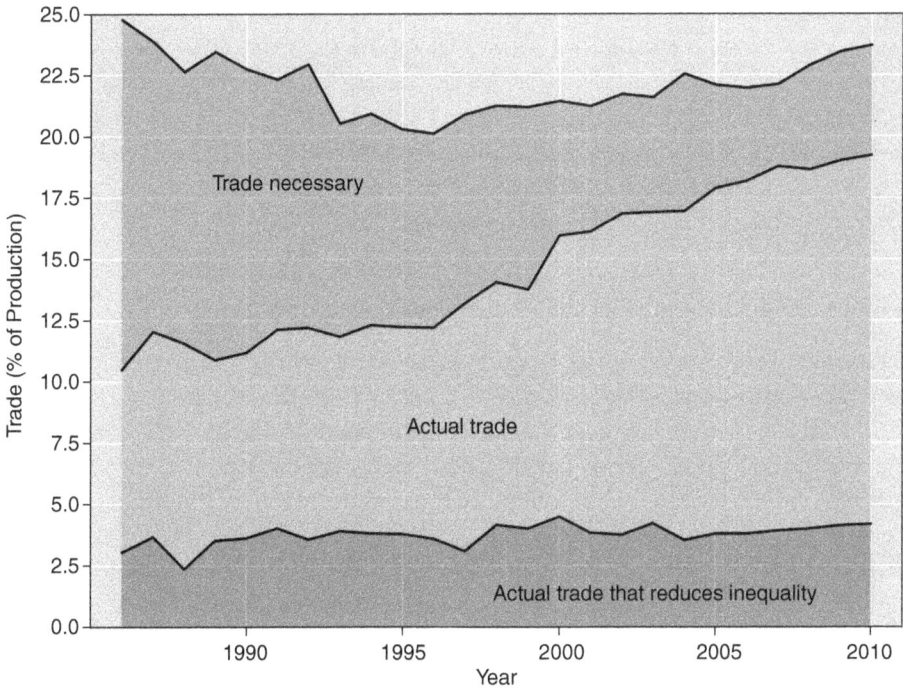

Figure 6.5 The trade necessary to fully redress inequality between population distribution and food production.

The United Nations has declared that adequate access to food and water are human rights. In the context of this chapter (i.e., the water–food nexus) we concentrate on water for food production. Therefore, because of the human right to food (UN, 1948) we invoke a right to water for food production.

Moral and political philosophers often distinguish two broad classes of human rights: negative and positive (Fried, 1978). Negative rights (e.g., the right to freedom of the press) can be enjoyed without requiring any action from the state or other parties. They just entail the obligation not to act against the press. Conversely, positive rights (e.g., the right of shelter) require that some resources be provided and, therefore, entail an obligation to act in such a way that those rights are met (e.g., Dasgupta, 1993; Pogge 2008). The fulfillment of positive rights requires a redistribution of resources in order to give everybody access to a bundle of "first tier" goods to meet their most basic vital needs. This reasoning allows us to evaluate the relationship between inequality and justice. While inequality is a fact of nature and is not necessarily unjust, it may leave some people without access to enough food to be free from hunger. Because violation of human rights is by definition unjust, both natural and man-made inequalities are unjust when the most basic vital needs are not met. Also, the right to food and its extension to the right to water for food production are positive rights (Carr et al. 2015); therefore, they entail a positive obligation. When some individuals, social groups or entire countries are not free from hunger, adequate institutions are needed to ensure that rights to food are met. As noted, trade, which accounts also for food aid, overall improves access to water, increases equality and reduces the number of people who are not free from hunger (Carr et al. 2015).

Nevertheless, it cannot be claimed that the virtual water trade is necessarily just. In fact, it could be argued that trade allows some avoidable deficits of human rights to persist. Moreover, even though *on-average* trade improves equality, in some cases it might act against the rights of some countries or individuals.

Virtues and Vices of Virtual Water Trade

Based on the above discussion, we can recognize some virtues in virtual water trade. It increases water equality (Carr et al. 2015), is associated with water savings (Dalin et al. 2012) and allows some societies to meet their water and food needs without engendering mass migrations, famine or conflict (Allan 1998). For instance, during the drought that affected the Sahel in the last three decades of the twentieth century the international community was able to respond to the associated decline in agricultural yields by increasing crop production in the United States, which reduced the societal effects of this climate anomaly. Food aid is itself a form of water solidarity associated with virtual water transfers among countries.

What are the downsides of virtual water trade? By making food and associated virtual water available to regions with chronic food deficits, trade allows some populations to grow unsustainably in conditions of dependency on resources they do not control. Such dependency has a negative impact on their food security and societal resilience (D'Odorico et al. 2010). It has been shown (Suweis et al. 2013) that the long-term trajectories of demographic growth in trade-dependent countries exhibit a dependency both on domestic resources and on virtual water available through trade; conversely the long-term growth of self-sufficient populations seems to rely on resources that are presently committed to exports to trade-dependent countries. These results indicate that both country groups will rely in the long run on the same pool of resources, and that there is an unbalance between populations and the resources they depend on. It is expected that in the long run net exporters will start reducing their exports, as happened with the export bans in the course of the recent food crises (e.g., Fader et al. 2012). The globalization of food through trade can be considered (D'Odorico and Rulli 2013) as a third food revolution (after the Industrial Revolution and Green Revolution). Its effect has been not to increase production but to redistribute virtual water from areas of excess to areas of deficit. Clearly, given the finite resources of the planet, this is not a long-term solution to assuring global water and food security, especially as its main impact is to maximize the use of agricultural resources and erode those safety margins (i.e., "redundancy") that historically have been used to give resilience to the global food system (D'Odorico et al. 2010). More recently, other studies have looked at the coupled food–population dynamics and shown how the increasing globalization of water and food by international trade is reducing the resilience of the global food system and increasing its instability and susceptibility to global instability and vulnerability to shocks (Puma et al. 2014; Suweis et al. 2015).

At the same time, by creating a disconnect between consumers and the land that supports them, international trade has the effect of undermining the ethic of environmental stewardship (Carr et al. 2012; O'Bannon et al. 2014). Good stewards of the land are typically people who live close to it. They care for its sustainable use and preservation for the generations to come. They also have the traditional knowledge that allows them to sustainably manage the land (Chapin et al. 2009).

As a result of the disconnect between people and the environment they rely on, consumers are not directly affected by the environmental impacts of their decisions. For instance, commercial farming typically entails pollution costs associated with the release of fertilizers and pesticides. The global patterns of gray water trade can be used to evaluate the pollution of the producer's land and water resources, an important externality of food production. In other words, importers virtually export environmental degradation to the producing countries (O'Bannon et al. 2014). This global displacement of land use has been recognized as a consequence of the globalization of food through trade (Meyfroidt et al. 2013).

It has also been shown that virtual water trade is not a long-term solution for water scarcity and food insecurity. The long-term trajectories of demographic growth show that the populations of the trade-dependent countries are increasingly relying on the virtual water available to them through imports. At the same time, the long-term projections of population growth in net exporting countries seem to indicate that these countries will soon self-rely on the water resources they are currently exporting (Suweis et al. 2013). Thus, we expect that in the long run exporters will reduce their exports, leaving food-insecure countries in a state of water scarcity, similar to what happened during the food crises of 2008 and 2011, when some countries banned exportation of food (Fader et al. 2013).

In conclusion, global food security has increasingly come to depend on virtual water trade. About 24 percent of the water used for food production is available to human societies through trade. Virtual water trade enhances equality and is an effective short-term solution to conditions of crisis but, in the long run, it consolidates the dependency of some societies on resources they do not control and disconnects the people from the environmental resources they rely on.

References

Allan, John. A. 1998. "Virtual Water: A Strategic Resource." *Global Solutions to Regional Deficits. Groundwater*, 36: 545–46. doi:10.1111/j.1745–6584.1998.tb02825.x.

Biazin, Birhanu, Geert Sterk, Melesse Temesgen, Abdu Abdulkedir and Leo Stroosnijder. 2012. "Rainwater Harvesting and Management in Rainfed Agricultural Systems in Sub-Saharan Africa: A Review." *Physics and Chemistry of the Earth, Parts A/B/C* 47: 139–51.

Carr, Joel A., Paolo D'Odorico, Francesco Laio, and Luca Ridolfi. 2012. "On the Temporal Variability of the Virtual Water Network." *Geophysical Research Letters* 39(6).

———. 2013. "Recent History and Geography of Virtual Water Trade." *PLoS One* 8(2): e55825. doi:10.1371/journal.pone.0055825.

Carr, J. A., David A. Seekell and P. D'Odorico. 2015. "Inequality or Injustice in Water Use for Food?" *Environmental Research Letters* 10(2): 024013. doi:10.1088/1748–9326/10/2/024013.

Chow, Ven T., David R. Maidment and Larry W. Mays. 1988. *Applied Hydrology*. New York: McGraw-Hill.

Dalin, Carole, Megan Konar, Naota Hanasaki, Andrea Rinaldo and Ignacio Rodriguez-Iturbe. 2012. "Evolution of the Global Virtual Water Trade Network." *Proceedings of the National Academy of Sciences* 109(16): 5989–94.

Dasgupta, Partha. 1995. "An Inquiry into Well-being and Destitution." *OUP Catalogue*. Oxford: Oxford University Press.

Debaere, Peter Marcel. 2012. "The Global Economics of Water: Is Water a Source of Comparative Advantage?" *American Economic Journal: Applied Economics* 6(2) 32–48.

D'Odorico, Paolo, Joel Carr, Francesco Laio and Luca Ridolfi. 2012. "Spatial Organization and Drivers of the Virtual Water Trade: A Community-tructure Analysis." *Environmental Research Letters* 7(3): 034007. doi:10.1088/1748–9326/7/3/034007.

D'Odorico, Paolo, Joel A. Carr, Francesco Laio, Luca Ridolfi and Stefano Vandoni. 2014. "Feeding Humanity Through Global Food Trade." *Earth's Future* 2(9): 458–69. doi:10.1002/2014EF000250, 2014.

Erisman, Jan Willem, Mark A. Sutton, James Galloway, Zbigniew Klimont and Wilfried Winiwarter. 2008. "How a Century of Ammonia Synthesis Changed the World." *Nature Geoscience* 1(10): 636–39.

UNION, PEAN. 2009. "Directive 2009/65/EC of the European Parliament and of the Council." *Official Journal of the European Union L* 302: 33.

Fader, Marianela, Dieter Gerten, Michael Krause, Wolfgang Lucht and Wolfgang Cramer. 2013. "Spatial Decoupling of Agricultural Production and Consumption: Quantifying Dependences of Countries on Food Imports Due to Domestic Land and Water Constraints." *Environmental Research Letters* 8(1): 014046.

Faostat, F. A. O. 2015. "Statistical Databases." *Food and Agriculture Organization of the United Nations*.

Falkenmark, Malin and Johan Rockström. 2004. *Balancing Water for Humans and Nature: The New Approach in Ecohydrology*. Earthscan.

Falkenmark, Malin and Johan Rockström. 2006. "The New Blue and Green Water Paradigm: Breaking New Ground for Water Resources Planning and Management." *Journal of Water Resources Planning and Management* 132(3): 129–32.

Fried, Charles. 1978. *Right and Wrong* (vol. 153). Harvard University Press.

Gephart, Jessica A., Michael L. Pace and Paolo D'Odorico. 2014. "Freshwater Savings from Marine Protein Consumption." *Environmental Research Letters* 9(1): 014005.

Gleick, Peter. 1993. *Water Crisis: A Guide to the World's Freshwater Resources*. New York: Oxford University Press.

Hoekstra, Arjen. 2008. "Globalization of Water." *Water Encyclopedia*.

Konikow, Leonard F. 2011. "Contribution of Global Groundwater Depletion Since 1900 to Sea-level Rise." *Geophysical Research Letters* 38(17): L17401.

Margat, J. and E. Custodio, "Economic Aspects of Groundwater Use." In: Groundwater Resources of the World: and Their Use. Edited by Igor S Zektser and Everett Lorne. *IhP Series on Groundwater*, no. 6. UNESCO.

Mekonnen, Mesfin M. and Arjen Y. Hoekstra. 2010. "The Green, Blue, and Grey Water Footprint of Crops and Derived Crop Products." vol. 1: Main Report. (http://waterfootprint.org/media/downloads/Report47-WaterFootprintCrops-Vol1.pdf).

———. 2011. National Water Footprint Accounts: The Green, Blue and Grey Water Footprint of Production and Consumption, vol. 2: Appendices. Value of Water Research Report Series No. 50, UNESCO-IHE, Delft, The Netherlands http://www.waterfootprint.org/Reports/Report50-NationalWaterFootprints-Vol2.pdf.

Meyfroidt, Patrick, Eric F. Lambin, Karl-Heinz Erb and Thomas W. Hertel. 2013. "Globalization of Land Use: Distant Drivers of Land Change and Geographic Displacement of Land Use." *Current Opinion in Environmental Sustainability* 5(5): 438–44.

O'Bannon, Clark, Joel Carr, David A. Seekell and Paolo D'Odorico. "Globalization of Agricultural Pollution Due to International Trade." *Hydrology and Earth System Sciences* 18(2): 503–10.

Pogge, Thomas W. 2008. *World Poverty and Human Rights*. Boston: Polity.

Puma, Michael J., Satyajit Bose, So Young Chon and Benjamin I. Cook. "Assessing the Evolving Fragility of the Global Food System." *Environmental Research Letters* 10(2): 024007.

Rawls, John. 2009. *A Theory of Justice*. Cambridge, MA: Harvard University Press.

Richter, Brian. 2014. *Chasing Water: A Guide for Moving from Scarcity to Sustainability*. Island Press.

Rockström, Johan, Malin Falkenmark, Mats Lannerstad and Louise Karlberg. 2012. "The Planetary Water Drama: Dual Task of Feeding Humanity and Curbing Climate Change." *Geophysical Research Letters* 39(15). doi:10.1029/2012GL051688.

Rockström, Johan, Jennie Barron and Patrick Fox. "Rainwater Management for Increased Productivity among Small-Holder Farmers in Drought Prone Environments." *Physics and Chemistry of the Earth, Parts A/B/C* 27(11): 949–59.

Rosegrant, Mark W., Claudia Ringler and Tingju Zhu. "Water for Agriculture: Maintaining Food Security under Growing Scarcity." *Annual Review of Environment and Resources* 34(1): 205. doi:10.1146/annurev.environ.030308.090351.

Rulli, Maria Cristina, Antonio Savior and Paolo D'Odorico. 2013. "Global Land and Water Grabbing." *Proceedings of the National Academy of Sciences* 110(3): 892–97.

Seekell, David A., Paolo D'Odorico and Michael L. Pace. 2011. "Virtual Water Transfers Unlikely to Redress Inequality in Global Water Use." *Environmental Research Letters* 6(2): 024017.

Shah, Tushaar, Jacob Burke, Karen Villholth, M. Angelica, E. Custodio, F. Daibes, J. Hoogesteger et al. 2007. *Groundwater: A Global Assessment of Scale and Significance*. No. H040203. International Water Management Institute.

Siebert, Stefan, Jacob Burke, Jean-Marc Faures, Karen Frenken, Jippe Hoogeveen, Petra Döll and Felix Theodor Portmann. 2010. "Groundwater Use for Irrigation: A Global Inventory." *Hydrology and Earth System Sciences* 14(10): 1863–80.

Suweis, Samir, Megan Konar, Carol Dalin, Naota Hanasaki, Andrea Rinaldo and Ignacio Rodriguez-Iturbe. 2011. "Structure and Controls of the Global Virtual Water Trade Network." *Geophysical Research Letters* 38(10): L10403

Suweis, Samir, Andrea Rinaldo, Amos Maritan and Paolo D'Odorico. 2013. "Water-controlled Wealth of Nations." *Proceedings of the National Academy of Sciences* 110(11): 4230–33.

Suweis, Samir, Joel A. Carr, Amos Maritan, Andrea Rinaldo and Paolo D'Odorico. 2015. "Resilience and Reactivity of Global Food Security." *Proceedings of the National Academy of Sciences* 112(22): 6902–07. doi: 10.1073/pnas.1507366112.

Tamea, Stefania, J. A. Carr, Francesco Laio and Luca Ridolfi. 2014. "Drivers of the Virtual Water Trade." *Water Resources Research* 50(1): 17–28. doi:10.1002/2013WR014707.

UN General Assembly. 1948. "Universal Declaration of Human Rights." *UN General Assembly*.

US Congress, Independence, Energy. 2007. "Security Act of 2007." *Public Law* 110(140): 19.

Wada, Yoshihide, Ludovicus P. H. van Beek, Cheryl M. van Kempen, Josef W. T. M. Reckman, Slavek Vasak and Marc F. P. Bierkens. 2010. "Global Depletion of Groundwater Resources." *Geophysical Research Letters* 37(20): L20402.

Chapter Seven

A HYBRID ANALYTICAL APPROACH FOR MODELING THE DYNAMICS OF INTERACTIONS FOR COMPLEX WATER PROBLEMS

Yosif Ibrahim and Shafiqul Islam

Abstract

Contemporary water management problems involve a wide range of variables, processes, actors and institutions with competing and often conflicting constraints. These water management problems often cross different physical, jurisdictional and political boundaries. Traditional systems engineering approaches—with known equilibrium principles and well defined system boundaries and causal mechanisms—are not well suited for such boundary crossing problems. When natural and societal systems are coupled—as in the case of many complex water management problems—underlying assumptions of systems engineering are often violated. There is a need for changing the water management paradigm and practices from traditional systems engineering to complex adaptive management. A new domain of knowledge—known as complexity science—is emerging to address these complex management problems where crossing of boundaries, interconnectedness, nonlinearity, feedback, and sensitivity to timing and small changes can profoundly affect system performance. Despite general awareness about these complexities, there is a gap in available analytical tools needed to critically examine the nature and implication of different adaptive water management options. Using network theory and agent based modeling, this study presents a simple network model to address this gap between theory and practice of complex water resources management. Using a hypothetical transboundary water case study with a range of competing and conflicting agenda for varying number of agents, preliminary analyses of different scenarios and simulations suggest that (a) for sustainable and effective evolution and implementation of transboundary water negotiations, it is important to create opportunities for 4 to 8 agents to form alliance with shared objectives and mutual gains; (b) irrespective of initial positions of competing agents, there appears to be opportunities for resolution(s) to emerge and prevail for a well structured negotiation process; (c) a balance between rational response, contextual difference, and randomness in agent's behavior allows system to evolve adaptively and follow a sustainable trajectory; and

(d) an optimal ratio between competing agendas and availability of resources is needed to achieve and sustain mutual gains.

Context and Motivation

Historically, water resources management addressed well-defined problems and traditional systems engineering techniques were used to solve these problems. Such approaches work well in addressing simple and complicated problems where there is a clear cause-and-effect relationship and the system has a well-defined boundary. While such approaches are intended to be objective, their performances and findings become highly subjective because a range of simplifying assumptions and unaccounted-for uncertainties need to be made in order to make them mathematically tractable.

On the other hand, complex problems are ill-defined, with ambiguous cause-effect relationships, and are often associated with strong ethical, political and professional values and issues. For complex water problems, the certainty of solutions and degree of consensus vary widely (Islam and Susskind 2012). In fact, there is often little consensus about what the problem is, let alone how to resolve it. Furthermore, complex problems are constantly changing because of interactions among the natural, societal and political processes involved. Addressing these problems requires a differentiation between systems thinking and systems engineering. Systems engineering requires a clear understanding of the relationships among different components of the system, its boundaries, its interactions at those boundaries and the ways in which perturbing parts of the system will change outcomes. For complex systems, is nearly impossible to enumerate such a clear delineation of system components, boundaries and interactions and represent the system within a mathematically tractable framework.

Complex water problems cross multiple domains (natural, societal, political) at different scales (e.g., time, space, jurisdictional, institutional), and aspects of these problems occur at different levels within each scale (e.g., short-term or long-term duration; household, city, county, state, etc.). For example, a watershed is usually a common unit of analysis for assessing basin-level resource characterization and management. However, a watershed may not always be aligned with the "problem-shed" and "policy-shed" for effective and efficient management for a complex water problem (Islam and Susskind 2012). Cohen and Davidson (2011) argue that there are at least five recognized challenges associated with the watershed approach to water resource management: the difficulties of boundary choice, accountability, public participation and asymmetries with "problem-sheds" and "policy-sheds." The interactions between domains or scales at various levels contribute to the complexity of a water problem and preclude solutions that were applicable for simple and complicated systems. Hence, there is a need for changing the management paradigm and practices from traditional systems engineering to complex adaptive management.

Many natural, societal and political systems are characterized by behavior that emerges as a result of often-nonlinear interactions and feedback among elements of the system at different scales. These systems are coming to be known as a complex adaptive system (CAS) in many disciplines (e.g., Gell Mann 1994; Holland 1998; Kaufmann 1995; Bar-Yam 1997). A CAS is a dynamical system capable of adapting to and evolving with a changing environment; consequently, it is important to appreciate that the system and its environment are not easily separable, and each affects, as well as changes, the other. A CAS is characterized by non-linearity and interdependency, in which small changes can create large impacts. Complex behaviors from these systems can emerge from simple rules applied locally within a CAS. Managing a CAS requires co-production of knowledge because co-evolution is inherent in the system through the process of emergent possibilities. A CAS approach for managing complex water problems is more appropriate than a typical systems engineering approach (De Weck et al. 2011; Miller and Page 2007; Rammel et al. 2007). The analysis and management of CAS require a combination of theoretical, experimental and applied methods (see Maguire et al. 2006; Allen et al. 2011).

Knowledge Gaps and Research Questions

The study of CAS is one of the fastest-growing areas of research in the natural and social sciences (Allen et al. 2011). In fact, management of natural resource systems—where natural, societal, and political processes are intricately coupled with nonlinear feedback—has begun to adapt interdisciplinary perspectives and tools available in managing CAS (Arthur et al. 1997; Levin, 1998; Janssen 1998; Ramos-Martin 2003; Rammel et al. 2007). There is increasing awareness that management of water resources systems—in the midst of often competing and conflicting needs and demands—can benefit from a CAS approach (Pahl-Wostl 2006). Islam and Susskind (2012) suggested the use of a CAS—using network theory to guide the implementation of their Water Diplomacy Framework (WDF)—to manage complex water networks. While WDF offers a systematic approach—based on complexity science and mutual gains—there is a gap between the theoretical construct of WDF and practice of WDF while treating water management as a CAS.

The present study is an attempt to begin addressing this gap by arguing that complex water resources systems are characterized by the interdependencies and feedback of elements from various scales and are driven by interactions among elements from natural, societal and political systems. We argue that sustainable water management requires explicit appreciation of uncertainty and feedbacks among interrelated elements of interacting systems. To operationalize such an appreciation and broaden our analytical scope, we consider a CAS perspective to provide an appropriate conceptual umbrella for addressing key aspects of adaptive complex water management problems. However, using a CAS perspective does not necessarily mean the development of new methodology; rather, it emphasizes the need for a novel and interdisciplinary combination of existing tools (such as agent-based system modeling, network-theory-based analysis, institutional analysis and development framework, multi-criteria appraisal, among others) to capture co-evolutionary interactions and complex dynamics of elements/agents active for a given water resource management problem. Using network theory and agent-based modeling, this present study develops a simple network model to address this gap between theory and practice of complex water resources management.

A Hybrid Network Model for a Coupled Natural and Societal System

We present a hybrid analytical approach, building on ideas from complexity science, network theory and traditional systems engineering. As discussed earlier, many water problems are complex because they arise when natural, societal and political forces interact and cross boundaries and scales. Water resources might be more effectively managed if we paid more attention to the interactions among the natural, societal and political forces that constitute water networks. These networks can best be understood as an interconnected set of nodes and links representing socio-ecological variables and processes set in political reality. The type and flow of information among the nodes define the level of complexity of water networks and water problems. The challenge is to understand the type and flow of information among the nodes and links in a water network and use that information to enhance the operational skills that decision makers and others in positions of responsibility need to manage these interactions.

In our proposed water network representation, agents are represented as nodes, and their relationships are modeled as links. An agent here represents an entity that influences the decision-making process. Such an entity could be a project or a group of actors/stakeholders involved in influencing the water management scenario. In such a representation, agents may compete, cooperate or remain neutral at different times. Consequently, such a network can evolve depending on the dynamics of interactions between agents and their links. Agents are capable of looking for situational conditions to maximize their position in the "fitness landscape" of their "inner environment" or local network. Fitness landscape was introduced by Kauffman (1995) within the context of an evolutionary network to visualize the relationship between genetic make-up of a cell (agent), and its reproductive success; fitness represents the "height" of the landscape and cells (agents) which are similar and said to be "close" to each other. In addition to its influence within the inner environment, an agent in a network can influence, and be influenced by, the "outer environment" representing the larger network. Stability is found when each agent fits successfully into a local network of alliance, and the alliance fits successfully into the larger network. If the outer environment (larger network) changes, then the alliance as well as agents within the alliance need to evolve and adapt to meet the nature of the changing landscape. This dynamic interaction of agents within their local and larger networks allows the proposed modeling framework to represent the adaptive nature of a complex water management situation. The proposed hybrid analytical approach introduced here is inspired by the generalized version of the simple random interaction model as introduced by Allen (2001) in studying the collective emergent behavior and the dynamic of interaction among agents. Table 7.1 provides a summary description of variables and parameters used in formulating the proposed model.

To model the dynamics of interactions among different competing elements/agents for a water resource management problem, consider first the simple interactions between

Table 7.1 Description of parameters and state variables for the proposed model.

Designation	Description	Value
m	Parameter which represents the number of agents (nodes) considered in the modeling system	m is varied to take values of 10, 16 and 20
N	Parameter that limit the resource availability and provide control to growth of individual agents, so that it cannot grow indefinitely	N is varied to take values of 20 and 30 (Table 6)
k	Agent connectivity parameter which measure how many nodes the agents is connected with	Varies between 0 and 20. $k=0$ means no connection; $k=20$ means agents is connected to 20 agents or nodes
ϕ	The inclination angle is state variable that measure the degree of cooperation (conflict) between two agents i and j	$0°\leq\phi\leq180°$
$X_j(t)$	State variable measure the growth of interest with time in pursuing agenda of agent j	Varies between 0 and X_{max} (Table 6)
$y_j(t)$	State variable that normalize the growth of interest for each agent given by Equation (7)	Varies between 0 and 1
ρ	Parameter that represent intensity of competition	Varied to have values of 10, 15 and 20 (Table 6)
by	Parameter describing the proportionality of growth rate for each agent j as a function of X_j	Varies from 0.15 to 0.23 depending on the agent (Table 6)
ψ_j	Parameter representing the average death rate or life-span of certain agent j	Kept constant for all agents in the system at a value of 0.01 (Table 6)
$Z_j(t)$	State variable measure the agent interaction described by Equation (8)	Varies between -1 and 1
$ZA_j(t)$	State variable represent the rational agent behavior described by Equation (9)	Varies between 0 and 1
$ZC_j(t)$	State variable represent deviation or perturbation around rational agent behavior to include subjectivity element in an agent behavior approximated by Equation (10)	Varies between 0 and 1
$ZR_j(t)$	State variable represent randomness in agent behavior associated with other unaccounted uncertainties or externalities approximated by power law distribution (Equation (10))	Varies between 0 and 1
f_r	Represent the strength or intensity of interaction	Varies between 0 and 1
$NI_j(t)$	State variable represent the net effect of interactions on agent j at time t (Equation (2))	Varies from 0 to $\sum x_j$
$CR_j(t)$	State variable represent the competition effect of other agents on agent j at time t (Equation 3)	Varies from 0 to $\sum x_j$

two agents i and j competing for resources N (Figure 7.1). The outcome of the interaction process—cooperative, neutral or conflictive—is described by Equation (1) below:

$$\text{Interaction } (i, j) = f_r.\mathcal{Z} = f_r.\cos\phi \tag{1}$$

where Interaction (i,j) represents the interaction of agent i on j; f_r represents the strength or intensity of interaction; $\mathcal{Z}$ is the interaction effect of agent i on j which varies between $(-1,1)$; and ϕ is the inclination angle between the two entities i and j (Figure 7.1). The inclination angle is a parameter that measures the degree of cooperation (conflict) between two agents i and j. When the inclination angle ϕ is in the range of $0°$ to $90°$, cooperative behavior is anticipated; when ϕ is within the range of $90° \leq \phi \leq 180°$ negative and conflictive interaction is expected. With m agents interacting with each other in a network, the net effects of other $(m-1)$ agents present in the system in creating influence on interest of agent j can be described by Equation (2):

$$NI_j(t) = \sum_{i=1}^{m-1} X_i(t).Cos\left[\varphi(t) + \omega_i(t)\right] = \sum_{i=1}^{m-1} f_r.\mathcal{Z}_{i,j}(t) \tag{2}$$

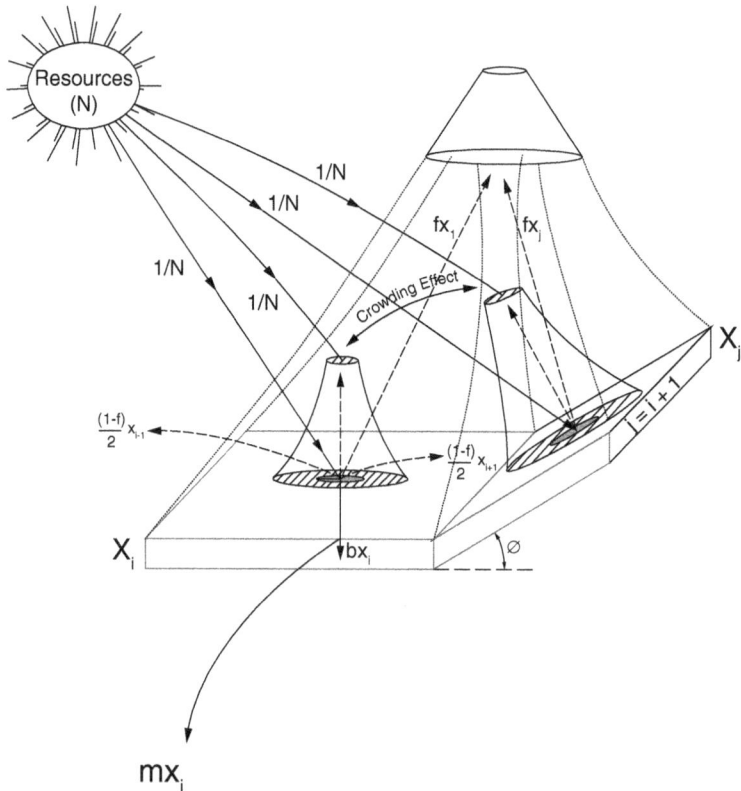

Figure 7.1 Conceptualizing the dynamic of interaction using the analogy of solar system.

where $NI_j(t)$ represents the net effect of interactions on agent j at time t; $X_i(t)$ is a state variable that measures the growth of interest in pursuing the agenda of agent i. For example, consider j as a hydropower dam and other agents of the network as different stakeholders affected by the dam. Each agent i is likely to consider the impact of the dam project (agent j) differently. Assume the group of stakeholders agrees to certain scoring criteria; $[Xj(t)]$ in this case represents the cumulative score and measures the growth of interest in pursuing the agenda of agent j; not all agents will have the same influence, and they are likely to perceive agent j differently. $\omega_i(t)$ is the lag in the phase angle $\phi(t)$ at time t; and the sum over i includes all the $m\text{-}1$ agents. The competition among different agents in promoting their interests as part of the cooperative agenda such as adhering to certain guiding principles (e.g., equitable and sustainable use of water) could be described by Equation (3):

$$CR_j(t) = \sum_{i=1}^{m-1} \left(\frac{X_i(t)}{1 + \rho D_{j,i}(t)} \right) \tag{3}$$

where $CR_j(t)$ represents the competition of other agents; ρ is a parameter representing intensity of competition and $D_{j,i}(t)$ is a distance variable that defines the behavior of two agents sharing a similar niche in the network and competing for the same resource. For example, within the context of the Nile Basin countries and the Cooperative Framework Agreement (CFA), all the upper riparian states in the Nile, apart from Egypt and Sudan, have smaller distance variables $D_{j,i}(t)$ between them, as they share similar niches and commonality in the sense that they have been historically disadvantaged from equitable utilization of the Nile's water. This distance variable $D_{j,i}(t)$ measures the proximity of the two agents $\boldsymbol{i}$ and $\boldsymbol{j}$ in their interest and could by approximated by Equation (4):

$$D_{j,i} = Z_i(t) - Z_j(t) = \cos\left[\phi_{i,j}(t)\right] \tag{4}$$

The prospect for the hydropower dam to emerge as a chosen project depends on the ability of the project to interact and compete with other project proposals available within the network of agents. We can describe the rate of change in interest in the hydropower dam that includes the positive and negative effects of the influence of other agents present in the network as:

$$\left[\frac{dX_j(t)}{dt}\right] = \left\{b_j.\Delta^K\left[X_j(t)\right]\right\}.\left\{\left[1 + NI_j(t)\right]\left[1 - \frac{CR_j(t)}{N}\right]\right\} - \left\{\Psi X_j(t)\right\} \tag{5}$$

where ψ is a parameter representing the average life-span of a certain agent's agenda or interest; b_j is a parameter describing proportionality of growth rate as a function of X_j. The parameter k describes the network connectivity among system agents. The operator Δ^k is described by the following finite difference polynomial approximation:

$$\Delta^k\left[X_j(t)\right] = \begin{cases} f_j(t)X_j(t) + \dfrac{(1-f_j(t))}{2}X_{j-1}(t) + \dfrac{(1-f_j(t))}{2}\ X_{j+1}(t) \quad k \le 1 \\[4mm] f_j(t)X_j(t) + \left[\dfrac{(1-f_j(t))}{2^{k-1}}\right]X_{j-1}(t) + \\[4mm] \left[\dfrac{(k-2)!(1-f_j(t))}{k!}\right]\displaystyle\sum_{m=2}^{k}(k-m)X_{j-m}(t) \quad k \ge 2 \end{cases} \tag{6}$$

The parameter f measure changes in interests, values, and norms of agents as defined in Equation (7):

$$f_j(t) = \left[1-y_j(t)\right]e^{-\frac{1}{m-k}\left[1-y_j(t)\right]}; \qquad y_j(t) = \left[\frac{X_j(t)}{X_j(\max)}\right] \tag{7}$$

The parameter $f_j(t)$ in Equation (7), is formulated in such a way that as the time-dependent variate $y_j(t)$ increases and approaches a value of 1, $f_j(t)$ asymptotically approaches to zero.

Accounting for Subjectivity and Non-linearity in a System's Agent Behavior

Complex water networks are usually characterized by a high degree of uncertainty and non-linearity. Small perturbations in initial conditions could lead to significantly different outcomes. Effects of non-linearity and uncertainty in shaping the outcome of complex systems are well documented. To address the effects of uncertainty, randomness and feedback in agent's behavior, we use three components—rational agent behavior; perturbation around rational behavior; and random behavior—$Z_j(t)$, for any agent j in the following form:

$$Z_j(t) = 2\left[\theta_1 ZA_j(t) + \theta_2 ZC_j(t) + \theta_3 ZR_j(t)\right] - 0.5 \tag{8}$$

where $ZA_j(t)$ represents rational agent behavior; $ZC_j(t)$ is the perturbation in agent behavior around rational behavior; $ZR_j(t)$ is the random noise element in agent behavior; Θ_1, Θ_2 and Θ_3 are the weights assigned for each behavior type. The average agent behavior, $ZA_j(t)$, may be characterized as rational and can potentially be estimated on the basis of previous observations and interactions. In most cases, this is the behavior that is accounted for in traditional systems engineering approaches and may described by Equation (9):

$$ZA_j(t) = f_j(t)y_j(t) + \left[\frac{1-f_j(t)}{2^{k-1}}\right]y_{j-1}(t) + \left[\frac{(k-2)!(1-f_j(t))}{k!}\right]\sum_{i=1}^{k-1}(k-i)y_{j-i} \tag{9}$$

The second component, $ZC_j(t)$, represents deviations from rational behavior and includes a subjectivity element in an agent's behavior. This component allows for certain degrees of freedom for an agent to act in a somewhat autonomous fashion (Dooley and Van de Ven, 1999). Within the context of WDF, this component allows an agent to explore alternative options that may deviate, for example, from a purely rational cost-benefit behavioral response represented by $ZA_j(t)$. We represent dynamics of $ZC_j(t)$ using a Bessel function as shown in Equation (10).

$$ZC_j(t) = \frac{1}{2}\left[1 - \sum_{m=1}^{k} \frac{(-1)^m \, y_j^{2m}}{m!}\right] \tag{10}$$

Figure 7.2 illustrates the oscillations and perturbations of agent behavior around the average for different k values. As shown in Figure 7.2, the choice of Bessel function allows us to account for high damping effects when the connectivity parameter is within the range $(3 \leq K \leq 20)$. This signifies the role of engaging disputed parties (in this case, more than three parties need to be engaged) in dialogue that could potentially lead to convergence and cooperative behavior. The third component, $ZR_j(t)$, represents issues of randomness in agent behavior associated with other unaccounted externalities. We use the following equation to represent this component:

$$ZR_i(t) = \text{RND}[0,1] \tag{11}$$

where RND[0,1] is a random number between 0 and 1 generated using power law distribution.

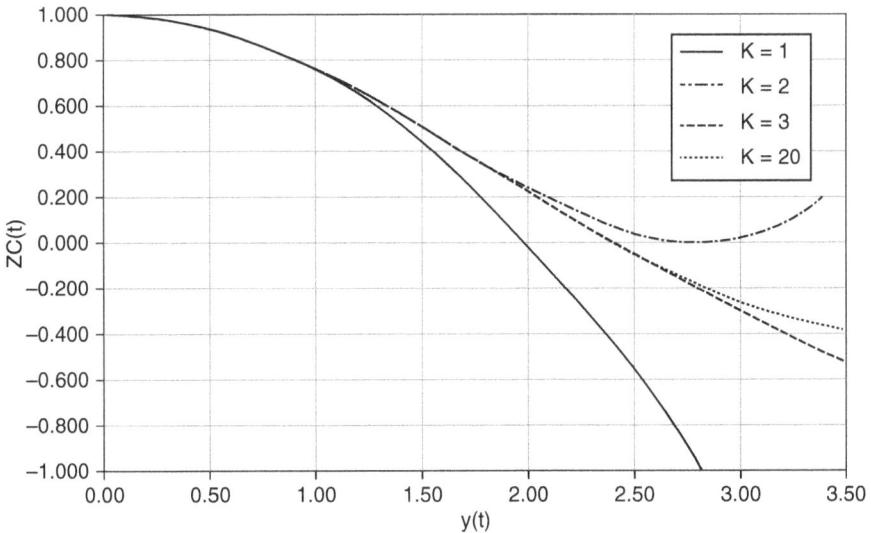

Figure 7.2 Numerical plot of Bessel's Function for modeling perturbed agent behavior.

Negotiating Transboundary Water Issues:
Application of the Network Model

We will use a hypothetical transboundary water basin to show how the proposed network model can be used to develop insight about the dynamics of competing and conflicting issues among multiple countries. The "Indopotamia" water management fable was first introduced by Islam and Susskind (2012) as a fictional—yet realistic—story to describe the typical issues and challenges relevant to transboundary water management. This fable was used to develop a negotiation simulation game—known as "Indopotamia"—as a nine-party negotiation game involving disputes over allocation of land and water resources shared by three countries in an international river basin. The game provides opportunities to discuss natural, societal and political dimensions of science-intensive policy disputes where high levels of uncertainty are involved.

The basin is shared by three countries as shown in Figure 7.3. These are Alpha, Beta and Gamma. Table 7.2 summarizes the socioeconomic indicators for the riparian states in the basin. Alpha is the largest of the three riparian states in the basin. Alpha is economically and politically dominant and has long monopolized access to the river. The rapid population growth and the associated increased demand for food and energy to satisfy domestic consumption seems never to leave enough water for Beta and Gamma. As it has continued to devote all of its resources to its own political

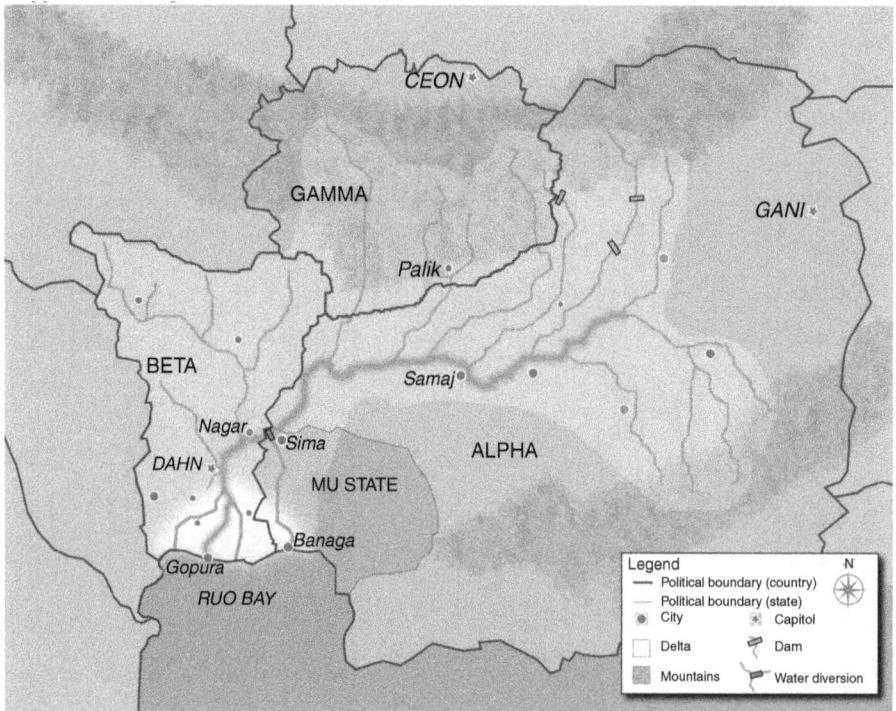

Figure 7.3 Map of Indopotamia River Basin (Ashcraft 2011).

Table 7.2 Description of parameters and state variables in the proposed model.

Indicator	Alpha	Beta	Gamma
Average National GDP (Million US$)	466,682	048,906	015,560
Average Per Capita GDP (US$)	463.00	356.00	241.00
Percentage Contribution of Agriculture to National GDP	25	25	40
Percentage Contribution of Industry to National GDP	27	24	22
Percentage Contribution of Services to National GDP	48	51	37
Percentage Population Workforce Employed in Agriculture Sector	58	52	93
Total Energy Use (MWhr)	410,000	014,500	015,200
Hydropower Potential (MW)	084,000	000,700	225,000
Percentage Hydropower Potential Developed	38	30	16
Percentage Population with Access to Electricity	43	20	15

and economic development, Alpha has earned a reputation of being self-serving and uncompromising.

Beta, on the other hand, has plenty of farmland, but it does not have an adequate supply of labor. In recent years, it has tried to shift to less labor-intensive, energy-powered agriculture. This move has been difficult because Beta cannot afford the required gasoline or coal. Unless Beta can import cheap labor, or generate hydropower from the river, its irrigation development and economic growth will continue to suffer. Beta is suffering significantly from devastating floods, and a significant part of the national budget is directed to address flood-protection and water-pollution issues. In addition, Beta's goal to expand rice production could be achieved through a power trade agreement with Gamma through the implementation of the planned Gamma hydropower project (a regional power trade and interconnection project).

Periodic droughts and forest fires plague Gamma. The lack of storage infrastructures undermines its ability to deal with these problems and has slowed its economic development. Much of Gamma's groundwater is contaminated with arsenic, forcing it to rely almost entirely on the river for its drinking water. After decades of civil war, Gamma's national plans for hydropower development and new power-sharing are a work in progress. Gamma made a commitment to its people to double the access to electricity and sewerage within 10 years. Gamma also intends to develop water resources to grow the national economy, improve food security and attract foreign investment. International financing for water projects and economic development is critical to Gamma's success. An international agreement would be a tremendous accomplishment for its government and would enhance its legitimacy in the eyes of the international community. Gamma is interested in building support for a multilateral agreement on a development approach—interlinking multiple issues—for the Indopotamia basin.

Mu—a state located in the southwestern part of Alpha—is the country's fifth most populous state, and fourth-largest state economy. Like other Alpha states, Mu is governed by a parliamentary system of representative democracy. Mu's governor is the strongest candidate running against the prime minister in next year's national election. Tension between the two candidates has been building recently, attracting media attention to their disagreements over how federal services are distributed. Mu's interest is to build support for a comprehensive, regional approach to development, which will bring real benefits to local communities.

Over sixty water professionals, decision makers, diplomats and non-governmental actors from the four Eastern Nile countries (Egypt, Ethiopia, Sudan and South Sudan) participated in an engaging workshop in Nazareth, Ethiopia, on 5–6 February, 2014, called Water Diplomacy Framework: From Theory to Practice—and organized by the Eastern Nile Technical Regional Office (ENTRO) and led by one of this book's co-authors (Shafiqul Islam). The workshop included a spirited water negotiation using the Indopotamia simulation game. Findings from the workshop are used to structure this case study.

Assumptions Related to the Network Model

To apply the proposed network model for the Indopotamia game, following assumptions are made:

- The decision makers in the Indopotamia basin have agreed to form a high-level committee and use the WDF as an approach for addressing their water management challenges and to negotiate a sustainable development plan. This development plan is expected to culminate into a Multilateral Development Agreement (MDA) for the Indopotamia basin. The MDA must be signed by all the member states of the basin, which includes Alpha, Beta, Gamma and Mu.
- To support the negotiation process, the committee works with a neutral facilitator and invites other development partners from the basin to participate in the negotiation process.
- The committee, with the support of a neutral facilitator, agrees to structure their negotiation process through several sessions or stages. The first session is devoted to understanding, aligning and integrating countries' national development plans into a regional investment program. Engagement in this session allows each member to discuss and reflect on its key interests and to understand each other's interest without making any commitment. This stage allows for creative exploration of options as well as for thinking about linking multiple options for mutual gain. The second stage is to develop a systematic procedure to engage in basin-wide Cooperative Regional Assessment (CRA) studies through joint fact finding, data sharing and monitoring. The third stage of negotiation is to agree on benefit sharing mechanisms and finalizing a comprehensive MDA.

Analysis and Findings

The primary objective of this study is to develop a systematic approach to identify and prioritize a set of policy options (e.g., basin-wide comprehensive study, regional approach to climate change) and projects (e.g., Gamma hydropower dam; Beta rice project) that make the development desirable for multiple stakeholders in a transboundary water negotiation. A key goal for this analysis is to provide information and engage decision makers to develop a sustainable basin development plan and an investment program by exploring and linking multiple projects and policy options through a process that focuses on identifying and ensuring mutual gain for all parties.

Performance Indicators

Five performance indicators are chosen to evaluate the outcome and impact of the development plan. These performance indicators are selected based on the principles of sustainable development, equitable uses and participatory decision making approaches.

Average Annual Growth Rate (AVR)

This indicator is used to measure the performance of the system in achieving certain targeted goals, such as ensuring a minimum targeted level of average annual growth. The AVR indicator is calculated using the simple compound interest formula as shown below:

$$\text{AVR} = \left[\left(\frac{\text{TP}_{\text{desired}}}{\text{TP}_{\text{initial}}} \right)^{1/n} - 1 \right] . 100\% \tag{12}$$

where AVR is the average annual growth rate (%); $\text{TP}_{\text{desired}}$ is the targeted or desired growth after n years and $\text{TP}_{\text{initial}}$ is the initial or current regional growth level.

Overall System Performance (OSP)

This indicator is used to measure Overall System Performance (OSP), which is the sum of the total growth generated at time t:

$$OSP(t) = \sum_{j=1}^{m} X_j(t) \tag{13}$$

Synergistic Response Per Agent (SPI)

This indicator is used to measure the average of positive interactions or intersection in the cooperative agenda per agent at any time t:

$$SPI(t) = \frac{\sum_{j=1}^{m} \text{NI}_j(t)}{m} \tag{14}$$

Competitive Response Per Agent (CPI)

This indicator measures the average competition per agent at any time t. This indicator reflects the competition between different water uses and users in the basin.

$$\text{CPI}(t) = \frac{\sum_{j=1}^{m} CR_j(t)}{m} \tag{15}$$

Equity in Benefit Sharing (EBS)

This indicator measures the equitable distribution of costs and benefits among the riparian states and is measured using a weighted method of maximizing the sum of the ratio of the targeted interest achieved by each state. Different weights δ can be assigned to each member state according to an agreed upon criterion. For this study, an equal weight of $\delta = 0.25$ is assigned for each member state

$$\text{EBS}(t) = \delta_1 \left(\frac{X_4(t)}{X_4(\text{max})} \right) + \delta_2 \left(\frac{X_9(t)}{X_9(\text{max})} \right) + \delta_3 \left(\frac{X_{12}(t)}{X_{12}(\text{max})} \right) + \delta_4 \left(\frac{X_{15}(t)}{X_{15}(\text{max})} \right) \tag{16}$$

where:
$X_4(\text{max})$, $X_9(\text{max})$, $X_{12}(\text{max})$ and $X_{15}(\text{max})$ represent the maximum desired interests of Gamma, Beta, Mu and Alpha, respectively, in the regional investment program.

Dynamics of Interactions among Agents

This case study builds upon the conversations and findings from the Water Diplomacy Workshop organized by the Eastern Nile Technical Regional Office (ENTRO) in Nazareth, Ethiopia, in 2014. The outcome of the group discussion from the Indopotamia simulation game was used to inform the order of interaction among agents as shown in Tables 7.3, 7.4 and 7.5. The system under consideration is represented by a maximum of 20 agents, where each agent has the capability of influencing decision making during the negotiation process. For example, for the 20-agent configuration of the network (Table 7.5), Agent 1 represents Gamma hydropower dam, which shares similar niche in the system with the following five agents: signing of the MDA, power interconnection and trade, Gamma national interest, regional approach to climate change and basin-wide comprehensive study. The rational agent behavior for each agent is estimated based on the outcome from the Nazareth workshop group discussion. In addition, our formulation of agent behavior also includes a provision to represent contextual and qualitative factors as well as random components (Equation 8). A set of 11 scenarios (Table 7.6) was designed by systematically varying the basic model parameters $(b_j, f_j, \psi_j, X_{\text{max}})$ to better understand the dynamics of interactions among agents and identify critical parameters that may trigger changes.

Scenario Analyses and Findings

Analysis of Optimal System Connectivity

This part of the analysis examines the evolution of network configuration (i.e., connectivity among different agents) that enables the network to evolve in a sustainable manner. Three different network configurations were tested. The first configuration represented a network of 10 agents/nodes (Table 7.3), where only large-scale investment projects, riparian states' national interest and the multilateral development agreement were represented as nodes or agents in the system. The second network configuration (Table 7.4) with the total number of nodes ($m = 16$) represented a more extensive network, where the fast track project was added as part of the investment portfolio to include projects such as flood management, watershed management, salinity in Alpha, etc. The third network configuration ($m = 20$, Table 7.5) represents the most extensive investment portfolio, where soft policy intervention such as monitoring, mainstreaming climate change issues are also included.

Table 7.3 List of agents and their order of interactions (number of nodes n=10).

i	Agent	k=0	k=1	k=2	k=3	k=4	k=5	k=6	k=7	k=8	k=9
1	Gamma Hydropower Dam	1	20	2	4	5	14	9	7	13	15
2	Power Interconnection and Trade	2	20	1	4	5	14	9	7	13	15
4	Gamma National Interest	4	1	2	20	5	14	9	7	13	15
5	Beta Rice Project	5	20	2	9	1	13	4	14	7	15
7	Beta IWLM Project	7	20	9	5	1	2	4	13	14	15
9	Beta National Interest	9	5	7	2	13	20	1	14	4	15
13	Alpha Food Security	13	20	15	14	5	2	9	4	7	1
14	Alpha Energy Security	14	20	15	13	2	5	9	4	7	1
15	Alpha National Interest	15	13	14	5	2	20	9	4	7	1
20	Multilateral Agreement	20	1	2	14	5	7	4	9	13	15

(a) K = 1 (b) K = 2 (c) K = 4

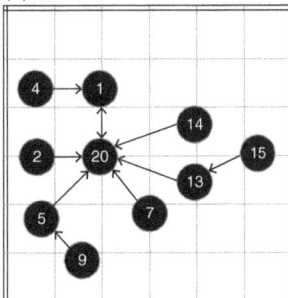
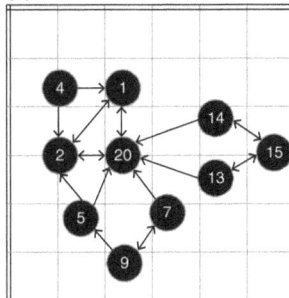
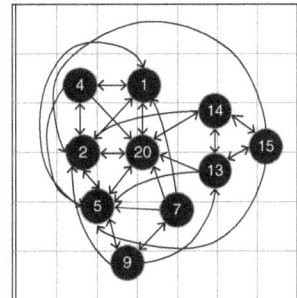

Table 7.4 List of agents and their order of interactions (number of nodes n=16).

i	Agent	k=0	k=1	k=2	k=3	k=4	k=5	k=6	k=7	k=8	k=9	k=10	K=11	K=12	K=13	K=14	K=15
1	Gamma Hydropower Dam	1	20	2	4	5	14	3	9	12	6	7	8	10	11	13	15
2	Power Interconnection and Trade	2	20	1	4	5	14	3	9	12	6	7	8	10	11	13	15
3	Gamma WSM Project	3	20	1	2	4	7	6	9	12	8	5	14	10	11	13	15
4	Gamma National Interest	4	1	2	3	20	5	14	10	9	12	6	7	8	11	13	15
5	Beta Rice Project	5	20	2	9	1	13	4	12	14	6	7	8	11	3	10	15
6	Beta Flood Management Project	6	20	9	7	8	5	3	10	11	1	2	4	12	13	14	15
7	Beta IWLM Project	7	20	9	5	6	8	3	10	11	1	2	4	12	13	14	15
8	Beta GWM Project	8	20	9	5	6	7	3	10	11	1	2	4	12	13	14	15
9	Beta National Interest	9	5	6	7	8	2	13	20	1	3	14	4	12	10	11	15
10	MU IWRDM Project	10	20	12	11	13	14	2	1	3	4	5	6	7	8	9	15
11	Alpha Salinity Management Project	11	20	15	13	14	12	10	5	2	3	4	9	6	7	8	1
12	MU State Interest	12	10	2	5	1	11	14	20	9	4	3	6	7	8	13	15
13	Alpha Food Security	13	20	15	14	11	5	2	9	10	12	4	6	7	8	3	1
14	Alpha Energy Security	14	20	15	13	11	2	5	9	10	12	4	6	7	8	3	1
15	Alpha National Interest	15	13	14	11	5	2	20	9	12	10	4	3	6	7	8	1
20	Multilateral Agreement	20	1	2	14	5	6	3	7	8	10	11	4	9	12	13	15

Table 7.5 List of agents and their order of interactions (number of nodes n=20).

i	Agent	Matrix Showing the Ranking the Order and connection of Agents According to System Connectivity Parameter k																			
		k=0	k=1	k=2	k=3	k=4	k=5	k=6	k=7	k=8	k=9	k=10	k=11	k=12	k=13	k=14	k=15	k=16	k=17	k=18	k=19
1	Gamma Hydropower Dam	1	20	2	4	16	17	19	18	5	14	3	9	12	6	7	8	10	11	13	15
2	Power Interconnection and Trade	2	20	1	4	5	14	17	19	16	18	3	9	12	6	7	8	10	11	13	15
3	Gamma WSM Project	3	20	1	2	4	7	6	16	17	19	18	9	12	8	5	14	10	11	13	15
4	Gamma National Interest	4	1	2	3	20	5	14	16	17	19	18	10	9	12	6	7	8	11	13	15
5	Beta Rice Project	5	20	2	9	1	13	16	17	19	18	4	12	14	6	7	8	11	3	10	15
6	Beta Flood Management Project	6	20	9	7	8	5	3	10	11	16	17	19	18	1	2	4	12	13	14	15
7	Beta IWLM Project	7	20	9	5	6	8	3	10	11	16	17	18	19	1	2	4	12	13	14	15
8	Beta GMW Project	8	20	9	5	6	7	3	10	11	16	17	18	19	1	2	4	12	13	14	15
9	Beta National Interest	9	5	6	7	8	2	13	20	1	3	14	4	12	16	17	18	19	10	11	15
10	Mu IWRDM Project	10	20	12	11	13	14	2	1	3	4	5	6	7	8	9	16	17	18	19	15
11	Alpha Salinity Management Project	11	20	15	13	14	12	10	5	2	16	17	18	19	3	4	9	6	7	8	1
12	Mu State Interest	12	10	2	5	1	11	14	20	16	17	18	19	9	4	3	6	7	8	13	15
13	Alpha Food Security	13	20	15	14	11	16	17	18	19	5	2	9	10	12	4	6	7	8	3	1
14	Alpha Energy Security	14	20	15	13	11	2	5	16	17	18	19	9	10	12	4	6	7	8	3	1
15	Alpha National Interest	15	13	14	11	5	2	16	17	18	19	20	9	12	10	4	3	6	7	8	1
16	Regional Approach to CC	16	17	19	18	20	1	2	14	5	6	3	7	8	10	11	4	9	12	13	15
17	Basin-Wide Comprehensive Study	17	16	19	18	20	1	2	14	5	6	3	7	8	10	11	4	9	12	13	15
18	Basin-Wide Monitoring Program	18	17	16	19	20	1	2	14	5	6	3	7	8	10	11	4	9	12	13	15
19	Level of SH Engagement (PJFF)	19	17	16	18	20	1	2	14	5	6	3	7	8	10	11	4	9	12	13	15
20	Multilateral Agreement	20	17	19	16	18	1	2	14	5	6	3	7	8	10	11	4	9	12	13	15

Table 7.6 Model input data, design of scenarios and initial conditions.

i	Agent	b_i	f_i	ψ_i	X_{max}	SC No. 1	SC No. 2	SC No. 3	SC No. 4	SC No.5	SC No. 6	SC No.7	SC No. 8	SC No. 8a	SC No. 9	SC No. 10
1	Gamma Hydropower Dam	0.15	1.00	0.01	9.00	0.50	9.00	0.30	0.30	0.90	0.90	0.90	0.90	0.90	0.90	0.90
2	Power Interconnection and Trade	0.15	1.00	0.01	9.00	0.50	0.30	0.30	0.30	0.90	0.90	0.90	0.90	0.90	0.90	0.90
3	Gamma WSM Project	0.23	1.00	0.01	3.00	0.50	0.30	0.30	0.30	0.30	0.30	0.30	0.30	0.30	0.30	0.30
4	Gamma National Interest	0.15	1.00	0.01	14.70	0.50	0.30	0.30	0.30	1.47	1.47	1.47	1.47	1.47	1.47	1.47
5	Beta Rice Project	0.15	1.00	0.01	9.00	0.50	0.30	9.00	0.30	0.90	0.90	0.90	0.90	0.90	0.90	0.90
6	Beta Flood Management Project	0.23	1.00	0.01	3.00	0.50	0.30	0.30	0.30	0.30	0.30	0.30	0.30	0.30	0.30	0.30
7	Beta IWLM Project	0.23	1.00	0.01	3.00	0.50	0.30	0.30	0.30	0.30	0.30	0.30	0.30	0.30	0.30	0.30
8	Beta GMW Project	0.23	1.00	0.01	3.00	0.50	0.30	0.30	0.30	0.30	0.30	0.30	0.30	0.30	0.30	0.30
9	Beta National Interest	0.15	1.00	0.01	14.40	0.50	0.30	0.30	0.30	1.44	1.44	1.44	1.44	1.44	1.44	1.44
10	Mu IWRDM Project	0.23	1.00	0.01	4.00	0.50	0.30	0.30	0.30	0.40	0.40	0.40	0.40	0.40	0.40	0.40
11	Alpha Salinity Management Project	0.23	1.00	0.01	5.00	0.50	0.30	0.30	1.30	0.50	0.50	0.50	0.50	0.50	0.50	0.50

	1	2	3	4	5	6	7	8	9	10	11	12	13	14	15	16
12 Mu State Interest	0.15	1.00	0.01	4.90	0.50	0.30	0.30	0.30	0.49	0.49	0.49	0.49	0.49	0.49	0.49	0.49
13 Alpha Food Security (JIBR)	0.15	1.00	0.01	9.00	0.50	0.30	0.30	4.00	0.90	0.90	0.90	0.90	0.90	0.90	0.90	0.90
14 Alpha Energy Security (GPT)	0.15	1.00	0.01	9.00	0.50	0.30	0.30	4.00	0.90	0.90	0.90	0.90	0.90	0.90	0.90	0.90
15 Alpha National Interest	0.15	1.00	0.01	16.50	0.50	0.30	0.30	0.30	1.65	1.65	1.65	1.65	1.65	1.65	1.65	1.65
16 Regional Approach to CC	0.23	1.00	0.01	2.00	0.50	0.30	0.30	0.30	0.50	0.50	0.50	0.50	0.50	0.50	0.50	0.50
17 Basin-Wide Comprehensive Study	0.23	1.00	0.01	2.00	0.50	0.30	0.30	0.30	0.50	0.50	0.50	0.50	0.50	0.50	0.50	0.50
18 Basin-Wide Monitoring Program	0.23	1.00	0.01	2.00	0.50	0.30	0.30	0.30	0.50	0.50	0.50	0.50	0.50	0.50	0.50	0.50
19 Level of SH Engagement (PJFF)	0.23	1.00	0.01	2.00	0.50	0.30	0.30	0.30	0.50	0.50	0.50	0.50	0.50	0.50	0.50	0.50
20 Multilateral Agreement	0.23	1.00	0.01	8.00	0.50	0.30	0.30	0.30	0.70	0.70	0.70	0.70	0.70	0.70	0.70	0.70
Weight for Average Agent Behavior: Θ_1					1.00	1.00	1.00	1.00	1.00	1.00	0.90	0.80	0.60	0.50	0.90	0.90
Weight for Departure from Average Agent Behavior: Θ_2					0.00	0.00	0.00	0.00	0.00	0.00	0.05	0.10	0.20	0.25	0.05	0.05
Weight for Random Noise in Agent Behavior: Θ_3					0.00	0.00	0.00	0.00	0.00	0.00	0.05	0.10	0.20	0.25	0.05	0.05
Model Competition Crowding Parameter ρ					10.0	10.0	10.0	10.0	10.0	10.0	10.0	10.0	10.0	10.0	15.0	20.0
Model Crowding Parameter N: Total Available Underlying Resources					20.0	20.0	20.0	20.0	20.0	20.0	20.0	20.0	20.0	20.0	20.0	30.0

We have designed scenarios to represent each network configuration (m=10, 16, and 20) with different sets of initial conditions to determine the optimum system connectivity parameters **k**. For k=0 represents an isolated network and agents are not connected. For k=1, each agent or node in the system is connected to only one other node and flow of information between the nodes is in the form of a chain of actions. For k=19, this means each node or agent in the system is connected to all other nodes and agent. A total of 230 simulations were carried out and the results are presented in Table 7.7 and Figure 7.4. Results presented in Figures and the Table suggest that—regardless of the number of nodes and the set of initial conditions considered—the optimum connectivity falls within the range of $4 \leq k \leq 8$.

Figure 7.5 shows the evolution of total system growth—which measures the evolution of aggregated interest in the investment package—as a function of time for different values of system connectivity parameter k (varied between 0 and 19 in Figure 7.5). The contour plots reveal consistent and systematic patterns in growth with time for different sets of

Table 7.7 Total system growth for different network configurations.

k	Number of Nodes =10					Number of Nodes =16					Number of Nodes=20				
	SC1	SC2	SC3	SC4	SC5	SC1	SC2	SC3	SC4	SC5	SC1	SC2	SC3	SC4	SC5
0	57	35	34	26	28	52	36	43	28	28	42	41	41	46	67
1	53	48	31	39	39	49	53	38	40	38	66	70	69	71	73
2	63	48	43	59	59	76	64	63	51	46	73	74	75	76	76
3	65	59	44	64	64	76	70	64	74	71	73	74	74	75	76
4	65	61	52	64	64	77	72	73	74	72	75	80	79	80	82
5	65	61	62	65	65	78	72	74	74	72	78	78	83	83	84
6	66	62	54	65	65	78	72	73	73	72	80	79	85	85	85
7	66	63	53	65	65	77	72	72	73	76	81	81	85	84	86
8	67	63	51	65	65	77	71	71	72	69	82	82	84	84	86
9	67	64	50	65	65	77	71	69	71	67	82	82	83	84	86
10						78	71	68	70	65	82	82	83	83	86
11						78	71	67	69	63	82	82	82	83	85
12						78	71	65	68	61	81	82	82	83	85
13						78	71	64	67	60	81	81	82	83	85
14						78	71	64	67	59	81	81	82	83	84
15						78	71	63	66	59	81	81	82	83	84
16											81	81	82	83	84
17											81	81	82	83	84
18											81	81	82	83	84
19											81	81	82	83	84

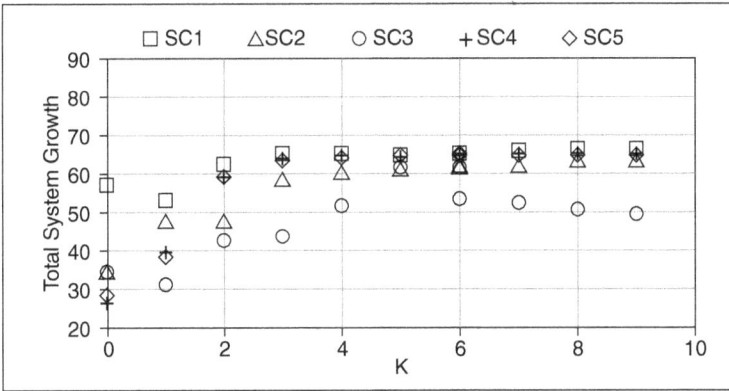

(a) Total Number of Nodes = 10

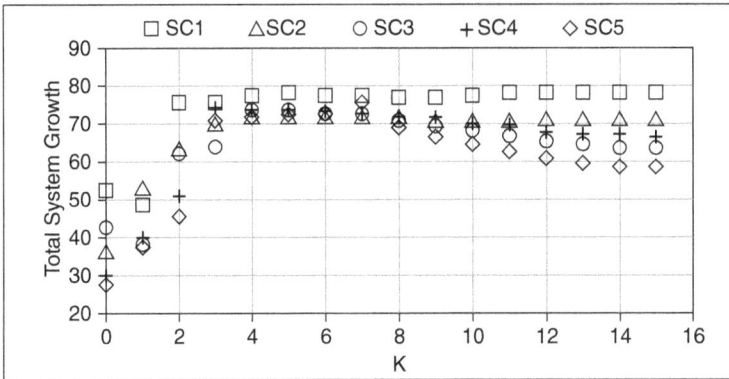

(b) Total Number of Nodes = 16

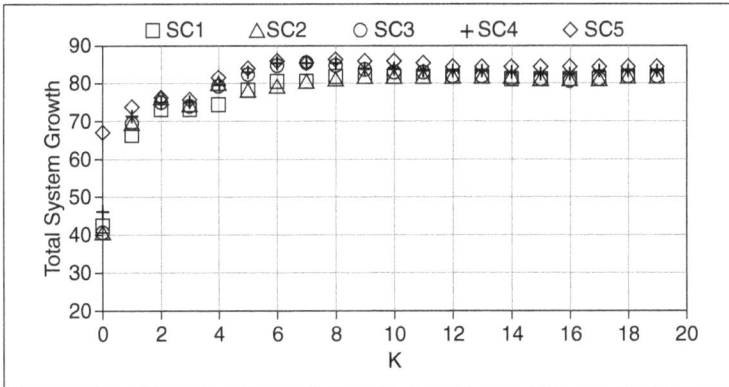

(c) Total Number of Nodes = 20

Figure 7.4 Total system growth for different network configurations (m=10, 16, and 20).

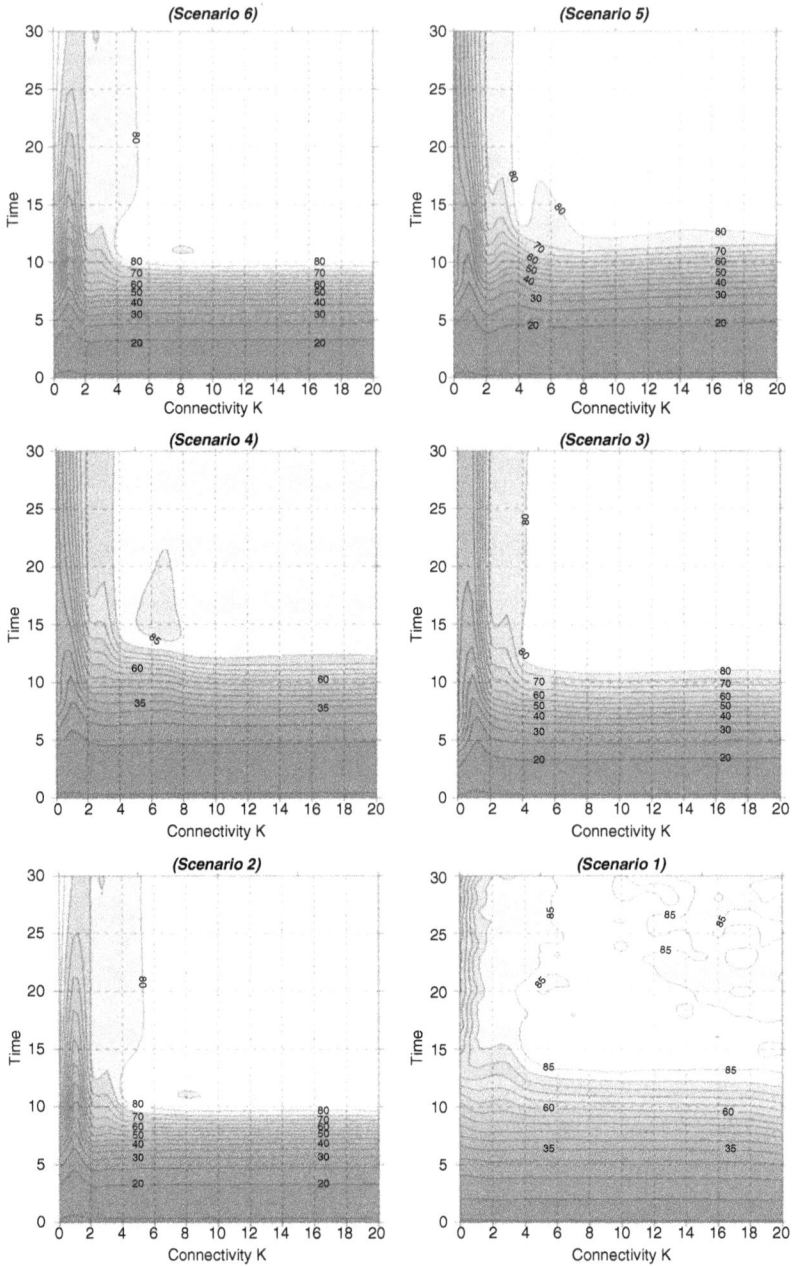

Figure 7.5 Evolution of total system growth for different scenarios and model connectivity K.

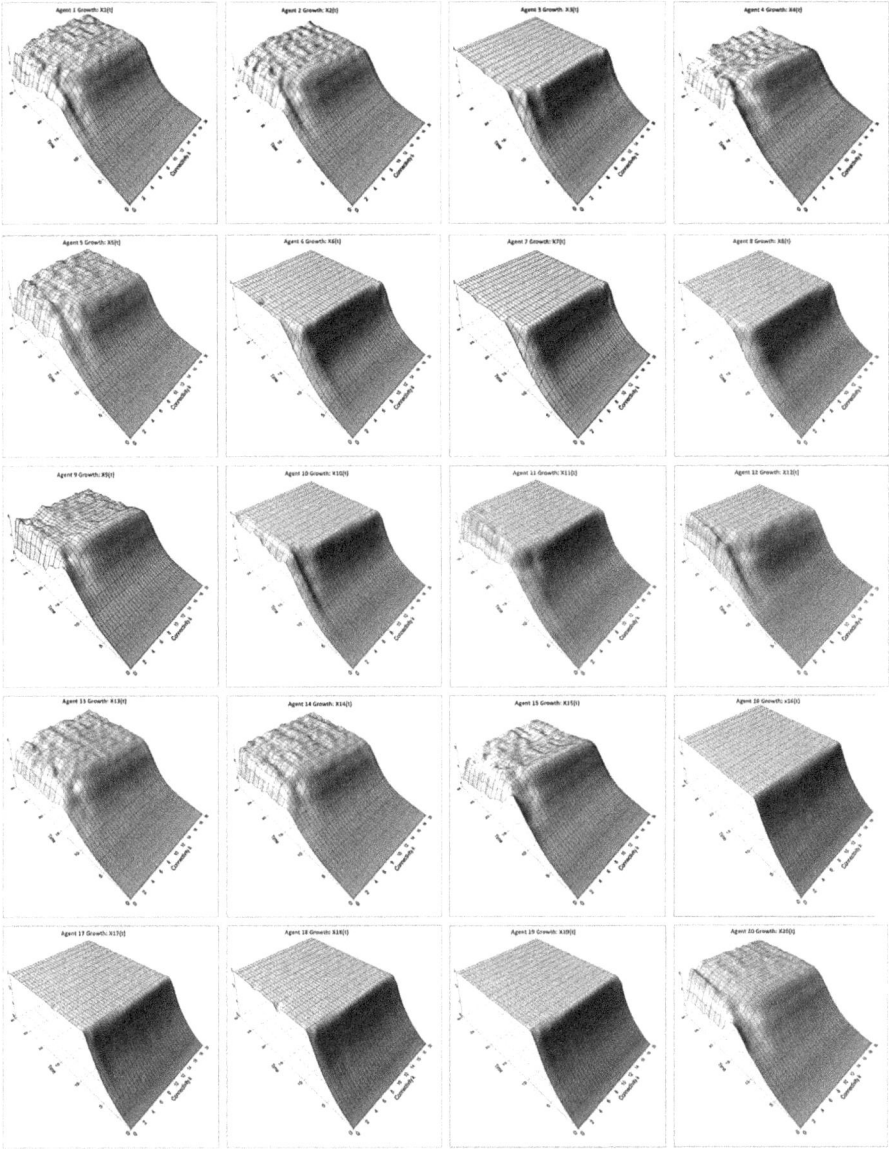

Figure 7.6 Evolution of agent-level growth as a function of system connectivity for Scenario 6.

initial conditions (scenarios 1 to 6). From Figures 7.4 and 7.5, it appears that the optimum system connectivity falls between the range $4 \leq k \leq 8$. For $k < 4$, different parts of the network evolve as an isolated group of agents with limited ability to communicate and influence overall system growth. On the other hand, for $k > 8$, the complexity of the network increases without any appreciable improvement in the overall system growth. Figure 7.6, represents the growth in satisfying interest for each of the 20 agents for scenario 6. The plots show similar evolution at the agent level in terms of optimal system connectivity (i.e., $4 \leq k \leq 8$). Thus, both at the agent level (Figure 7.6) and system level (Figure 7.5), it appears to be efficient and optimal to maintain a connectivity within the range $4 \leq k \leq 8$, with $k=6$ providing a reasonable estimate of optimal system connectivity. These findings suggest that for a sustainable and effective evolution of a transboundary water negotiation, it is important to create opportunities for four to eight agents to form an alliance with shared objectives and mutual gains.

Emergence of Structural Attractors and the System Self-Organization Property

A key puzzle for any complex negotiation scenario is to understand the impact of initial position of each agent in influencing the evolution of the system. Use of traditional optimization techniques or game theory based approaches to quantify the impact of initial position usually leads to narrow solution space akin to zero-sum approaches. Our proposed network model with evolving configuration, on the other hand, provides an opportunity for the network to emerge through exploration and enhancement of mutual gains. To analyze such emergent properties of the proposed network model, simulations were carried out for 11 scenarios and results are summarized in Table 7.8. Table 7.8 provides detailed results of the network simulations for the case of m=20 and under different sets of initial conditions with k=6.

Figure 7.7 shows the evolution of each agent's growth with time for scenarios 1 to 5. It appears that, irrespective of initial position (scenarios 1 through 5 starts at different position for different agents), there is a consistent pattern of convergence in system response over time. These results suggest that initial positions of competing agents may not matter and shall not discourage efforts to initiate a negotiation process. Irrespective of initial positions, there are opportunities for solution(s) to emerge and prevail for a well-structured negotiation process.

Sensitivity analysis of agent behavior and associated system response

To address the effects of uncertainty, randomness and feedback in agent's behavior, we have introduced three components—rational agent behavior; perturbation around rational behaviour; and random behavior—as shown in Equation (8). Relative importance of these three components is determined by their associated weights Θ_1, Θ_2 and Θ_3 in Equation (8). The sensitivity analysis is carried out by varying the weights for different agent's response. Sensitivity analyses for five different scenarios (5, 6, 7, 8 and 8a) under optimal connectivity $(k = 6)$, are summarized in Table 7.8 and presented in Figure 7.8. The overall system performance shows a systematic and consistent improvement in growth when comparing, for example, Scenario 5 (OSP = 85.5: Θ_1 = 1.0, Θ_2 = 0.0 and Θ_3 = 0.0) with scenario 8 (OSP = 95: Θ_1 = 0.6, Θ_2 = 0.2 and Θ_3 = 0.2). Here, Scenario 5 represents a rational agent response with no subjectivity or random components while Scenario 8 includes certain degree of flexibility in agent's behavior by including subjectivity (Θ_2 = 0.2) and random components (Θ_3 = 0.2). The agent interaction phase

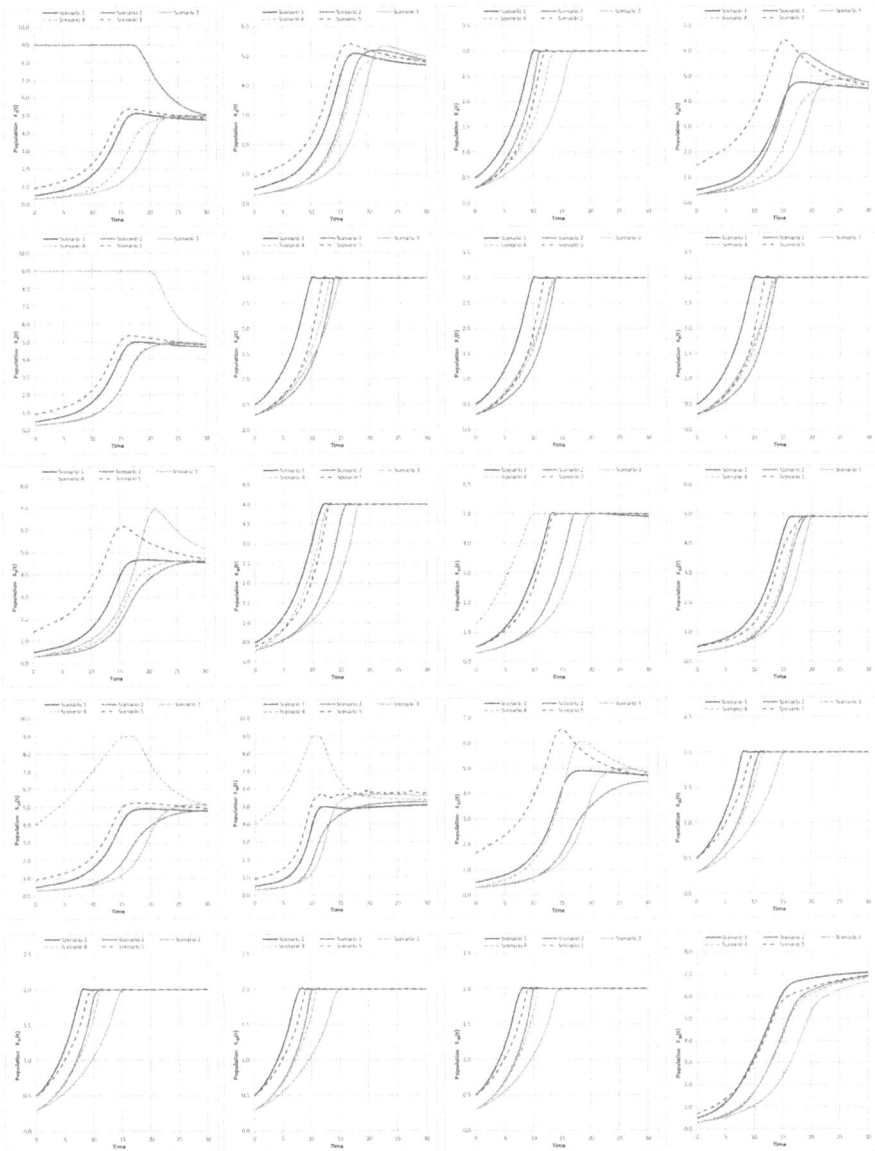

Figure 7.7 Evolution of agent-level growth with time for different initial positions (Scenarios 1 to 5).

diagrams—particularly Scenarios 6, 7 and 8 in Figure 7.8—show the situation of all agents positively interacting with each other ($\phi \leq 90^\circ$ and $\zeta \geq 0$). This indicates that as some degree of flexibility and randomness is allowed, the system performance improves (OSP = 87.8 for Scenario 6; OSP = 90.8 for Scenario 7; and OSP = 95 for Scenario 8). When Comparing Scenario 8 with Scenario 8a, it can be realized that the overall performance of the model slightly drop (OSP = 95 for Scenario 8 compared with OSP 94.8 for scenario 8a) as we further relax control on agent behavior and the interaction phase diagram for scenario 8a show antagonistic behavior for some of the agents (Figure 7.8). This suggests that there may be a need to balance control and degree of freedom in an agent's behavior. A value of $\Theta_I \leq 0.6$ might pose a risk in disrupting the

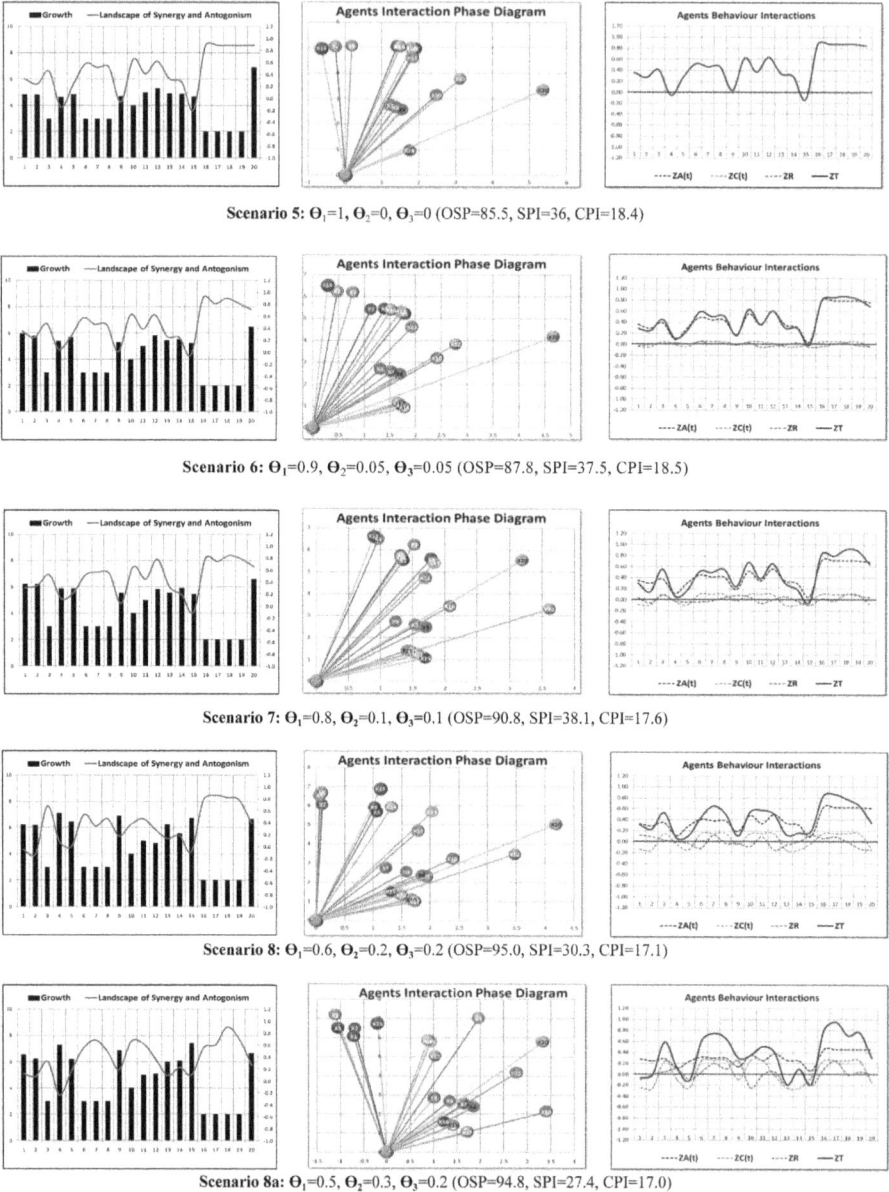

Scenario 5: Θ_1=1, Θ_2=0, Θ_3=0 (OSP=85.5, SPI=36, CPI=18.4)

Scenario 6: Θ_1=0.9, Θ_2=0.05, Θ_3=0.05 (OSP=87.8, SPI=37.5, CPI=18.5)

Scenario 7: Θ_1=0.8, Θ_2=0.1, Θ_3=0.1 (OSP=90.8, SPI=38.1, CPI=17.6)

Scenario 8: Θ_1=0.6, Θ_2=0.2, Θ_3=0.2 (OSP=95.0, SPI=30.3, CPI=17.1)

Scenario 8a: Θ_1=0.5, Θ_2=0.3, Θ_3=0.2 (OSP=94.8, SPI=27.4, CPI=17.0)

Figure 7.8 Sensitivity analyses for different agents' behavior for different scenarios.

rational logic of interaction among agents. Hence, it may be prudent to keep subjectivity component as $\Theta_2 \leq 0.20$ and random components $\Theta_3 \leq 0.20$.

Sensitivity of System Competition and Resource Parameters

This part of the analysis explores the significance of the competition parameter ρ and the resource parameter N on the overall system performance. The competition parameter ρ (Equation 3) represents the relative strength, position and concentration of the individual agent within the system environment. The resource parameter N (Equation 5) represents the physical limitation in available natural resources (e.g., water or land availability for agricultural development or labor availability for agriculture development).

Scenario 9 is designed to study the sensitivity of model performance to the competition parameter ρ where the value of the model parameter is varied: $\rho = 10$ *for scenario 8* to $\rho = 15$ *for scenario 9* as shown in Table 7.6. When comparing the results of scenario 9 with scenario 8 (Table 7.8), there is an increase in total system performance or growth from (*OSP* from *95* to 97.29) and increases in the Synergistic Response per agent (*SPI* from *30.34* to 52.82). This increase in the system's ability to evolve could be attributed to the effect of relaxing the control in the rules of competitions among system agents to have the opportunity to compete for limited underlying resources. Hence, the effect of the crowding parameter ρ is to partially control the rule of interactions and competitions and, as ρ increases, there may be an increase in the overall growth. However, such an increase in growth may come at the cost of possible delays in the early realization of benefits with some reduction in the average annual growth over the first ten-year horizon (AVR reduced from 17.72 percent to 16.36 percent), when comparing scenarios 9 with 8 as shown in Table 7.8.

Scenario 10 is designed to study the effect of resource-availability model parameter N, on the overall system performance. Since the system under consideration is sensitive to the combined effect of parameters ρ and N, there is a need to couple the effect of change in parameter N with possible change in parameter ρ. This is achieved by keeping the ratio of (ρ / N) in the range of *0.5* to *1.0*. Hence, the coupled effect of the competition parameters is studied by varying the model resource parameter N from ($N = 20$ to $N = 30$), and the competition parameter ρ from ($\rho = 15$, to $\rho = 20$), while keeping other model parameters unchanged. By comparing the results of scenario 10 with the results from other sets of scenarios, maximum overall system growth (OSP $=116.3$) appears to occur for $\rho / N = 0.67$ (Scenario 10) while for $\rho / N = 0.75$ (Scenario 9, OSP $= 97.29$) system growth diminishes. Thus, it is important to consider the effects of crowding parameter and resource availability jointly while exploring creative option packages for mutual gain.

It must be emphasized that these are preliminary findings, and the outcome of these numerical experiments are intended to provide general guidance to demonstrate the utility of a hybrid approach to explore the dynamics of the complex water network of range of agents with different types and levels of interdependencies and feedback. There is a need for further refinement of this network model and for sensitivity analysis to generalize these preliminary findings. We hope the reflective water community will use this hybrid approach to identify, characterize and analyze water case studies—from

Table 7.8 Evaluation of model performance for different indicators and system connectivity parameter, $k=6$.

Performance Indicators	SC 1	SC 2	SC 3	SC 4	SC 5	SC 6	SC 7	SC 8	SC 8a	SC 9	SC 10
Overall System Performance (OSP)	79.99	79.33	84.62	84.61	85.53	87.78	90.83	95.00	94.8	97.29	116.3
Synergistic Response Per Agent (SPI)	34.35	34.40	25.49	36.21	35.95	37.46	38.11	30.34	27.4	52.82	85.19
Competitive Response Per Agent (CPI)	18.70	18.90	18.40	18.40	18.40	18.50	17.60	17.10	17.0	18.80	26.90
Equity in Benefit Sharing (EBS)	00.63	00.64	00.65	00.64	00.64	00.68	00.70	00.71	0.71	00.70	00.83
10-Years Average Annual Growth	16.98	9.920	07.68	12.85	14.20	15.65	16.48	17.72	17.67	16.36	17.93
15-Years Average Annual Growth	22.74	16.05	13.30	17.85	19.54	19.85	19.98	20.81	20.45	20.64	22.65
20-Years Average Annual Growth	23.02	18.24	18.92	19.18	19.06	19.60	20.00	20.59	20.61	20.44	22.42
30-Years Average Annual Growth	22.84	18.31	18.60	18.65	18.62	19.72	19.86	19.91	20.73	19.99	22.27
10-Years AVR in Regional Power Trade	12.59	12.03	09.33	11.59	09.84	11.14	12.27	13.92	13.96	11.77	13.06
15-Years AVR in Regional Power Trade	23.82	24.58	17.29	23.20	19.06	20.01	20.84	19.92	21.06	21.61	24.36
20-Years AVR in Regional Power Trade	25.90	32.95	30.51	32.02	19.22	20.50	20.51	21.79	21.89	21.90	25.89
30-Years AVR in Regional Power Trade	25.11	32.11	32.44	32.04	18.20	20.31	20.82	21.98	22.05	21.93	25.60
10-Years AVR in Joint Rice Production	12.43	10.84	00.00	11.51	09.76	11.05	12.01	13.80	13.75	11.67	12.99
15-Years AVR in Joint Rice Production	23.67	22.81	00.00	23.01	18.92	19.83	20.49	21.20	21.08	21.75	25.86
20-Years AVR in Joint Rice Production	25.80	31.66	00.00	31.75	19.28	20.05	20.76	21.54	21.31	21.91	25.51
30-Years AVR in Joint Rice Production	25.15	32.09	−5.19	32.04	18.28	19.42	20.61	20.88	20.78	20.83	24.56
Time to Total System Peak (Years)	17.00	25.00	21.00	18.00	15.00	15.00	16.00	15.00	15.00	14.00	12.00
Alpha Share in Benefits	08.19	08.37	08.61	08.37	08.33	08.99	09.21	09.27	09.42	09.24	11.30
Beta Share in Benefits	08.82	08.91	09.11	08.90	08.89	09.41	09.59	09.60	09.83	09.60	11.40
Gamma Share in Benefits	09.15	09.45	09.48	09.31	09.27	10.66	11.03	11.53	11.51	11.08	14.08
Mu Share in Benefits	04.47	04.49	04.50	04.48	04.48	04.58	04.62	04.67	04.59	04.63	04.90
10-Years AVR in MDA Growth	20.52	18.74	13.61	18.84	15.97	17.75	19.28	21.18	21.22	18.59	20.65
15-Years AVR in MDA Populations	28.59	30.73	23.07	30.35	23.67	24.31	24.53	25.72	25.45	25.83	27.58
20-Years AVR in MDA Populations	29.82	35.52	33.24	35.19	24.82	24.77	25.23	25.27	25.31	26.27	27.58

different places, settings, and scales—and build an actionable approach of managing complex water problems guided by globally emerging normative anchors of sustainability and equity that is deeply informed by practice and complexity of local and situational conditions.

Acknowledgments

The work presented in this paper was partially supported by the Nile Cooperation for Results (NCORE) Project, under the multi-donor Nile Basin Trust Fund of the Nile Basin Initiatives administered by the World Bank—in particular, Component 3 of the Eastern Nile Technical Regional Office (ENTRO). The study benefited from the Water Diplomacy Workshop organized by ENTRO in Nazareth, Ethiopia, during 5–6 February 2014 and the outcome of the group discussions around the case study of Indopotamia River Basin. This work was also supported, in part, by two grants from the US National Science Foundation (RCN-SEES 1140163 and NSF-IGERT 0966093).

References

Allen, Peter, 2000. "Knowledge, Learning, and Ignorance." *Emergence* 2(4): 78–103.
———. 2001. "What Is Complexity Science: Knowledge of the Limits to Knowledge." *Emergence* 3(1): 24–42.
Allen Craig, Joseph J. Fontaine, Kevin L. Pope and Ahjond S. Garmestani. 2011. "Adaptive Management for a Turbulent Future." *Journal of Environmental Management* 92: 1339–45.
Arthur, W. Brian, Steven Durlauf and David A. Lane. 1997. *The Economy as an Evolving, Complex System II*. Reading, MA: Addison-Wesley.
Ashcraft, Catherine M. 2011. "Indopotamia: Negotiating Boundary Crossing Water Conflicts" (Game). Water Diplomacy Workshop and Program on Negotiation at Harvard Law School.
Bar-Yam, Yaneer. 1997. *Dynamics of Complex Systems*. Reading, MA: Addison-Wesley.
Cohen, Alice and Seanna Davidson. 2011. An Examination of the Watershed Approach: Challenges, Antecedents, and the Transition from Technical Tool to Governance Unit. *Water Alternatives* 4(1): 1–14.
deWeck Olivier, Daniel Roos and Christopher Magee. 2011. *Engineering Systems: Meeting Human Needs in a Complex Technological World*. Cambridge, MA: MIT Press.
Dooley, Kevin, and Andrew Van de Ven. 1999. "Explaining Complex Organizational Dynamics." *Organizational Science* 10, 3: 338–72.
Gell-Mann, Murray. 1994. *The Quark and the Jaguar: Adventures in the Simple and the Complex*. New York: W. H. Freeman.
Holland, John B. 1998. *Emergence: From Chaos to Order*. New York: Perseus.
Islam, Shafiqul and Lawrence Susskind. 2012. *Water Diplomacy: Negotiated Approach to Managing Complex Water Networks*. New York: The RFF Press Water Policy Series, Routledge.
Janssen, Marco. 1998. "Use of Complex Adaptive Systems for the Modeling of Global Change." *Ecosystems* 1: 457–63.
Kauffman, Stuart A. 1995. *At Home in the Universe: The Search for the Laws of Self Organization and Complexity*. York: Oxford University Press.
Lansing, J. Stephen. 2003. "Complex Adaptive Systems." *Annual Reviews in Anthropology* 32: 183–204.
Levin, Simon. 1998. "Ecosystems and the Biosphere as Complex Adaptive Systems." *Ecosystems* 1: 431–36.

Maguire, Steven, Brian Mckelvey, Laurent Mirabeau and Nail Oztas. 2006. "Organizational Complexity Science." In: *The Handbook of Organizational Studies*, 2nd edn., edited by Stewart Clegg, Cynthia Hardy, Walter Nord and Tom Lawrence, 165–214. Thousand Oaks, CA: Sage.

Miller, John and Scott E. Page. 2007. *Complex Adaptive Systems: An Introduction to Computational Models of Social Life*. Princeton: Princeton University Press.

Pahl-Wostl, Claudia. 2006. "The Importance of Social Learning in Restoring the Multifunctionality of Rivers and Floodplains." *Ecology and Society* 11(1): 10.

Ramos-Martin, Jesus. 2003. "Empiricism in Ecological Economics: A Perspective from Complex Systems Theory." *Ecological Economics* 46: 387–98.

Rammel, C., Sigrad Stagel and Harald Wilfing. 2007. "Managing Complex Adaptive System: A Co-evolutionary Perspective on Natural Resource Management." *Ecological Economics* 63: 9–21.

Chapter Eight

A CALL FOR CAPACITY DEVELOPMENT FOR IMPROVED WATER DIPLOMACY

Dena Marshall, Léna Salamé and
Aaron T. Wolf

Abstract

Capacity development is a critical component of water diplomacy, navigating the complexities of transboundary water conflicts. Performed at the institutional and personal levels, capacity development can include initiatives to improve the practitioner's technical capabilities in water resources management, understanding of international water law, interpersonal communications and dispute resolution. One example of capacity development that interweaves society, institutional and individual capacities is the effective use of methods of public participation, by which communities that are impacted by transboundary water policy become critical voices in the decision-making processes. In this chapter we discuss the individual, institutional and societal levels of capacity development and their basin-wide implications for improved outcomes in water diplomacy.

Introduction

Anything else you're interested in is not going to happen if you can't breathe the air and drink the water. Don't sit this one out. Do something.

(Carl Sagan)

Water management is conflict management. There is no such thing as managing water for a single purpose because all water management is multi-objective and based on navigating competing interests. These interests include domestic users, agriculturists and environmentalists—any two of which are regularly at odds—and the chances of finding mutually acceptable solutions drop exponentially as more stakeholders are involved. Add international boundaries and the chances decrease exponentially yet again. (MacQuarrie et al. 2008, 176) Within each international basin, demands from environmental, domestic and economic users increases annually, while the amount of freshwater in the world remains roughly the same, as it has been throughout history. Given the scope of the problems and the resources available to address them, avoiding water conflict is vital. In international waters, the promising trend reveals a record of acute conflict over international water resources overwhelmed by a record of cooperation. Despite the tensions inherent in the international setting, riparians have shown tremendous creativity in approaching regional development, often through preventive diplomacy, and the creation of "baskets of benefits," which allow for positive-sum, integrative allocations of joint gains (MacQuarrie et al. 2008, 177). It is in preventive diplomacy and cultivating the creativity to develop baskets of benefits where capacity building becomes an invaluable tool for improved water diplomacy.

The complexities of our water problems and the pressure they exert on our lives and future generations mean that we, as individuals, institutions and societies must shift the way we invest in our capacities to address them. Water is an interdisciplinary resource that draws on our expertise and consensus-oriented approaches in hydrology, geology, climatology, economics, engineering, politics, sociology, psychology, religion and much more, bringing people together in a way that other resources cannot. To meet transboundary water's broad reaches, our diplomatic initiatives should be equally interdisciplinary and also consensus driven. Diplomatic efforts to address water conflicts as, or before, they arise should, therefore, be built upon at least basic understandings of law, science, cross-cultural communication, psychology and economics. With the understanding that our current educational and management institutions are generally not specifically organized to address such complexity, we can still begin to stretch beyond the usual frames. We begin by acknowledging the many calls for capacity building at several points along the diplomatic spectrum, and in several ways we encourage institutions to invest in capacity building for water diplomacy. For example, to fulfill the 2003 sustainability requirements of the United Nations' Millennium Development Goals (MDGs), member countries agreed that Africa would need an estimated 300 percent increase in the number of trained water professionals, while Asia would need a 200 percent increase, and Latin America and the Caribbean a 50 percent increase, in all disciplines. (World Water Development Assessment Program 2003) To date, these goals have not yet been met while requests for capacity building continue. Leading up to the 2004 Zambezi Watercourse Commission Agreement (ZAMCOM), Zambia requested assistance with building skills in international waters and negotiations (Subramanian et al. 2014, 832). And as a method of risk reduction in the Niger basin, riparian neighbors have invested in capacity development and knowledge expansion to improve their skills in negotiations and institution building (Subramanian et al. 2014, 837–38). At the 2015 Knowledge Exchange in International Waters conference held in Beijing, international participants representing Asian countries and African countries requested capacity building training in international water law and water conflict management as one avenue for increasing their individual and institutional competencies in water diplomacy. (Knowledge Exchange in International Waters 2015).

It is time to reach beyond standard approaches in water diplomacy and encourage strong investments in capacity development for the individuals, institutions and societies that make up our shared waters communities. In particular, we promote the use of training and collaborative workshops at the stakeholder level to improve the way the riparian neighbors monitor, predict and preempt transboundary conflicts, promoting human and environmental security in international river basins—regardless of the scale at which they occur.

Water Diplomacy

For our purposes, the term water diplomacy refers to all positive interactions among official and unofficial actors concerned with the management of shared water resources and aiming to achieve peaceful and sound management of such resources. Actors (i.e., stakeholders) range from international, regional and national governmental and intergovernmental institutions to representatives of indigenous tribes living on opposite shores of a river, including non-official representatives of a given state, local and spiritual leaders, grassroots groups, opponents to a state's position, and any kind of non-governmental organization. In its classic interpretation, however, water diplomacy refers to processes led by state actors and their representatives to advance national interests related to the management of shared water resources. State actors have the formal obligation to respect, protect and fulfill their citizens' right to water, while other official actors of international public law (i.e., intergovernmental organizations) can also be involved in providing third-party support such as good offices or mediation.

The diplomacy process itself is a nonlinear iterative process in which the formal and informal processes can often be complementary and mutually reinforcing. Formal processes of water diplomacy include the tracks in which official agreements are taken and legal obligations and rights are endorsed at the highest level. Informal water diplomacy often occurs in parallel where actors tend to behave more freely; they can be more creative and resourceful in developing options for potential mutual agreements at formal state level. Advances in informal water diplomacy processes may help to open channels through the constraints and obstacles of formal processes. Working along the informal diplomacy channels could be a place for capacity development.

Capacity Development

> For joint bodies to be effective, their institutional and human capacities are crucial. Staffs of joint bodies should have a broad competence and skills that bridge disciplines. The capacities of managers, especially at the national and local levels, should be strengthened, not only to raise understanding of the complexities of managing shared water resources but also to derive the benefits made possible through cooperation. Negotiation, diplomacy and conflict resolution skills need to be developed and improved. The capacity to develop and implement policies and laws as well as the relevant enforcement mechanisms is vital, and should be developed accordingly, as is setting up funding arrangements, both internal and external. (UN Water 2008, 6–7)

Capacity development, capacity building and capacity enhancement are often used interchangeably (United Nations Committee of Experts on Public Administration 2006). For purposes of this chapter we use the United Nations Development Program (UNDP) definition for capacity development: "The process through which institutions, individuals and societies obtain, strengthen and maintain the capabilities to set and achieve their own development objectives over time" (UNDP 2009). Capacity development can therefore take shape in various ways, such as through specialized training in international water law and the development of regional standards, training in dispute resolution, communications and negotiation, the gathering and sharing of hydrological and GIS data and monitoring and assessing water resources. All of these capacity development efforts can complement each other to enhance water diplomacy (see, e.g.: Grey and Sadoff 2007).

In its 2008 thematic paper, the United Nations Task Force on Transboundary Waters emphasizes the importance of having appropriate regional, transboundary and national institutional structures, and strongly encourages capacity development at the personal, societal and institutional levels. This concept is echoed in the UNDP International Network for Capacity Building in Integrated Water Resources Management *Conflict Resolution and Negotiation Skills for IWRM Training Manual*: "To ensure positive sum outcomes, it is essential that stakeholders have appropriate knowledge and capacity to collectively move away from potentially destructive actions and behaviours" towards collaborative actions (Salame, Swatuk and van der Zaag 2009). We see this call for capacity building prove its worth in shared waters conflicts where stakeholders navigate the dialogue with technical fluency and also demonstrate a personal inclination to commit to the collaborative process. When a stakeholder, formal or informal, maintains a solid understanding of the technical, scientific and legal aspects of the water issues under consideration, *and also* exhibits her or his commitment to the collaborative process, she or he is more likely to navigate the issue with credibility and grace. To illustrate this, we draw on a current example of the transboundary dialogues that have evolved through the Columbia River Treaty Review process in our regional backyard, the Columbia River Basin of the Pacific West of the United States and Canada.

Columbia River Treaty Review Transboundary Dialogue: A Case Study in Capacity Development for Water Diplomacy

The Columbia River Basin is a complex system characterized by a highly engineered main stem and detailed operating agreements for flood control and hydropower; it is also susceptible to large-scale variations due to climate change, energy market changes, and hydro-meteorological fluctuations. At 672,100 square kilometers, the Columbia River Basin is the fourth largest river basin in North America. It has been home to over thirty indigenous Tribes and First Nations, for whom the annual migrations of salmon and deer, and seasonal gathering of roots and berries have defined the spiritual, cultural, geographic and economic identity for thousands of years. The native people of the Columbia Basin are considered the River People or the Salmon People. During the 1850s federal expansion efforts to assume possession of territories belonging to the tribes, the United States entered into several treaties, appropriating most Native lands but preserving "the right of taking fish at all usual and accustomed places, in common with citizens of the Territory, and of erecting temporary buildings for curing them; together with the privilege of hunting, gathering roots and berries, and pasturing their horses and cattle upon open and unclaimed land" (United States and Yakama Nation 1855). Since then, through a series of federal and state motions, those rights have been continuously challenged, thus limiting Native Americans' access to their ancestral lands. Meanwhile, the Columbia River basin has grown significantly in population, straining the region's natural resources. In 1961, the United States and Canada entered into the Columbia River Treaty, which provided for the building of four dams and reservoirs in Canada and one in the United States to produce a consistent flow of hydropower and ensure protection against major flooding along the main stem of the Columbia. It was ratified in 1964 through a treaty protocol and a Canada–British Columbia agreement that limited and clarified many treaty provisions, defined rights and obligations between the British Columbia and Canadian governments, and allowed for the sale of power, the Canadian Entitlement, to downstream US users. The treaty does not provide for other uses, such as water for native fish migration, irrigation or navigation. (Cosens 2012, 79) To implement the treaty, Canada and the United States designated official entities to manage the technical and political aspects by assigning British Columbia Hydro on the Canadian side and the United States Army Corps of Engineers with Bonneville Power Administration on the American side as the official treaty entities, and by establishing the bi-national International Joint Commission, and the Permanent Engineering Board to manage the legal and technical aspects of treaty implementation. The Columbia River Treaty has no official end date, but either the United States or Canada may deliver a notice of intent to terminate with ten years anticipation at any point 60 years after ratification: the earliest notification date would have been in September 2014. By a measure of strong diplomatic relations among the treaty parties, implementation of the treaty provisions, and ongoing management of the river itself, the Columbia River Treaty is considered one of the most successful transboundary river treaties in the world.

In 2009 a formal treaty review process was launched and officially guided by the designated United States and Canadian entities. While the entities were tasked with their independent review processes and mandates to report to their national bodies (US Department of State and Canadian Crown Government), they engaged in a series of formal review committee meetings and highly structured public information sessions designed to inform basin residents of the treaty and process of review. In Canada, a parallel series of community open houses throughout the province of British Columbia provided an informal consultative forum for basin residents to learn about the treaty review process but, perhaps more important, to exchange information between basin residents and policymakers. These community open houses were designed and conducted by the Columbia Basin Trust (an organization founded to connect basin communities impacted by the treaty), and often produced emotional testimonies from people who had been displaced when reservoirs were built to accommodate the four treaty dams. Testimonies often included stories of houses destroyed and towns inundated, sacred native burial grounds desecrated by water-level fluctuations, and concerns for climate change adaptations and ecological restoration.

Meanwhile, on both sides of the boundary an interdisciplinary informal basin-wide water diplomacy process began to emerge in the Columbia Basin among the universities in the basin. The Universities Consortium on Columbia River Governance (UCCRG) was comprised of universities in each basin state—including the University of British Columbia, University of Montana, University of Idaho, University of Washington and Oregon State University—and had as its initial and evolving objective to inform or influence in some way the formal treaty review process. To this end, the Northwest Power and Conservation Council in the United States, and the Columbia River Trust in British Columbia, contributed support to the universities to host a series of annual transboundary dialogues. As the UCCRG dialogues evolved from an informal core of interested academics intending to influence the formal process to become an open forum for all basin residents to engage in transboundary discussion on a multitude of Columbia Basin issues, the public engagement process deepened, broadened and grew into a unique mechanism independent of the Columbia River Treaty Review process and the treaty itself. In addition, opportunities for capacity development began to emerge. Lay participants began to study the treaty and organized field trips to treaty dams and the affected communities; regional policymakers sought to understand technical aspects of the treaty; institutional leaders studied international law, hydropower development, climate change and salmon ecology; tribal leaders developed methods of transboundary cooperation and, for the first time in history, published an official statement reflecting their shared indigenous common vision; basin residents who had only geography in common began talking with each other, hearing each other, and learning from each other. Participants in the informal dialogue processes often expressed a sense of forward progress toward a deeper understanding of the river basin communities with their shared interests and dynamics.[1] Eventually, the energy and creativity generated from the informal dialogue processes began to influence discussions at the formal levels. Over time some of the concepts emerging from the UCCRG and related informal dialogues were reflected in official regional communications. The regional recommendation delivered to the US Department of State in September 2014 included recommendations to modernize the Columbia River Treaty by naming a third entity—the Native American Tribes and First Nations of the Basin—by incorporating ecological function and climate change considerations and developing mechanisms to promote anadromous fish passage.

The Columbia River Basin demonstrates a unique case study. It is a river basin situated in one of the most stable parts of the world, with no water-scarcity issues, shared between two countries with strong diplomatic relations and a well-developed upstream–downstream relationship. Water diplomacy through the formal Columbia River Treaty review process and the informal transboundary dialogue has rich applications at many levels: international, domestic, regional and basin-wide. Considering the multiple levels of diplomatic activities, the size of the basin and wide range of stakeholders engaged in the process, the Columbia Basin process could be considered a simple problem in which cause–effect relationships are relatively clear, a limited range of solutions is possible for a given management intervention and there generally is a high degree of public input in decision-making processes. However, even in this relatively, or apparently, "easy" case, cooperation was established and is sustained only because various types of investments are made in formal and informal capacities' development at various levels. And this becomes even truer as the rate of complexities of a case rises.

Capacity Development for Improved Water Diplomacy

Why go to war over water? For the price of one week's fighting, you could build five desalination plants. No loss of life, no international pressure, and a reliable supply you don't have to defend in hostile territory. (Israeli Defense Forces analyst. quoted in Wolf 1998).

Developing the capacity to monitor, predict and preempt transboundary water conflicts is key to promoting human and environmental security in international river basins. (MacQuarrie et al. 2008, 177) These capacities can be strengthened through strategic and well-formulated efforts that combine practical application with skills-building methods. Taking the example of the Columbia Basin: to gain a deeper understanding of the issues stakeholders have actively sought to learn more about the principles of international water law (e.g., McCaffery 2008, Wouters 2013), water conflict management and the broad spectrum of designing public participation processes. In negotiations over international waters, the stakeholders, institutions and our societies must be able to flex with the ever-changing dialogue while also holding firm to the core principles of collaboration, inclusiveness and transparency. Because each case carries with it a unique level of complexity with its own geographic, cultural, political and socioeconomic specificities, there is therefore no standard methodology but rather indefinite alternatives for stakeholders to explore, test, stop, reflect upon and retract throughout the negotiation process.

We briefly discuss capacity development at the individual, institutional and societal levels.

Individual Capacities[2]

Perhaps the most sensitive of perceptions is the perception of insufficiency. As humans, we need to feel competent in our affairs, a need that is heightened for those imbued with representational or leadership authority. Individual capacities refer to the abilities and competencies that we carry with us, such as communication or negotiation skills, or possessing technical expertise in engineering, law or science. Water managers, political leaders and negotiators entering the negotiation room rely deeply on their own abilities to navigate the technical, legal, political and interpersonal complexities of water diplomacy. They need to know how to manage cultural sensitivities and, above all, they need solid skills in conflict management and negotiation. (Unver et al. 2010) They must have the soft skills and a personal inclination to commit to the collaborative process, build trust and transparency and engage a wide range of stakeholders to support the negotiated outcomes. Those abilities can be strengthened through individual capacity development training in negotiation, communication and water conflict management among other skills (Stockholm International Water Institute 2009).

Individual Capacity Development Training in Water Conflict Management

The first type of training encompasses introductory training in transboundary water conflict prevention, management and cooperation and is described in MacQuarrie et al.

(2008). This training provides a broad overview of transboundary water issues; basic knowledge, principles and practices in cooperation, and prevention and management of conflict. "This type of training is designed to build awareness and improve the general understanding of conflict prevention and management practices for several target stake-holder groups within a basin" (MacQuarrie et al. 2008). It is designed to enhance regional cooperation through three levels of learning objectives. Short courses of this type are offered at Oregon State University in Water Conflict Management and Transformation, Tufts University in Water Diplomacy, UNESCO-IHE Institute for Water Education in Water Conflict Management, and the University of East Anglia in Water Security (Jarvis 2014). Also, the International Union for the Conservation of Nature (IUCN), Stockholm International Water Institute (SIWI) and University of Dundee have offered similar programs. In addition, the United Nations Economic and Social Commission for Western Asia (ESCWA) engages individual capacity building training in negotiation skills in Jordan, Lebanon, Palestine and the Syrian Arab Republic (UN-Water 2008, 15).

Demand for such training modules and material is increasing. Every year new capacity-building programs emerge to meet the needs and interests of the growing pool of students and professionals in water diplomacy processes.

Institutional Capacities

Institutional capacities refer to the programmatic competencies contained with an orga-nization, such as data collection and monitoring, research and advocacy, outreach and communications, or providing a neutral forum for dialogue and collaboration. Research conducted through Oregon State University's Program in Water Conflict Management and Transformation (PWCMT) suggests that institutional capacity is key to success-ful and enduring cooperation in shared river basins. Results indicate that conflict in a basin is more likely if (a) there are rapid political or physical changes in the basin, and (b) basin institutions are unable to absorb and manage those conditions. Therefore, insti-tutional capacity development is critical to support water diplomacy in complex sys-tems. The need for institutional capacity development is shared across the globe with the least-developed countries and post-conflict countries having particularly urgent needs, followed by developing countries and those in the global north. When supported well, international river basin institutions can absorb and manage major changes in a river basin through a number of instruments such as treaties, cooperative arrangements, the creation and distribution of technical data, promoting stakeholder involvement in poli-cymaking and the distribution of costs and benefits (Wolf et al. 2003a, 2003b, 2005).

Since its launch in 2000, From Potential Conflict to Cooperation Potential (PCCP)—a program of the United Nations Educational Scientific and Cultural Organization—tries to respond to this challenge by facilitating multilevel and interdisciplinary dialogues to foster peace, cooperation and development related to the management of transboundary water resources. The PCCP program produces capacity-development tools that can be used by individuals but mostly serve at the institutional level. This program has devel-oped region-specific and global educational materials and offered multidisciplinary train-ing courses around the world. It has organized collaborative research undertaken jointly by various parties concerned with the resources of the same basin and developed new knowledge about such basins. It has also designed and sustained innovative water diplomacy processes and enhanced cooperation techniques, trust and dialogue among official stakeholders within a number of under-studied basins.

Institutional Capacity Development in Facilitation and Mediation

A second type of capacity building focuses on building skills in facilitation and mediation at a national or regional level. It brings together actors concerned with a specific water body or common regional transboundary issues and offers them the necessary skills to address their challenges. In the Mekong region for example, tailor-made training offered in 2007, 2008, 2010 and 2012 through Oregon State University in collaboration with the Mekong River Commission Secretariat prepared national Mekong committees and line agencies to assist with potential transboundary issues or differences. Their responsibilities included information sharing, meeting facilitation, and activities requiring basic conflict management skills in a shared waters environment. In this setting, the training was designed to assist staff to carry out these tasks, employing exercises and practices aimed at developing practical skills rather than concepts. Examples of skills emphasized in this area include communication, negotiation, facilitation and mediation. This type of training has been offered for the Mekong River committees, and organizations focused on the Nile River Basinamd Jordan River Basin, as well as for China, Central Asia and the United States Bureau of Reclamation.

Society's Capacities–Public Participation

In the Columbia Basin example, public participation successes have been a keystone of the diplomatic process. Societal capacities refer to the competencies of our communities to engage in decision-making processes that impact them, and also the competencies of political leadership to seek, listen to, and incorporate community priorities into public decisions in shared waters. To be meaningful, public participation processes should be designed to encourage involvement among the full range of stakeholder interests, including underserved communities such as women, the poor, and indigenous communities that are or would be directly or indirectly impacted by a project. Communities historically excluded from public processes must be brought into the public-participation process, and direct investments in their capacity development should be made to ensure their meaningful participation. Despite being historically unheard in the public discourse of water negotiation, these communities are key stakeholders in water diplomacy and must be engaged in meaningful ways at every possible opportunity.[3] The Columbia Basin example demonstrates a broad reach of public efforts to engage indigenous, youth and basin residents.

Public participation has been acknowledged as a key part of negotiation processes for decades. The first international convention to address the importance of public participation in water management was the 1992 Dublin International Conference on Water and Environment, declaring, "Water development and management should be based on a participatory approach, involving users, planners and policy makers at all levels."[4] In the same year, the Rio Declaration and the Helsinki Convention (UNECE Convention on the Protection and Use of Transboundary Watercourses and International Lakes, 1992) reconfirmed the value of information sharing and public participation in environmental decision making, stating, "Environmental issues are best handled with participation from all concerned citizens, at the relevant level."[5]

The benefits of productive public participation efforts exhibit in many ways, including broader support for public leaders and the policies they shepherd, the possibility of achieving creative and flexible solutions to complex problems and, in economic terms, more sound investment choices and diminished social costs in the achievement of water security (Grey and Sadoff 2007, 569). Accountability through meaningful public participation processes allows organizations and systems to monitor, learn, self-regulate and adjust their behavior accordingly (Southern African Development Community 2010). It provides legitimacy to decision making, increases transparency and responsiveness and helps reduce the influence of vested interests (UNDP 2009, 14).

Meaningful public participation also creates a sense of ownership in and support for the decisions and policies derived from the negotiation process. It is a mechanism for gaining a better or common understanding between the various stakeholders on the nature of a given problem and its desired outcomes. Stakeholder participation strengthens integration, thereby contributing to conflict prevention and risk reduction (UN-Water 2008, 9). If done right, effective public participation can also motivate improvements in institutional arrangements, leadership, knowledge and accountability.[6] These themes are reflected in the recent work of the Open Working Group of the UN General Assembly on Sustainable Development (United Nations General Assembly 2014a)[7] and the recent secretary-general report on the post-2015 sustainable development agenda.[8] Understanding all this, however, political considerations in a shared waters environment can have profound impacts on the outcomes of a project, and should not be ignored.

The Four Worlds of Conflict Transformation Applied to Capacity Development

The process of capacity development implies a type of transformation, bringing with it a shift in perceptions, relationships, and understandings. When actors can apply a framework for understanding this transformation, the inevitable changes are less surprising and, perhaps, even embraced. Incorporated into capacity development initiatives, orientation as to the four worlds of conflict transformation can provide a valuable tool for actors in water diplomacy. Wolf's four worlds rubric defines a transformational spectrum from conflict to cooperation, as experienced through four discreet yet interdependent stages. The phases of integration progress from the intrapersonal experience (the internal, personal experience that we each feel as we engage in conflict) to the interpersonal experience (the feeling of engaging in conflict with one or two other people), to the communal experience (the feeling of an energy shift in the room as trust begins to build and underlying interests are revealed and shared), to the systems experience (the

Table 8.1 Summary of the four phases of water conflict transformation.

Geographic Scope	Collaborative Skills	Common Water Claims	Negotiation Stage
Nations	Trust building	Rights	Adversarial
Watersheds	Skills building	Needs	Reflexive
"Benefit-sheds"	Consensus building	Benefits	Integrative
Region	Capacity building	Equity	Action

Source: Wolf (2008), fn 58 citing Rothman (1989) and Kaufman (2002).

incorporation of a highly collaborative, structured new way of doing business, as applied to complex systems such as institutions guided by principles of intentionality or the precautionary principle. (Wolf 2012, 78–79). Thus strengthened, institutional arrangements are nourished by the communal experience of accountability measures and interpersonal leadership, which benefits from a deepened intra-personal understanding.

The four phases of water conflict transformation are progressive and parallel, depicted by Wolf (2012)[9] and summarized in Table 8.1. Applied to a collaborative skills-building scenario in water diplomacy, the four stages correspond to a progression of capacity development to consensus building to skills building to trust building. To improve water diplomacy, stakeholders, institutions and society should invest heavily in their capacities to *build trust, develop negotiation and communication skills, work toward consensus*, and *build interdisciplinary capacities in the technical, scientific, legal and sociological aspects of shared waters*. With such interdisciplinary investments in developing the soft and hard skills required to navigate the diplomatic process in a shared-waters environment, institutions build resiliency to avoid or manage complex water conflicts into the future.

Conclusions and Recommendations

Water is, by its nature, an interdisciplinary resource—the attendant disputes can only be resolved through active dialogue among disciplines (Wolf 1998, 257). Capacity development for meaningful outcomes in water diplomacy can take shape in three ways: (a) through enhancing individual skills in negotiations, communications and water-conflict management; (b) through building institutional knowledge capacity in the technical, scientific and legal aspects of shared waters; and (c) through community capacity development in public-participation methodology for the meaningful engagement of all stakeholders in a shared-waters environment.

With today's water issues increasingly more complex, and the related crisis exponentially quicker in expanding and becoming entrenched, we must improve our capacities in water diplomacy (Unver et al. 2010). Actors engaged in water matters need to master the skills of water diplomacy to meet expectations—water managed without cooperation is water managed inefficiently. The classical single disciplinary approach is not enough; our complex shared-waters problems demand multi-disciplinary action and negotiators with broad-spectrum skillsets. To be effective, today's negotiators need strong technological knowledge as well as skills in negotiation, communications, and water-conflict management.

The speed at which diplomacy is undertaken must also be evaluated. Given the galloping global changes that we are facing, and pace with which they impact the hydrological cycle and our water resources, the international community needs to act faster. If we want to preempt and anticipate those changes instead of addressing them after a crisis has inflicted its damages, the international community must invest in immediate, more frequent, regular and sustained capacity-development efforts related to water diplomacy. Only by acting fast will it be possible to achieve any kind of development goal: catching-up (at a first stage) with the delays and then anticipating and preventing impacts of a water crisis on our current environment and that of future generations.

Finally, the international community, as a whole, has the responsibility to invest in capacity development or enhancement (depending on the case) for water diplomacy. Some actors may be givers and receivers of capacity development/enhancement efforts, while others may only have one of these roles. One thing is certain, though: Each member of this community has a role to play and a responsibility to fulfill. To do so, we should not just think outside the box, but act as if there were no box at all.

Notes

1 Participants in the Columbia River Basin Symposium, personal communication with the authors. Spokane, WA, October 2014.
2 See, e.g. http://www.unesco.org/new/en/natural-sciences/environment/water/ihp/ihp-programmes/pccp/
3 To accomplish equity objectives public processes are designed to include: (a) Meeting times and locations that work for working families. (b) The provision of childcare and meals that reflect local cultural norms. (c) Travel assistance in the form of subsidies or shuttle transportation to and from meeting locations. (d) Written materials in colorful, easy-to-understand text and images, engaging audio and visual displays. (e) Onsite language translation in all local languages. (f) Follow through with open lines of communication, an identified point of contact and notice of future opportunities and avenues to communicate with project personnel via telephone, email, mail, walk-in office hours or follow-up visits to the project site and/or impacted communities.
4 Dublin Statement 1992. Accessible at: http://www.wmo.int/pages/prog/hwrp/documents/english/icwedece.html. In relevant part, "**Capacity building:** All actions identified in the Dublin Conference Report require well-trained and qualified personnel. Countries should identify, as part of national development plans, training needs for water-resources assessment and management, and take steps internally and, if necessary with technical co-operation agencies, to provide the required training, and working conditions which help to retain the trained personnel. Governments must also assess their capacity to equip their water and other specialists to implement the full range of activities for integrated water-resources management. This requires provision of an enabling environment in terms of institutional and legal arrangements, including those for effective water-demand management. Awareness raising is a vital part of a participatory approach to water resources management. Information, education and communication support programmes must be an integral part of the development process."
5 Rio Declaration A/CONF.151/26 (Vol. I) Principle 10: "Environmental issues are best handled with the participation of all concerned citizens, at the relevant level. At the national level, each individual shall have appropriate access to information concerning the environment that is held by public authorities, including information on hazardous materials and activities in

their communities, and the opportunity to participate in decision-making processes. States shall facilitate and encourage public awareness and participation by making information widely available. Effective access to judicial and administrative proceedings, including redress and remedy, shall be provided."

6 UNGA A/69/700 par. 128. "Strategies will also have to be reviewed and implemented at the local level, with the full engagement of local authorities. [...] Institutional and human capacities will, in many cases, need to be strengthened for effective implementation and monitoring. This includes bolstering capacities to assess needs, collect data and formulate responses across sectors and institutions."

7 A/68/970, Goal 6.a,b. Available at http://www.undocs.org/A/68/970. The Working Group has established as its Goal 6 by 2030 to "expand international cooperation and capacity building support developing countries in water and sanitation related activities and programs, including water harvesting, desalination, water efficiency, recycling and reuse technologies [...] and to support and strengthen the participation of local communities in improving water and sanitation management."

8 UNGA A/69/700, par. 127. United Nations General Assembly synthesis report of the Secretary-General on the post-2015 sustainable development agenda emphasizes the importance of capacity building and meaningful stakeholder engagement at multiple levels. At the national government level, in "consultation with all stakeholders," governments and their institutions are called upon to "develop the skills and capacities to deliver sustainable development."

9 The four stages of water conflict transformation are built primarily on the work of Jay Rothman, who initially described his stages as adversarial, reflexive, integrative and action. The skill set in each stage is drawn from the work of Edy Kaufman.

References

Cosens, Barbara. 2012. "The River People and the Importance of Salmon." In *Columbia River Treaty Revisited* edited by B. Cosens, Corvalis, OR: Oregon State University Press.

Jarvis, Todd. 2014. *Contesting Hidden Waters: Conflict Resolution for Groundwater Aquifers.* New York: Routledge.

MacQuarrie, Patrick, V. Viriyasakultorn and Aaron Wolf. 2008. "Promoting Cooperation in the Mekong Region through Water Conflict Management, Regional Collaboration, and Capacity Building." *GMSARN International Journal* 2: 175–84.

McCaffrey, S. 2003. *The Law of International Watercourses: Non-Navigational Uses* Oxford Monographs in International Law. Oxford: Oxford University Press, 112–74.

Salame, Lena, Larry Swatuk and Pieter van der Zaag. 2009. "Developing Capacity for Conflict Resolution Applied to Water Issues." In *Capacity Development for Improved Water Management,* edited by Maarten Blokland, Guy Alaerts, Judith Kaspersma and Matt Hare, 103–20. Delft: UNESCO-IHE.

Southern African Development Community. 2010. Guidelines for Strengthening River Basin Organisations: Stakeholder Participation. SADC Secretariat: Gaborone, Botswana. Online: http://www.sadc.int/files/2313/5333/8274/SADC_guideline_stakeholder.pdf

Subramanian, Ashok, Bridget Brown and Aaron Wolf et al. 2014. "Understanding and Overcoming Risks to Cooperation along Transboundary Rivers." *Water Policy* 16: 824–43.

UK Department for International Development (DFID). 2015. "Knowledge Exchange in International Waters." Phase I Report, on file with the authors.

United Nations Committee of Experts on Public Administration. 2006. "Definition of Basic Concepts and Terminologies in Governance and Public Administration." United Nations Economic and Social Council. March 27–31. http://unpan1.un.org/intradoc/groups/public/documents/un/unpan022332.pdf

United Nations Development Programme (UNDP). 2009. *Capacity Development, a UNDP Primer*. New York: UNDP. http://www.undp.org/content/dam/aplaws/publication/en/publications/capacity-development/capacity-development-a-undp-primer/CDG_PrimerReport_final_web.pdf

United Nations General Assembly (UNGA). 2014a. *The Road to Dignity by 2030: Ending Poverty, Transforming All Lives and Protecting the Planet: Synthesis Report of the Secretary-General on the Post-2015 Sustainable Development Agenda*. Report A/69/700. December 4. http://www.un.org/ga/search/view_doc.asp?symbol=A/69/700

———. 2014b. *Report of the Open Working Group of the General Assembly on Sustainable Development Goals*. Report A/68/970. August 12. http://www.un.org/ga/search/view_doc.asp?symbol=A/68/970

United Nations Water (UN-Water). 2008. "Transboundary Waters: Sharing Benefits, Sharing Responsibilities." Thematic Paper.

United States and Yakama Nation. 1855. "Treaty between the United States and the Yakama Nation of Indians, concluded at Camp Stevens, Walla Walla Valley, 9 June 1855," Washington, DC.

Unver, Olcay., Lena Salamé and Thomas Etitia. 2010. "Mejores prácticas en la gestión del agua transfronteriza (Best Practices in Transboundary Water Management)" (original in Spanish). *Ingeniera y Territorio*, Las Naciones Unidas y el Auga (91): 28–35. Online: http://www.ciccp.es/revistaIT/portada/index.asp?id=563

Wolf, Aaron. 1998. "Conflict and Cooperation along International Waterways," *Water Policy* 1(2): 251–65.

———. 2008. "Shared Waters: Conflict and Cooperation," *Annual Review of Environment and Resources* 32: 3.1–3.29, cited in Patrick MacQuarrie, Vitoon Viriyasakultorn and Aaron T. Wolf, "Promoting Cooperation in the Mekong Through Water Conflict Management, Regional Collaboration and Capacity Building," *GMSARN International Journal*: 175–84.

———. 2012. "Spiritual Understandings of Conflict and Transformation and Their Contribution to Water Dialogue." *Water Policy* 14: 73–88.

Wolf, Aaron T., Shira. B. Yoffe and Mark Giordano. 2003. "International Waters: Identifying Basins at Risk." *Water Policy* 5(1): 29–60.

Wolf, Aaron T., Kerstin Stahl and Marcia F. Macomber. 2003. *Conflict and Cooperation within International River Basins: The Importance of Institutional Capacity*. Water Resources Update 125. Universities Council on Water Resources.

Wolf, Aaron T., Annika Kramer, Alexander Carius and Geoffrey Dabelko. 2005. "Managing Water Conflict and Cooperation." In: E. Assadourian et al., *State of the World 2005: Redefining Global Security*, 80–95. Washington, DC: The World Watch Institute.

Wouters, Patricia. 2013. "Dynamic Cooperation in International Law and the Shadow of State Sovereignty in the Context of Transboundary Waters" (Part 2), *Environmental Liability* 4: 141.

World Water Assessment Programme (Editor). 2003. *United Nations World Water Development Report 1: Water for People—Water for Life*. Paris and Oxford: UNESCO and Berghahn Books.

Chapter Nine

WATER COMPLEXITY AND PHYSICS-GUIDED DATA MINING

Udit Bhatia, Devashish Kumar, Evan Kodra
and Auroop R. Ganguly

Abstract

Climate projections, especially at decadal to century scales, rely on physics-based computer models. While the models have generated useful information about global warming and hydrology, "the sad truth of climate science is that the most crucial information is the least credible" (Schiermeier 2010). The possible reasons include intrinsic variability of the climate system as well as our lack of understanding of the physics and the inability to include known physics within the current generation of computer models. Data sciences, ranging from statistics and signal processing to machine learning and nonlinear dynamics, continue to help fill some of the crucial gaps in climate. We hypothesize that these data science solutions can be improved if they are driven by physical knowledge, especially when this knowledge cannot be incorporated in current climate models because they are incomplete or incompatible. We have called this paradigm Physics-Guided Data Mining (PGDM) (Ganguly et al. 2014). In addition to motivating and introducing PGDM, this chapter presents three case studies on precipitation, based on our prior work. Statistical downscaling, which generates higher-resolution projections from lower-resolution model simulations, benefits from a blend of sparse learning techniques with physically motivated covariates (Das et al. 2014; 2015). Multimodal uncertainty quantification shows the potential to improve when physical relations with ancillary variables are considered together with historical skills and future consensus, within a Bayesian framework (Ganguly et al. 2013; Kodra 2014a; Smith et al. 2009). Characterization of internal variability and associated model performance benefit from data-driven analysis of multi-initial condition ensembles, combined with physical understanding of oceanic indices and their initializations (Kodra et al. 2012). The proposed PGDM paradigm, illustrated through our prior publications, shows the potential to bridge crucial knowledge gaps in climate science and help in translation to water resources impacts.

The Grand Water Challenge

Our planet and society continue to be critically shaped by water. In the 21st century and beyond, water is without question among the major clear and present challenges facing the world (Hall et al. 2014). Water in the atmospheric column is perhaps the most important of all the greenhouse gases and causes the most significant uncertainties in our understanding and projections of climate variability and change. Water in the oceans is critical for the survival of the majority of species while water towers in glaciers sustain riverine systems. Water in the rivers and in groundwater is important for direct human consumption, including drinking, as well as for agriculture and for energy production. While water as a resource is critical to our survival and well being, inadequate water causes droughts, contaminated water causes disease and excess water leads to floods and flash floods, all of which may continue to result in severe loss of lives and property.

Hydrologists, hydro-meteorologists, hydraulic engineers and other interdisciplinary scientists and engineers have studied or discussed water sustainability (Vörösmarty et al. 2000) and security (Vörösmarty et al. 2010), freshwater variability impacts (Hall et al. 2014), heavy precipitation (Xie et al. 2010) and coastal floods or droughts (Griffin and Anchukaitis 2014), as well as impacts on agriculture (Foley 2011) and energy and the water–food–energy nexus (Le et al. 2011). The impacts of nonstationarity on water policies (Milly et al. 2008) have been debated, while the significance of water on interconnected sustainable systems has been discussed (Bagley et al. 2013).

Water is intimately related to climate change and weather or hydrologic extremes at disparate scales through multiple feedback channels. Water directly impacts food and energy security and, together with climate, act as threat multipliers for the energy and food crises, besides indirectly influencing governance and conflicts. The water challenge is broad but can perhaps be categorized into three interrelated groups: (a) water resources including water quantity and quality; (b) water-related hazards such as floods and droughts, along with their interactions with critical infrastructures and lifelines; and (c) water-influenced nexus among the food and energy systems and interactions with terrestrial or marine ecosystems. Water resources, hazards and the nexus, as defined above, are stressed by changes in population and lifestyle choices, climate variability and change, as well as regional land use and urbanization, resulting in complex feedback among the coupled natural–engineered–human systems across spatial and temporal scales. A 2014 Pew survey at the World Economic Forum ranked the top ten global risks: Water crises were ranked third after fiscal crises and unemployment, closely followed by the water-related or water-influences issues (Table 9.1: left). The US National Academy of Engineering (NAE) ranks the need to *provide access to clean water* as fifth among the 14 grand challenges in engineering. The top four challenges relate indirectly to sectors that are deeply connected to water—specifically, energy (solar and fusion), climate (carbon sequestration) and food (nitrogen cycle). The final grand challenge listed is to *engineer the tools of scientific discovery*, which in turn is a crucial aspect explored later in this chapter (Table 9.2).

Computational data sciences, or data-intensive discovery in the sciences, has been termed as the fourth paradigm of scientific discovery (Tolle et al. 2011). The earth and

Table 9.1 Water as a global risk of the highest concern animating engineering grand challenges.

America's Grand Challenges	Global Risks of High Concern in 2014
Make solar energy economical	Fiscal crises in key economies
Provide energy from fusion	Structurally high unemployment
Develop carbon sequestration methods	**Water Crises**
Manage the nitrogen cycle	Severe income disparity
Provide access to clean water	*Failure of climate change mitigation and adaptation*
Restore and improve urban infrastructure	*Greater incidence of extreme weather events*
Advance health informatics	Food crises
Engineer better medicines	Failure of major financial mechanism/institute
Reverse-engineer the brain	Profound political and social inability
Secure cyberspace	
Enhance virtual reality	
Advance personalized learning	
Engineer the tools of scientific discovery	
Prevent nuclear terror	

The top ten global risks of highest concern as reported by the Pew Research Center based on their Global Perception Survey 2013–2014 at the World Economic Forum (WEF). Water-related and water-influenced hazards are shown in italics.
Fourteen grand challenges in engineering as recognized by the US National Academy of Engineers (NAE). The physics-guided data mining (PGDM) approach discussed in this chapter exemplifies the final challenge on the NAE list: engineer the tools of scientific discovery.

environmental sciences or engineering have rapidly transformed from data-poor to data-rich sciences over the span of the last several decades with the advent of remote sensors, advances in large-scale computational models as well as developments in data storage and movement in a high performance computing environment. Observed data from sensors, whether remote or in-situ, as well as archived simulations from complex computational models, which encapsulate our understanding of the physics, need to be mined for actionable insights. However, major challenges remain both in capturing the physics and in extracting the data-driven insights.

Climate Stress on Water Systems

Water in atmosphere, oceans, land and biosphere does not just significantly impact climate variability and change, but remains among their largest sources of uncertainty (Vörösmarty et al. 2000; Vörösmarty et al. 2010). Conversely, climate variability and change impact regional hydrometeorology and the water balance, including availability and quality of surface and ground water, as well as water-related natural hazards such as floods and droughts (Van Vliet et al. 2012; Greve et al. 2014). Water availability and stream temperatures together influence freshwater and marine water quality, energy and food security as well as terrestrial and marine ecosystems (Walther 2010; Hall et al. 2014). Climate influences the probability of hazards such as floods, whether caused by heavy rain and snowmelt or hurricanes and storm surge as well as droughts, at multiple space–time resolutions. Adapting to hazards in turn requires the resilient design of critical infrastructures and effective management of key resources. However, climate change is one of many stressors impacting water systems.

Water resources including quantity and quality, water-related hazards such as floods and droughts, as well their impacts on built and natural infrastructures and lifelines, and the water-influenced nexus with food, energy and ecosystems, are stressed by changes in population and lifestyle choices, climate variability and change, resilience of infrastructures and societies and regional land use and urbanization, and ultimately impact the coupled natural-engineered human systems across scales (Aerts et al. 2014; Linkov et al. 2014). The influence of the stressors on water as well as the impacts of water on stressed systems is critically dependent on planning time horizons, spatiotemporal resolutions and the resilience of the coupled system (Milly et al. 2008; Hawkins and Sutton 2009; Hawkins and Sutton 2010). Sensitivity analyses of climate impacts on water systems may occasionally assume other stressors to be invariant (a rather strong assumption) or known functions of time, where the functions may be informed through scenario based projections. However, the feedback amongst various stressors and stressed systems need to be considered.

The need to understand the climate as a stressor on the water system leads to multiple challenges in both physics and data. Physics-guided data mining advances are motivated to address these challenges. Once developed, the PGDM solutions may be generalized across domains.

Water as a Data Challenge

Data science challenge discussed in this section for water challenges are primarily but perhaps not exclusively in context of climate as a stressor. The spectrum of water challenges in the resources, hazards and nexus, operate at multiple space and time scales, and across disparate planning horizons, over which scientific insights, projections, as well as policy insights, need to be generated (Figure 9.1).

The four time horizons to be considered are the following:

(1)   Near-real time to weeks
(2)   Seasonal to inter-annual
(3)   Decadal to mid-century
(4)   Multi-decadal to century and beyond.

The four time horizons have natural boundaries from multiple perspectives, specifically, natural systems and their predictability, engineered systems and their vulnerability, as

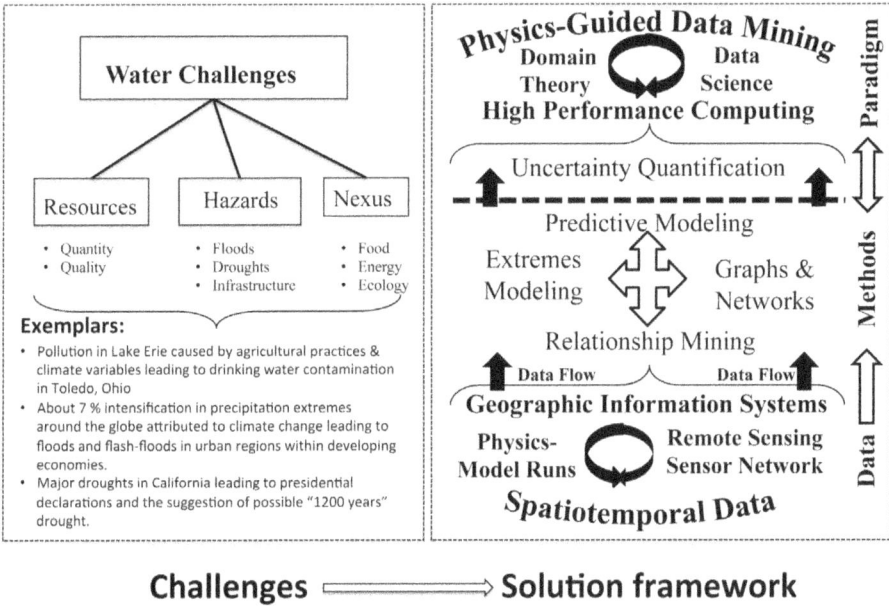

Figure 9.1 Grand challenges in water and the physics-guided data mining framework. The three categories of water challenges (left) are water resources (water quantity and quality), water-related hazards (floods, droughts and impacts on infrastructures) and the water-impacted nexus (food, energy and ecosystems). The exemplars include drinking water scarcity, floods and droughts. The physics-guided data mining (PGDM) paradigm is depicted (right) in both bottom-up and top-down formats, with the flow of data and uncertainty. The PGDM as a paradigm attempts to bring together the best available physics, some of which may be captured within high-performance physics-based computer models, along with mining massive and complex data for actionable insights. The data may be spatiotemporal in nature and embedded in geographical information systems. Data science methods such as relationship mining, predictive modeling, graphs and extreme value modeling need to be guided by physics and cognizant of deep uncertainties, including aleatory, epistemic and nonlinear dynamical.

well as human systems and their adaptability, as well as different tradeoffs in terms of what action may be necessary and indeed what insights may be actionable.

Near-real time to weekly time scales represent horizons over which high-resolution predictions are possible, while beyond these timescales limits to predictability in hydrological and meteorological systems creep in because of chaos or extreme sensitivity to initial conditions (Khan et al. 2005; 2007). Within these time frames, short-term monitoring and predictions may lead to urgent and immediate events and emergency management. While weather changes and short-term population movements may be important at these timescales, major changes in climate or demographics are not expected, while infrastructures, lifelines, economic resources and technological capabilities are expected to remain more or less unchanged.

Seasonal to interannual time scales produce changes in seasonal and interannual climate patterns such as the El Nino phenomenon where phenomena related to natural climate and water variability such as seasonal changes in monsoons in the Southwest United States or seasonal floods in Iowa or the severity of a blizzard in winter in the Northeast United States, or the possible water-quality degradation in lakes such as Erie leading to drinking water pathogens in Ohio, or seasonal droughts in California or the Southeast and the implications of these droughts on annual wine or orange production in California are of concern. While specific weather events are not predictable at these timescales, the average seasonal and annual patterns may be characterized.

Major demographic shifts are not expected although steady population growth and movement may occur (Vörösmarty et al. 2000; Arnell and Lloyd-Hughes 2013). Seasonal physics-based climate models have some predictive skills, as do certain hydrological models, but on the whole the physics at these scales is less well understood. Data science methods have been developed that attempted to creatively use limited physical understanding within statistical or artificial intelligence approaches (Das et al. 2014; Ganguly et al. 2014).

Decadal to mid-century time scales range from about 5 to 30 years in the future (and perhaps all the way to the 2050s), do not have weather or seasonal or even annual predictability for hydrological or hydro-meteorological processes. However, physics-based climate models can project climate trends and variability based on assumed emission scenarios, which in turn are based on storylines for population and technological or societal changes at aggregate scales. However, while non-stationary signals, which include trends in global warming and changes in the statistics of weather patterns and relations, are expected to become prominent at these horizons and the variability in mean and extreme climate is expected to be large and exhibit an increase. In fact, the uncertainty is expected to be comparable to the trend, and the intrinsic variability of the climate projections is expected to be a large component of the overall uncertainty (Hawkins and Sutton 2009). While climate change and associated uncertainty is expected to be a major factor in water challenges at these time horizons, population growth and movement may dominate with regard to water resources (Vörösmarty et al. 2000; Arnell and Lloyd-Hughes 2013). Water quality may be dominated by changes in industrial and agricultural discharge regulations and exacerbated by climate change. Floods and droughts may not be predictable on an event, seasonal or annual occurrence basis at these time horizons, but their attributes, such as intensity or severity, duration and frequency changes may be predictable. Design standards based on flood and precipitation intensity–duration–frequency curves may change and can be quantified based on climate models and observations, leading to changes in protective reinforcement around existing infrastructures and lifelines. The hydrologists and meteorologists may consider these time scales too far out for effective predictability at the space–time scales of familiarity, while the climate community may consider these too near term when the climate signal is not fully detectable given the dominance of variability. However, although the water, climate, energy and food communities have often stayed away from these time horizons, stakeholders ranging from city planners and infrastructure managers to resource managers and energy producers as well as agriculture sectors and private sectors, including insurance and real estate, are demanding actionable insights at these time horizons. Thus, these decadal horizons represent both a major societal challenge and a significant window of opportunity for data scientists, especially those who can deal with space–time data and handle nonlinear dynamics or predictability issues.

Beyond decadal to mid-century time scales, climate trends are expected to dominate over internal variability, but projections for population and human systems are not available and, indeed, are difficult to project other than as what-if scenarios. In addition, many stakeholder decision horizons do not extend to these scales, other than perhaps for long-term policymakers. Credible and actionable insights are possible only at aggregate scales, further restricting which of these insights may be actionable. The climate-change adaptation community, as well as the related integrated-assessments community, has been working at these models with a variety of interconnected but highly simplified models. The models range from being physically based to agent-based and often rely on system dynamics approaches. The community is so used to examining simple models that global-climate or Earth-system models in their full complexity were never used in these integrated assessment models (IAMs), although recent developments have attempted to incorporate the latest Earth system models. However, while integration with a single model run, or even a few ensemble runs, may be reasonable at these time scales in general, such integration may be inadequate when the full range of initial-condition variability needs to be considered, which is likely to make them difficult to use at the decadal scales discussed previously and, occasionally, even at the multi-decadal scales considered here. In addition, a single model (or a single parameter run from one model), and even a few ensemble model runs, may be inadequate to capture multi-model variability (Tebaldi and Knutti 2007; Knutti and Sedláček 2013). Thus, data science may be useful at these scales to comprehensively characterize predictability and uncertainty in the climate system, and to examine how they cascade across water systems, to include influencing the performance of system models of water, energy and food.

However, while data-driven approaches may augment physics-based models through parameter estimation and uncertainty assessment, pure data-driven models may be of limited use because of the underlying nonlinear dynamics and the inherent predictability issues. On the other hand, data-driven models may be very useful at these scales by helping to understand the science of climate change and associated extremes, developing enhanced projections with uncertainty and examining aggregate impacts in terms metrics relevant for adaptation and mitigation.

Thus, while the ability to bring together physics and data sciences is important across spatiotemporal scales and planning horizons, the specific solutions may need to be different. The section on case studies discusses three problems in climate-stressed water systems. One major challenge in statistical downscaling, where data science methods are devised to bring relatively low-resolution climate-model simulations to the scales at which they may be useful for water resources and water-related hazards and the nexus with food and energy. A second challenge is uncertainty quantification, specifically for uncertainties resulting from multiple climate and Earth system model ensembles, where historical skills or performance metrics are necessary but are not sufficient guides to the credibility of future projections. A third challenge is to delineate our lack of understanding of the climate system from intrinsic variability which, in turn, may be useful to inform design and planning principles for climate adaptation. Our case studies point to physics-guided data mining strategies, which may take us toward addressing these challenges.

Physics-Guided Data Mining

Computational modeling and simulation methods and tools have been transforming the Earth and environmental science and engineering communities by encapsulating best available physical understanding and data into high-performance computing environments. Thus, global climate or Earth system models, as well as watershed-scale hydrological models, may consider detailed information about topography and vegetation as well as mass, momentum and energy-conservation equations and empirical laws developed in the past decades. However, these computer models hide the complexities of the physics that is not too well understood through parameterizations, which in turn range from single coefficients to complex mechanistic schemes.

Given gaps in our understanding and legacy models, complex computer models may occasionally appear parametrically and even structurally constrained to comprehensively encapsulate all available understanding of physical processes. Archived or (near) real-time model simulations reflect both the strengths and weaknesses of computer models. Observations from remote and in situ sensors suffer from uncertainties in sensors and measurements and may occasionally lead to inconsistencies and large uncertainties. However, they represent the best available "truth"—hence, confronting models with observations as well as directly mining observations are expected to lead to realistic insights and even help improve physics-based models and our understanding of the science that can, in turn, drive engineering principles. Given that not all available physical understanding may be encapsulated within computational physics-based models, and considering that data mining without appropriate constraints and guidance may not respect physical principles, there is a need to bring together the best of data science methods with physics awareness to get the best possible insights from models and observations. Data mining has transformed business analytics and our everyday lives through embedded intelligence in search engines and the social media. Computational models have already revolutionized the sciences.

There is a clear and present need to bring the power of data science methods to the sciences where data sizes and complexity have been growing significantly, and where computational models and archived simulations thereof have been increasing rapidly as well. However, to be applicable to the sciences and engineering, the data sciences need to be aware of and guided by physical and engineering principles.

Premise of Physics-Guided Data Mining

Physics-based computational models, such as in hydrology, meteorology and climate, derive strengths and weaknesses from our a priori understanding of physical processes and from the structural form of the computational models. The physics in PGDM may refer to the physics already embedded in the physics-based computational models and, hence, captured in their simulations, or to the physics that may not be well captured in the models but could add value to the design of the data mining formulations.

In the context of short-term (0–6 hour) quantitative precipitation forecasting, numerical weather prediction models may provide credible forecasts at the relatively lower (compared to observations) model resolutions over relatively larger lead times but not at low resolutions and over short-forecast lead times. While certain relatively larger-scale processes may be better handled by the weather models, higher-resolution advection or evolution may be better modeled through data-driven techniques. The higher-resolution processes may use both lower-resolution parameters and higher-resolution data to develop forecasts, while the corresponding data-driven methods may need to bring together techniques developed in statistics and computer science. Physical understanding not only can provide a way to partition the overall problem into component processes and eventually bring them together, but also to develop data mining solutions that use lower-resolution model outputs to drive higher resolution change in state variables. Previous researchers have developed what could be called PGDM methods to tackle these challenges (Ganguly and Bras 2003)

Climate projections are based on computational models of climate and Earth systems, but there have been questions (Kumar et al. 2014) about the ability of the models to keep pace with the urgency of solutions required by stakeholders. In the context of climate projections at stakeholder-relevant resolutions and time horizons, both our physical understanding of global (Ganguly et al. 2009; Kodra and Ganguly 2014) and regional (Ghosh et al. 2012) extremes or change in statistical weather patterns and their impacts on urban (Mishra et al. 2015), freshwater and marine (Wang et al. 2015) systems, may benefit from PGDM. The work on temperature extremes and their relations to possible explanatory variables (Kodra and Ganguly 2014) may be viewed as one example of PGDM. The development of an uncertainty quantification framework (Kodra 2014b) that balances historical skills, model consensus and physical- or correlation-based intuition not captured in climate models may be viewed as another example of PGDM. Exemplars and future directions for PGDM in the context of climate extremes are provided elsewhere (Ganguly et al. 2013; 2014).

We note that while our references to the prior literature in this chapter is often limited to work in which the authors have been involved, we emphasize that this is more for reasons of focus and familiarity. A thorough review reveals a considerably rich literature that may serve as the foundations for PGDM of the future.

While not directly related to PGDM, our premise includes a couple of other considerations. First, the data mining aspect of PGDM is broadly interpreted to include computational statistics, signal processing, econometrics, information theory, machine learning, pattern recognition, network science, nonlinear dynamics and database mining, and such, which are all data-driven methods in search of consistent patterns or systematic relationships between variables. In other words, we focus on interdisciplinary data-driven solutions without regard to the specific data discipline of origin for any specific method or combination of methods. Second, the challenge of what may be called *big data – small data* is nearly ubiquitous in earth and environmental sciences and engineering, including but not limited to water and climate. The ability to deal with massive volumes of complex data is a necessity, but it is often the extreme values or the unusual patterns or the sudden or gradual changes that are of interest. The space–time nature of the data and problems as well as the complex dependence structure makes the design of solutions rather challenging. Third, there is a need to consider uncertainty propagation across data, models and processes and develop ways to capture comprehensive uncertainty across all components.

Earth and environmental systems are non-linear, dynamical, subject to multiple feedback mechanisms, exhibit low frequency and intermittence, as well as non-stationary behavior. Thus, purely data-driven extrapolation may not be adequate. However, physics-based computational models, while good at handling certain kinds of processes, cannot represent all processes well. In addition, certain processes are not understood well enough to embed within models but could still be encapsulated quantitatively with data-driven solutions. Physics-guided data mining (PGDM) provides an overarching framework to incorporate data mining tools with physical understanding that may or may not be already contained in state-of-the-art computational models—with the ultimate aim of generating novel and actionable insights at space–time scales and time horizons required by stakeholders. The physics could either help formulate or drive the data mining approach, help inform the strategy for covariate selection, or facilitate the generalization and interpretability of the results.

PGDM is closely related to data assimilation techniques (e.g., the ensemble Kalman filter), which explicitly update physics-based models with observations. However, while data assimilation may be considered a part of the overarching PGDM umbrella, here we explicitly consider those cases in which the projection horizons may preclude updates to state variables or model parameters. The ability to update models with observations does not carry forward in the context of long-range predictions. In addition, climate models often steer clear of explicit calibration strategies to avoid past conditions from overly biasing future projections (especially given expected changes under radiative forces). Stochastic partial differential equations and uncertainty methods such as polynomial chaos expansions are examples of other techniques that attempt to combine physics with data. However, while complementary, their nature and scope are usually different from PGDM approaches in the sense that they fundamentally rely on model structures and characterize uncertainty by introducing randomness in model parameters.

From a methods perspective, our PGDM paradigm is distinct from the other approaches mentioned above, although researchers and practitioners have previously combined physics intuition and data sciences in ways recommended by PGDM. Our PGDM concept may be described through two broadly construed principles: (a) The use of the best available hydroclimate physics, which are not included in models, to guide preprocessing or post-processing steps in data mining and machine learning as well as to inform the computational architecture; (b) the use of machine learning or statistical methods (such as sparse learning) and conceptual frameworks or architectures (such as mixed effects and mixture models) that best fit the climate and water challenge.

Case Studies

The essence of PGDM, as the name suggests, is to bring together physical understanding, which may or may not be encapsulated in computational models (either because the understanding is incomplete or is incompatible with the structure of the current generation of climate models), together with data-driven insights, where the data may include observations from remote or in-situ sensors or model simulations. The umbrella term is purposely left rather broad with a desire to be inclusive. Indeed, there may be claims that earth or environmental scientists and engineers, as well as their counterparts in other sciences and engineering, have developed and used variants of PGDM. We believe this is true. The purpose here is not necessarily to claim novelty, but to propose a way of thinking for data-intensive methods to make a difference in sciences and engineering.

This chapter elucidates three PGDM case studies in the context of water challenges, where climate acts as a stressor. Arguably, the two most important challenges for stakeholders and decision makers to generate actionable insights using multi-model ensembles of earth system models are: (a) downscaling, and (b) uncertainty characterization. Downscaling can be further classified into statistical and dynamical techniques (see Case Study 1 for details), while uncertainty in future climate projections arise primarily from model response and internal variability, other than scenario-based projections of changes in the greenhouse-gases emissions. Particularly, PGDM has added values to statistical downscaling (Ghosh and Mujumdar 2008; Das et al. 2014) and quantification of model response uncertainty (Kodra et al. 2012). In the following sections, we present two case studies highlighting the application of PGDM to statistical downscaling and uncertainty quantification. While data sciences, specifically from nonlinear dynamics and information theory, and physical insights can potentially help inform our characterization of intrinsic variability in climate, the field is less developed. Despite focused studies with data-driven metrics, such as decadal scale predictability (Branstator and Teng 2010) from ensembles of multiple initial condition runs using relative entropy, general theories and methods are lacking. The third case highlights how PGDM, broadly construed, may help address this challenge, and emphasizes the role of internal variability for risk-management decisions at spatiotemporal scales and planning horizons that are relevant to decision makers and stakeholders across multiple sectors.

Case Study I: Statistical Downscaling

Until and unless global circulation models (GCMs) are unable to provide credible information at the spatiotemporal scales relevant to policymakers and stakeholders, downscaling will continue to remain relevant to impact analysis and climate adaptation. Dependent upon GCM outputs, downscaling inherits many of its problems and generates massive volumes of additional data, thus augmenting the complexity and big-data challenge. While the role of dynamical downscaling for hypothesis examination cannot be denied, as outlined in Ganguly et al. (2014), the approach based on regional climate models (RCMs) is much more resource intensive, and parameterization of topology and higher-resolution process models often result in significant inter-model disagreement. On the other hand, statistical downscaling involves developing the quantitative relationship between large-scale atmospheric variables and local surface variables. Statistical downscaling approaches can be broadly classified under three headings, namely: (a) weather typing, including the method of analogs (Zorita and von Storch, 1999); (b) transfer function (Ghosh and Mujumdar 2008); and (c) weather generators (F. Giorgi et al. 2001).

The most popular methodology for statistical downscaling is based on the transfer function approach, which establishes a quantitative relationship between the predictor (local-scale climate variable or target variable) and the predictor variables containing large-scale climate information, often obtained from the GCM output. These downscaling methods are based on two fundamental assumptions: (a) the relationship between chosen predictors and predict and is physically meaningful; and (b) these relationships will hold true in the future. However, an implicit assumption that the statistical downscaling relationships will hold true or remain unaltered in non-stationary climate change may not hold true (Figure 9.2). Hence, design of experiments, which include calibration and validation periods to include range decadal to century climate indicators to test the falsifiability of these relationships can be used to gauge the performance of downscaling models (Salvi et al. 2015).

Secondly, there is a little consensus over the choice of most suitable predictors. As outlined in Jeong et al. (2011), the selection of predictors or covariates in any downscaling application requires a balanced combination of expert knowledge and data-driven techniques, including correlation analysis, principal-component decomposition, and automatic-variable selection (Ghosh and Mujumdar 2008; Ghosh 2010; Hammami et al. 2012). Also, dimensionality reduction and covariate selection purely based on data mining techniques may not be able to capture the "higher order" processes (O'Gorman and Schneider 2009), which are not directly simulated by the GCMs.

We hypothesize that inclusion of the processes, which are not directly simulated or cannot be incorporated in current GCMs, could potentially help in informing the better choice of predictors. One example of such process is the scaling relationship of precipitation extremes. O'Gorman and Schneider (2009) provided the physical basis for changes in precipitation extremes in a range of climate models in warming climate, and quantitatively related the changes in precipitation extremes to change in temperature and vertical velocity. This is in contrast to the earlier held belief intensity of precipitation extremes increase in the proportion of atmospheric water vapor content.

Mathematically, scaling of precipitation extremes at any latitude can be expressed as:

$$P_e \sim -\left\{ \omega_e \frac{dq_s}{dp} \Big|_{\theta^*,T_e} \right\} \tag{1}$$

Here, P_e is a high percentile of precipitation, ω_e is the corresponding upward vertical velocity, {.} is a mass-weighted integral over the troposphere, and the moist-adiabatic derivative of saturation specific humidity is evaluated at the conditional mean temperature T_e when extreme precipitation occurs. Hence, inclusion of physical processes, which are not captured by Earth System Models (ESMs), to inform the weather types and covariates could potentially help in more realistic projections at finer scales.

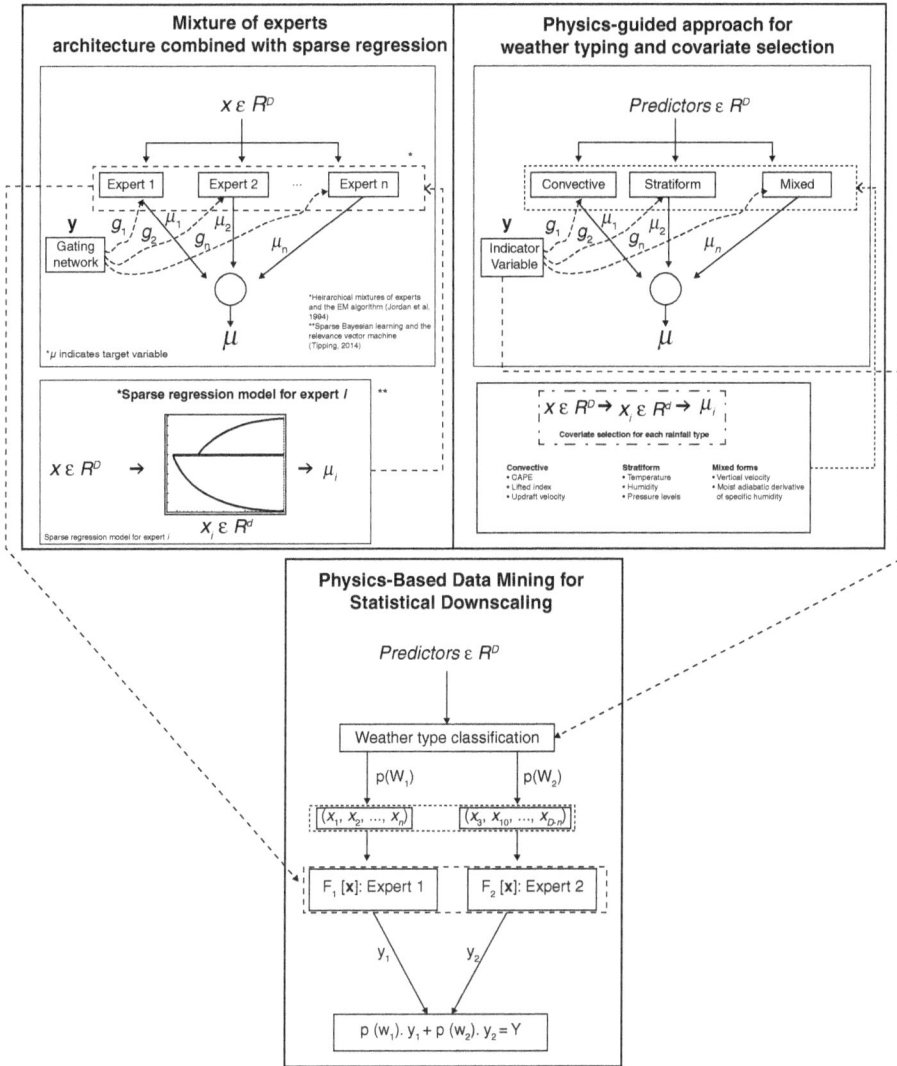

Figure 9.2 Physics-guided data mining for statistical downscaling. In the context of precipitation, the rainfall types can be broadly classified as (a) Convective; (b) Stratiform; and (c) Mixed form of the two. The indicator variables governing these processes may be used as a "gating function" to assign conditional probability of occurrence to the processes. These indicator variables may or may not form the subset of candidate predictors. The role of these variables is solely to assign the likelihood to all possible weather types.

As a specific case, we examine the work of Das et al. (2014), which brought together the best available approaches in physics and data. Specifically, weather typing and transfer function approaches were combined to develop the statistical downscaling approach, with both physics and data-guided choice of indicator variables used for weather typing (via clustering and classification) and covariate selection in transfer functions (via regression). The machine learning technique termed the "mixture of experts" architecture was further developed to include a Bayesian mixture of "sparse regressors." The latter leveraged newly developed methods in intelligent use of sparsity to enable high-dimensional regression. This PGDM approach can accommodate prior knowledge about "must link" constraints between pairs of datapoints, which can potentially lead to an approach for physics-guided feature selection. The approach suggests that instead of directly regressing on the variables of interest, regressors can be selected depending on weather or climate regimes, which in turn reflects the fact that different weather regimes may entail different relationships. The choice of different weather regimes may itself be a function of indicator variables, which in turn depends on inclusion of known physics governing the variable of interest.

In another line of work (Das et al. 2015), a sparse Bayesian framework to discover the covariates affecting the frequency of precipitation extremes at different spatial scales was introduced. The posteriors over regression coefficients indicating the dependence of extremes on the corresponding covariates were obtained. The candidate covariates represent the station variables (e.g., mean annual minimum temperature, mean seasonal temperatures, convective available potential energy (CAPE station)); regional variables (e.g., mean annual precipitation, mean seasonal temperatures and CAPE at the regional level); and large-scale climate indices and variables such as North Atlantic Oscillation, Global Mean Temperature Anomaly. The results indicate that the physics-guided data-driven approaches could potentially inform more "realistic" feature selection prior to statistical regression. Their results suggest that dependence of extreme frequency on covariates is the function of physical processes driving the precipitation processes at the given location. For example, their findings suggest a strong dependence of the rainfall extremes on large-scale climate indices in the central and southwest regions of United States. Similar links were found between precipitation extreme frequency in the southeast regions, with Tropical North and South Atlantic patterns and Nino indices.

Figure 9.3 illustrates an outline of a physics-guided framework for statistical downscaling where a mixture of experts (Jacobs et al. 1991; Jordan 1994) architecture can be used as an analogue to develop a physics-based approach for weather typing, covariate selection and pre-processing.

Case Study II: Uncertainty Quantification

Uncertainty quantification is the big challenge for water and climate sciences and for data sciences in general. In climate, different GCMs differ substantially in their projection; hence, combining multiple models in a single framework remains a challenge. Uncertainty, in general, is known to be higher in rainfall projections because climate models miss the adequate description of basic processes such as cloud formation, mist convection and upward velocities (O'Gorman and Schneider 2009).

Initial approaches to quantify uncertainty treat Multi-modal Ensembles (MMEs) as samples, despite the fact that the output of these models is not random in the true sense.

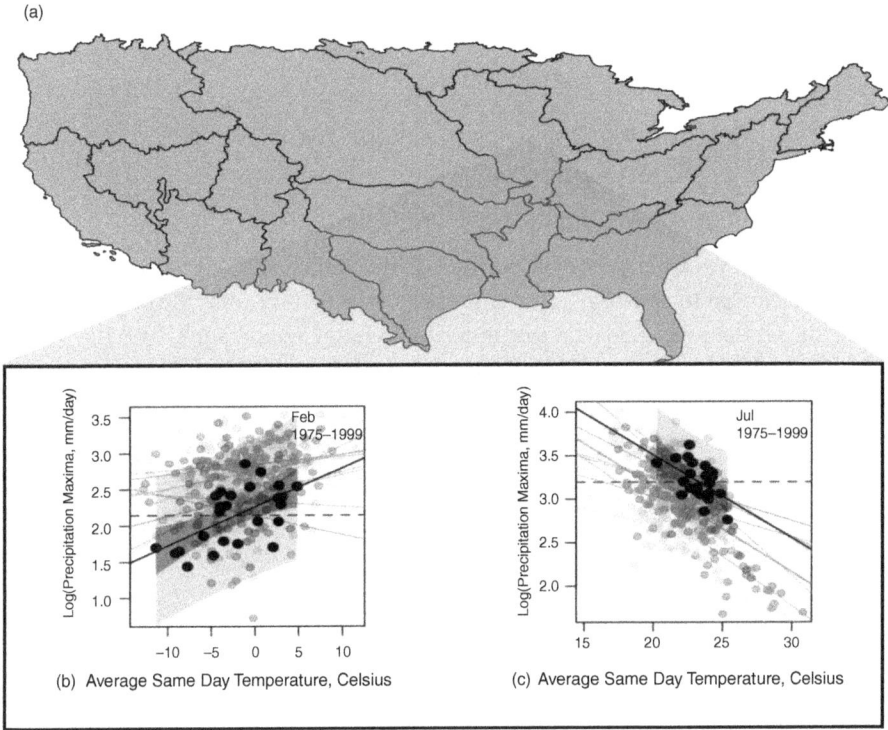

Figure 9.3 Clausius–Clapeyron scaling of precipitation extremes (a) watersheds in the Upper Mississippi Region of Continental USGS; (b) each point is a natural log transformed monthly precipitation maximum (Y-axis) and the corresponding same day average surface temperature (X-axis) from Februaries of 1975–1999. Black dots are observations, and black shading represents 95 percent regression confidence bounds for the fit between temperature and precipitation. The opaque black shading represents 95 percent prediction intervals. gray dots represent the same statistics from 15 CMIP5 GCMs, and corresponding colored lines are their respective regression fits. (c) The same but for July.

Since outputs from multiple ensembles are not independent of each other (F. Giorgi and Mearns 2003), multiple statistical frameworks (Tebaldi and Knutti 2007; Smith et al. 2009; Tebaldi et al. 2011) have been proposed to analyze these outputs. For example, Giorgi and Mearns (2002) introduced the concept of the reliability ensemble averaging method for calculating average, uncertainty range and a measure of reliability of simulated climate changes from ensembles of different atmosphere–ocean general circulation model. In this approach, the estimated change in the state variable, say temperature, is given by:

$$\delta T_{avg} = \frac{\sum_i^N R_i \delta T_i}{\sum R_i} \qquad (2)$$

where N it the total number of models, δT_i is the model simulated change of ith model and R_i is a model reliability factor given as a product of bias and skill:

$$R_i = \left[\left[\frac{\in}{\left| X_j - \mu \right|} \right]^m \left[\frac{\in}{\left| \delta T_j - (v - \mu) \right|} \right]^n \right]^{\frac{1}{m \times n}} \tag{3}$$

where $| X_j - \mu |$ was a measure of historical GCM simulation (X) bias from true historical average climate μ, and $| \delta T_j - (v - \mu) |$ was a measure of how far the projection from GCM j stayed from the change in true historical to future temperature, $(v - \mu)$.

Here, the first component, $[.]^m$ and the second component, $[..]^n$, account for skill and consensus respectively.

The notion of assigning skill and consensus formed the foundation of statistical frameworks (C. Tebaldi et al. 2004; 2005), which formalized them in a Bayesian model. According to their model, each GCM,X_i, is assigned weight according to the following conditional expectation:

$$E\left[\lambda_j | \{X_1, X_2, X_3, ..X_n\} \right] = \frac{a + 1}{b + 0.5\left((X_i - \mu)^2 + \theta (Y_i - v)^2 \right)} \tag{4}$$

Where a and b are prior parameters, X_i and Y_i are GCM simulations of the past and future state variable, respectively, $| X_j - \mu |$ is the measure of model bias (inverse of skill), and $\theta(Y_i - v)$ is the measure of degree of consensus with the future truth, v. Parameter θ plays a crucial role in assigning degree of importance to the consensus. As pointed in original studies as

One of the recent developments (Smith et al. 2009) also allowed for joint consideration of multiple regions, and handle the uncertainty estimation for multiple parameters at once.

As reported in the original studies and in other studies (Ganguly et al. 2013), the model suggested in Smith et al. (2009) assigns more weight to the consensus and less to the reliability and, hence, the bias criterion is not a good representation of the model skill. The idea of whether and how climate models should be weighed has been the topic of discussion in recent literature. However, assigning weights to MMEs and averaging the output to screen the model output may hard to justify for the physical processes that are well represented in climate models and are robust to model uncertainties (Santer et al. 2009).

We argue that inclusion of the weight for GCM based on their skills related to the physical process may result in meaningful reduction of the uncertainty. One such relationship is discussed in the context of Clausius Clapeyron (CC) scaling of precipitation extremes (Pall, Allen and Stone 2006).

The CC provides physical intuition for why temperature may contain useful information content for better constraining projections of precipitation extremes under climate change: as temperature increases, so does atmospheric moisture capacity and, as a consequence, potential precipitation. As suggested in O'Gorman and Schneider (2009) and Sugiyama et al. (2010), precipitation extremes are better described by a scaling process such as the vertical gradient of specific humidity with respect to pressure, updraft velocity and local temperatures. However, in comparison to the other variables that may control this scaling effect, temperature is consistently measured and relatively well projected by GCMs. This provides an opportunity for data-driven methods to harness the information content in temperature to at least quantify, and perhaps constrain, future regional uncertainty in precipitation extremes under global warming scenarios.

Figure 9.3 demonstrates that the Clausius-Clapeyron best serves as the guiding principle for this type of research, reinforcing the notion that other mechanisms that are less well understood may dominate in some regions and seasons. Opportunities exist to gauge the reliability of GCMs by measuring their ability to replicate the regional properties of an observational temperature–precipitation extremes relationship. This represents a prime example of where data-driven methods can be guided by our current understanding of physics. One possible solution framework is inclusion of a statistical transfer function from air temperature to precipitation extremes in the Bayesian framework which, in-turn, assigns the weight to GCMs based on three components: (a) skill based on past observations; (b) consensus; and (c) skill in modelling physical processes and ancillary variables (Figure 9.4).

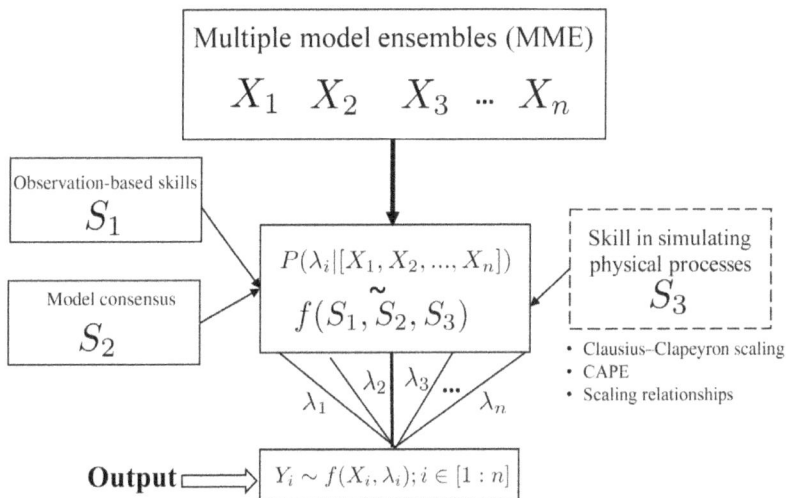

Figure 9.4 Extended Bayesian scheme for uncertainty quantification in multi-modal ensembles. Statistical frameworks to date have largely relied on skills of models relative to historical observations (S_1) and degree of agreement with future consensus (S_2) to assign weights to each model to quantify uncertainty in multi-model ensembles. Inclusion of physical process–related informationd information (S_3) to assign weights for GCMs may lead to our ability to meaningfully reduce uncertainty.

Case Study III: Role of Internal Variability in Climate Change

Long-term climate change projections are generally viewed as a boundary condition problem, unlike weather prediction, which is an initial value problem (Figure 9.5). However, as described by the IPCC (Kirtman et al. 2013), a recent paper in *Nature Climate Change* (Deser et al. 2012), and a rebuttal article in *Nature* (Hawkins et al. 2014), internal variability of the climate system and corresponding initial condition model ensembles remain important in climate projections. This is especially true for near-term projection lead times (~0 to 30 years), which incidentally is of importance to stakeholders across multiple sectors (Schindler and Hilborn 2015; Ganguly et al. 2015), especially at higher space–time resolutions. However, characterization of internal variability in climate-change assessment, particularly in the context of precipitation, has received disproportionately less attention in the literature relative to the stakeholder relevance. Here, we discuss two examples, specifically in the context of precipitation, where internal variability plays a pivotal role in generating reliable insights, and where PGDM may add value.

Ganguly et al. (2012) evaluated the model skills to assess the reliability of individual models and to guide the model selection strategies for forecasting India's monsoons. The results suggested that the Norwegian model, Bjerknes Centre for Climate Research (BCCR), outperformed the other 6 and multimodal average of 7 GCMs, selected from the third phase of the World Climate Research Program's Coupled Model Intercomparison Projects (CMIP3), in simulating the statistics for All-India Monsoonal Rainfall. The superior performance of BCCR can be explained on the basis of understanding of physical mechanisms relating the North Atlantic sea surface temperature anomalies and Indian Monsoon Rainfall (IMR). The Atlantic Multidecadal Oscillation (AMO), with a period

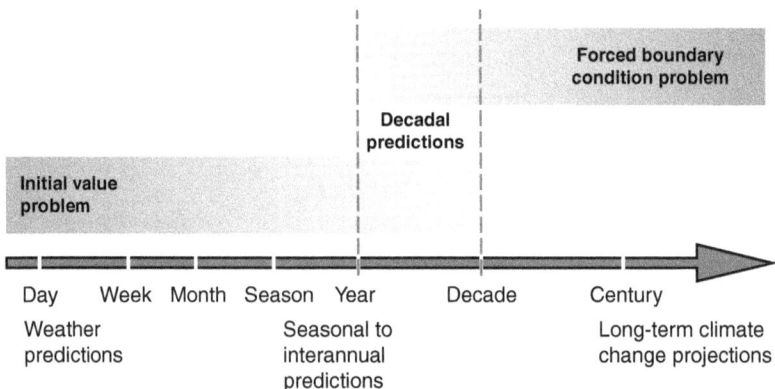

Figure 9.5 Climate predictions at different time scales. A schematic representation of the relative importance of initial conditions and external forcings for climate projections. Annual to decadal predictions occupy the middle ground between the two
Source: "Climate Change 2013: The Physical Science Basis." Contribution of working group I to the Fifth Assessment Report of the Intergovernmental Panel on Climate Change (http://www.ipcc.ch/report/graphics/images/Assessment%20Reports/AR5%20-%20WG1/Chapter%2011/FigBox11.1–2.jpg)

of 65–70 years, influences the meridional gradient of tropospheric temperature in the region, which in turn drives the Eurasian temperature and IMR with approximately same periodicity. While BCCR has been shown to reproduce the AMO relatively well (Ottera et al. 2010), Stoner, Hayhoe, and Wuebbles (2009) showed that the same model does not reproduce temporal and spatial aspects of AMO very realistically. Researchers suggested that patterns associated with AMO might be associated with elusive variability, which is only a few fractions of a degree variation in the ocean surface temperature. Hence, initializations using a single run may not capture this subtleness in variability. Kodra et al. (2012) also highlighted the sensitivity of results to choice of initial conditions. Significant within-model differences in periodic variability and phase shifts were observed from different initial condition runs of 4 GCMs. Hence, careful evaluation of individual model, multi-model ensembles, and multi-initial-condition ensembles, in addition to physical drivers, is imperative for regional projections from climate models.

Our preliminary analysis of 30 different initial condition runs of CESM-CAM5 suggests a complete change in sign for the mean annual precipitation over several regions, including Australia, parts of the United States and parts of Africa and Asia. Figure 9.6 shows the changes in annual mean precipitation patterns for the period 2010–2039 with respect to historical climatology (1961–1990) for three different initial conditions for the Australian region.

Caveats and Future Work

While knowledge-discovery techniques (e.g., data mining, machine learning) have been used in climate and water studies, concerns remain about the significance of discovered relationships (Caldwell et al. 2014) as well as about effective validation strategies (Salvi et al. 2015). One major issue of applying data mining to weather and climate is the unavailability of long enough time series to properly validate and test the methods. The issue is exacerbated in a climate context because of the need for the long-range projections. Field significance tests, appropriate incorporation of dependence structures, where possible, and creative design of experiments may help alleviate these challenges.

A critical scientific barrier is the issue of scale. Specifically, can we model physical phenomena at any scale? What is the appropriate scale for specific processes? How do potentially multiscale processes interact with each other? What to do if there is disparity between the scale of the physical processes versus the scales that are relevant to stakeholders? An argument could be made that we need to model at the scale of the process, which is where the physics is actually relevant. Unfortunately, the relevant scale may be synoptic or mesoscale when it comes to precipitation (and temperature for that matter) given the strong connection to dynamics. This means that we need information about the short-term characteristics of the atmospheric circulation. A key difficulty is that we need to reduce the high dimensionality of the atmospheric circulation without losing the information we need to model it. The challenge is identifying the relevant physics and the best physical representation of that physics at the proper scale. Physics-guided data mining methods of the future need to pay particular attention to these disparities across resolutions, especially the multiscale nature of the processes.

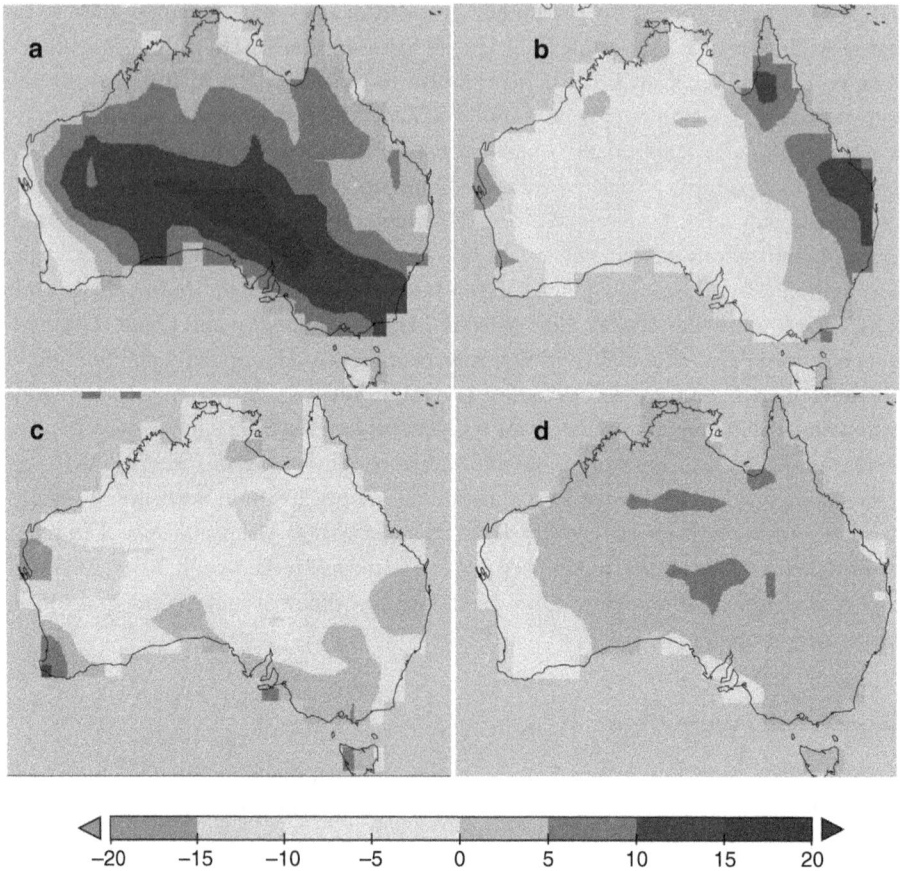

Figure 9.6 (a–c) Changes in spatial pattern for annual mean precipitation for 2010–2039 with respect to historical climatology (1961–1990) for (a–c) selected initial conditions; (d) multi-modal ensemble media from an ensemble of 30 initial conditions run from CESM-CAM5 model. A complete change in sign (dry versus wet patterns) for the mean annual precipitation is observed over Australia.

Acknowledgments

This work was funded by an NSF CISE Expeditions in Computing award: NSF CyberSEES award; NSF BIGDATA award; and Northeastern University's Office of the Provost.

References

Aerts, Jeroen C. J. H., W. J. Wouter Botzen, Kerry Emanuel, Ning Lin, Hans de Moel and Erwann O. Michel-Kerjan. 2014. "Evaluating Flood Resilience Strategies for Coastal Megacities." *Science* 344(6183): 473–75. doi:10.1126/science.1248222.

Arnell, Nigel W. and Ben Lloyd-Hughes. 2013. "The Global-Scale Impacts of Climate Change on Water Resources and Flooding under New Climate and Socio-Economic Scenarios." *Climatic Change* 122(1–2): 127–40. doi:10.1007/s10584-013-0948-4.

Bagley, Justin E., Ankur R. Desai, Keith J. Harding, Peter K. Snyder and Jonathan A. Foley. 2013. "Drought and Deforestation: Has Land Cover Change Influenced Recent Precipitation Extremes in the Amazon?" *Journal of Climate* 27(1): 345–61. doi:10.1175/JCLI-D-12-00369.1.

Branstator, Grant and Haiyan Teng. 2010. "Two Limits of Initial-Value Decadal Predictability in a CGCM." *Journal of Climate* 23(23): 6292–6311. doi:10.1175/2010JCLI3678.1.

Caldwell, Peter M., Christopher S. Bretherton, Mark D. Zelinka, Stephen A. Klein, Benjamin D. Santer and Benjamin M. Sanderson. 2014. "Statistical Significance of Climate Sensitivity Predictors Obtained by Data Mining." *Geophysical Research Letters* 41(5): 1803–08. doi:10.1002/2014GL059205.

Das, Debasish., J. Dy, Jennifer Ross, Zoran Obradovic and A. R. Ganguly. 2014. "Non-Parametric Bayesian Mixture of Sparse Regressions with Application towards Feature Selection for Statistical Downscaling." *Nonlinear Processes in Geophysics* 21(6): 1145–57. doi:10.5194/npg-21-1145-2014.

Das, D., Auroop R. Ganguly and Zoran Obradovic. 2015. "A Bayesian Sparse Generalized Linear Model with an Application to Multiscale Covariate Discovery for Observed Rainfall Extremes over the United States." *IEEE Transactions on Geoscience and Remote Sensing* PP(99): 1–14. doi:10.1109/TGRS.2015.2445911.

Deser, Clara, Reto Knutti, Susan Solomon and Adam S. Phillips. 2012. "Communication of the Role of Natural Variability in Future North American Climate." *Nature Climate Change* 2(11): 775–79. doi:10.1038/nclimate1562.

Foley, Jonathan A. 2011. "Can We Feed the World and Sustain the Planet?" *Scientific American* 305(5): 60–65. doi:10.1038/scientificamerican1111-60.

Ganguly, Auroop. R., Evan A. Kodra, Ankit Agrawal, Arindam Banerjee, S. Boriah, Snigdhasu Chatterjee, So Chatterjee et al. 2014. "Toward Enhanced Understanding and Projections of Climate Extremes Using Physics-Guided Data Mining Techniques." *Nonlinear Processes in Geophysics* 21(4): 777–95. doi:10.5194/npg-21-777-2014.

Ganguly, Auroop R., Evan. Kodra, Snigdhasu Chatterjee, Arindam Banerjee and Habib N. Najm. 2013. "Computational Data Sciences for Actionable Insights on Climate Extremes and Uncertainty." In: *Computational Intelligent Data Analysis for Sustainable Development*, 127–56. Chapman and Hall/CRC Data Mining and Knowledge Discovery Series. Chapman and Hall/CRC. http://www.crcnetbase.com/doi/abs/10.1201/b14799-8.

Ganguly, Auroop R., Devashish Kumar, Poulomi Ganguli, Geoffrey Short and James Klausner. 2015. "Non-Stationarity and Deep Uncertainty: Water Stress on Power Production" Accepted (to appear).

Ganguly, Auroop R. and Rafael L. Bras. 2003. "Distributed Quantitative Precipitation Forecasting Using Information from Radar and Numerical Weather Prediction Models." *Journal of Hydrometeorology* 4(6): 1168–80. doi:10.1175/1525–7541(2003)004<1168:DQPFUI>2.0.CO;2.

Ganguly, Auroop R., Karsten Steinhaeuser, David J. Erickson, Marcia Branstetter, Esther S. Parish, Nagendra Singh, John B. Drake and Lawrence Buja. 2009. "Higher Trends but Larger Uncertainty and Geographic Variability in 21st Century Temperature and Heat Waves." *Proceedings of the National Academy of Sciences* 106(37): 15555–59. doi:10.1073/pnas.0904495106.

Ghosh, Subimal. 2010. "SVM-PGSL Coupled Approach for Statistical Downscaling to Predict Rainfall from GCM Output." *Journal of Geophysical Research: Atmospheres* 115(D22): D22102. doi:10.1029/2009JD013548.

Ghosh, Subimal, Debasish Das, Shih-Chieh Kao and Auroop R. Ganguly. 2012. "Lack of Uniform Trends but Increasing Spatial Variability in Observed Indian Rainfall Extremes." *Nature Climate Change* 2(2): 86–91. doi:10.1038/nclimate1327.

Ghosh, Subimal and Pradeep P. Mujumdar. 2008. "Statistical Downscaling of GCM Simulations to Streamflow Using Relevance Vector Machine." *Advances in Water Resources* 31(1): 132–46. doi:10.1016/j.advwatres.2007.07.005.

Giorgi, Fillippo, J. Christensen, M. Hulme, H. von Storch, P. Whetton, R. Jones, L. Mearns, et al. 2001. *Regional Climate Information- Evaluation and Projections.* http://epic.awi.de/4973/.

Giorgi, Filippo and Linda O. Mearns. 2002. "Calculation of Average, Uncertainty Range, and Reliability of Regional Climate Changes from AOGCM Simulations via the 'Reliability Ensemble Averaging' (REA) Method." *Journal of Climate* 15(10): 1141–58. doi:10.1175/1520-0442(2002)015<1141:COAURA>2.0.CO;2.

Giorgi, Fillippo and Linda O. Mearns. 2003. "Probability of Regional Climate Change Based on the Reliability Ensemble Averaging (REA) Method." *Geophysical Research Letters* 30(12): 1629. doi:10.1029/2003GL017130.

Greve, Peter, Boris Orlowsky, Brigitte Mueller, Justin Sheffield, Markus Reichstein and Sonia I. Seneviratne. 2014. "Global Assessment of Trends in Wetting and Drying over Land." *Nature Geoscience* 7(10): 716–21. doi:10.1038/ngeo2247.

Griffin, Daniel and Kevin J. Anchukaitis. 2014. "How Unusual Is the 2012–2014 California Drought?" *Geophysical Research Letters* 41(24): 2014GL062433. doi:10.1002/2014GL062433.

Hall, Jim W., David Grey, Dustin Garrick, Fai Fung, Casey Brown, Simon J. Dadson and Claudia W. Sadoff. 2014. "Coping with the Curse of Freshwater Variability." *Science* 346(6208): 429–30. doi:10.1126/science.1257890.

Hammami, Dorra, Tae Sam Lee, Taha B. M. J. Ouarda and Jonghyun Lee. 2012. "Predictor Selection for Downscaling GCM Data with LASSO." *Journal of Geophysical Research: Atmospheres* 117(D17): D17116. doi:10.1029/2012JD017864.

Hawkins, Ed, Bruce Anderson, Noah Diffenbaugh, Irina Mahlstein, Richard Betts, Gabi Hegerl, Manoj Joshi et al. 2014. "Uncertainties in the Timing of Unprecedented Climates." *Nature* 511(7507): E3–5. doi:10.1038/nature13523.

Hawkins, Ed and Rowan Sutton. 2009. "The Potential to Narrow Uncertainty in Regional Climate Predictions." *Bulletin of the American Meteorological Society* 90(8): 1095–1107. doi:10.1175/2009BAMS2607.1.

———. 2010. "The Potential to Narrow Uncertainty in Projections of Regional Precipitation Change." *Climate Dynamics* 37(1–2): 407–18. doi:10.1007/s00382-010-0810-6.

Jacobs, Robert A., Michael I. Jordan, Steven J. Nowlan and Geoffrey E. Hinton. 1991. "Adaptive Mixtures of Local Experts." *Neural Computation* 3(1): 79–87. doi:10.1162/neco.1991.3.1.79.

Jeong, D. I., A. St-Hilaire, T. B. M. J. Ouarda and P. Gachon. 2011. "Comparison of Transfer Functions in Statistical Downscaling Models for Daily Temperature and Precipitation over Canada." *Stochastic Environmental Research and Risk Assessment* 26(5): 633–53. doi:10.1007/s00477-011-0523-3.

Jordan, Michael I. 1994. "Hierarchical Mixtures of Experts and the EM Algorithm." *Neural Computation* 6: 181–214.

Khan, Shiraj, Auroop R. Ganguly and Sunil Saigal. 2005. "Detection and Predictive Modeling of Chaos in Finite Hydrological Time Series." *Nonlinear Processes in Geophysics* 12(1): 41–53.

Khan, Shiraj, Sharba Bandyopadhyay, Auroop R. Ganguly, Sunil Saigal, David J. Erickson, Vladimir Protopopescu and George Ostrouchov. 2007. "Relative Performance of Mutual Information Estimation Methods for Quantifying the Dependence among Short and Noisy Data." *Physical Review E* 76(2): 026209. doi:10.1103/PhysRevE.76.026209.

Kirtman, B., S. B. Power, J. A. Adedoyin, G. J. Boer, R. Bojariu, I. Camilloni, F. J. Doblas-Reyes, et al. 2013. "Near-Term Climate Change: Projections and Predictability." In *Climate Change 2013: The Physical Science Basis. Contribution of Working Group I to the Fifth Assessment Report of the Intergovernmental Panel on Climate Change,* edited by T. F. Stocker, D. Qin, G.-K. Plattner, M. Tignor, S. K. Allen, J. Boschung, A. Nauels, Y. Xia, V. Bex, and P. M. Midgley, 953–1028. Cambridge and New York: Cambridge University Press. www.climatechange2013.org.

Knutti, Reto and Jan Sedláček. 2013. "Robustness and Uncertainties in the New CMIP5 Climate Model Projections." *Nature Climate Change* 3(4): 369–73. doi:10.1038/nclimate1716.

Kodra, Evan. 2014a. "Addressing Gaps in Climate Science: Hypothesis and Physics Guided Statistical Approaches." *Interdisciplinary Engineering Dissertations*, May. http://iris.lib.neu.edu/interdisc_eng_diss/4.

———. 2014b. "Addressing Gaps in Climate Science: Hypothesis and Physics Guided Statistical Approaches." *Interdisciplinary Engineering Dissertations*, May. http://iris.lib.neu.edu/interdisc_eng_diss/4.

Kodra, Evan and Auroop R. Ganguly. 2014. "Asymmetry of Projected Increases in Extreme Temperature Distributions." *Scientific Reports* 4(July). doi:10.1038/srep05884.

Kodra, Evan, Subimal Ghosh and Auroop R Ganguly. 2012. "Evaluation of Global Climate Models for Indian Monsoon Climatology." *Environmental Research Letters* 7(1): 014012. doi:10.1088/1748-9326/7/1/014012.

Kumar, Devashish, Evan Kodra and Auroop R. Ganguly. 2014. "Regional and Seasonal Intercomparison of CMIP3 and CMIP5 Climate Model Ensembles for Temperature and Precipitation." *Climate Dynamics* 43(9–10): 2491–2518. doi:10.1007/s00382-014-2070-3.

Le, Phong V. V., Praveen Kumar and Darren T. Drewry. 2011. "Implications for the Hydrologic Cycle under Climate Change due to the Expansion of Bioenergy Crops in the Midwestern United States." *Proceedings of the National Academy of Sciences* 108(37): 15085–90. doi:10.1073/pnas.1107177108.

Linkov, Igor, Todd Bridges, Felix Creutzig, Jennifer Decker, Cate Fox-Lent, Wolfgang Kröger, James H. Lambert et al. 2014. "Changing the Resilience Paradigm." *Nature Climate Change* 4(6): 407–9. doi:10.1038/nclimate2227.

Milly, P. Chris D., Julio Betancourt, Malin Falkenmark, Robert M. Hirsch, Zbigniew W. Kundzewicz, Dennis P. Lettenmaier and Ronald J. Stouffer. 2008. "Stationarity Is Dead: Whither Water Management?" *Science* 319(5863): 573–74. doi:10.1126/science.1151915.

Mishra, Vimal, Auroop R. Ganguly, Bart Nijssen and Dennis Lettenmaier. 2015. "Changes in Observed Climate Extremes in Global Urban Areas." *Environmental Research Letters* 10(2): 024005. doi:10.1088/1748–9326/10/2/024005.

O'Gorman, Paul and Tapio Schneider. 2009. "The Physical Basis for Increases in Precipitation Extremes in Simulations of 21st-Century Climate Change." *Proceedings of the National Academy of Sciences* 106(35): 14773–77. doi:10.1073/pnas.0907610106.

Ottera, Odd Helge, Mats Bentsen, Helge Drange and Lingling Suo. 2010. "External Forcing as a Metronome for Atlantic Multidecadal Variability." *Nature Geoscience* 3(10): 688–94. doi:10.1038/Ngeo955.

Pall, Pardeep, Myles Allen and Dáithí A. Stone. 2006. "Testing the Clausius–Clapeyron Constraint on Changes in Extreme Precipitation under CO2 Warming." *Climate Dynamics* 28(4): 351–63. doi:10.1007/s00382-006-0180-2.

Salvi, Kaustubh, Subimal Ghosh and Auroop R. Ganguly. 2015. "Credibility of Statistical Downscaling under Nonstationary Climate." *Climate Dynamics* (June): 1–33. doi:10.1007/s00382-015-2688-9.

Santer, Benjamin D., Karl Taylor, Peter J. Gleckler, Celine Bonfils, Tim P. Barnett, David Pierce, Tom Wigley et al. 2009. "Incorporating Model Quality Information in Climate Change Detection and Attribution Studies." *Proceedings of the National Academy of Sciences* 106(35): 14778–83. doi:10.1073/pnas.0901736106.

Schiermeier, Quirin. 2010. "The Real Holes in Climate Science." *Nature News* 463(7279): 284–87. doi:10.1038/463284a.

Schindler, Daniel E. and Ray Hilborn. 2015. "Prediction, Precaution, and Policy under Global Change." *Science* 347(6225): 953–54. doi:10.1126/science.1261824.

Smith, Richard L., Claudia Tebaldi, Doug Nychka and Linda O. Mearns. 2009. "Bayesian Modeling of Uncertainty in Ensembles of Climate Models." *Journal of the American Statistical Association* 104(485): 97–116. doi:10.1198/jasa.2009.0007.

Stoner, Anne Marie K., Katharine Hayhoe and Donald J. *Wuebbles*. 2009. "Assessing General Circulation Model Simulations of Atmospheric Teleconnection Patterns." *Journal of Climate* 22(16): 4348–72. doi:10.1175/2009JCLI2577.1.

Sugiyama, Masahiro, Hideo Shiogama and Seita Emori. 2010. "Precipitation Extreme Changes Exceeding Moisture Content Increases in MIROC and IPCC Climate Models." *Proceedings of the National Academy of Sciences* 107(2): 571–75. doi:10.1073/pnas.0903186107.

Tebaldi, Claudia, Julie Arblaster and Reto Knutti. 2011. "Mapping Model Agreement on Future Climate Projections." *Geophysical Research Letters* 38: L23701. doi:10.1029/2011gl049863.

Tebaldi, Claudia and Reto Knutti. 2007. "The Use of the Multi-Model Ensemble in Probabilistic Climate Projections." *Philosophical Transactions of the Royal Society of London A: Mathematical, Physical and Engineering Sciences* 365(1857): 2053–75. doi:10.1098/rsta.2007.2076.

Tebaldi, Claudia, Richard L. Smith, Doug Nychka and Linda O. Mearns. 2005. "Quantifying Uncertainty in Projections of Regional Climate Change: A Bayesian Approach to the Analysis of Multimodel Ensembles." *Journal of Climate* 18(10): 1524–40. doi:10.1175/JCLI3363.1.

Tebaldi, C., L. O. Mearns, D. Nychka and R. L. Smith. 2004. "Regional Probabilities of Precipitation Change: A Bayesian Analysis of Multimodel Simulations." *Geophysical Research Letters* 31(24): L24213. doi:10.1029/2004GL021276.

Tolle, K. M., D. Tansley and A. J. G. Hey. 2011. "The Fourth Paradigm: Data-Intensive Scientific Discovery [Point of View]." *Proceedings of the IEEE* 99(8): 1334–37. doi:10.1109/JPROC.2011.2155130.

van Vliet, Michelle T. H., John R. Yearsley, Fulco Ludwig, Stefan Vögele, Dennis P. Lettenmaier and Pavel Kabat. 2012. "Vulnerability of US and European Electricity Supply to Climate Change." *Nature Climate Change* 2(9): 676–81. doi:10.1038/nclimate1546.

Vörösmarty, Charles J., Pamela Green, Joeseph Salisbury and R. B. Lammers. 2000. "Global Water Resources: Vulnerability from Climate Change and Population Growth." *Science* 289(5477): 284–88.

Vörösmarty, Charles J., Peter B. McIntyre, Mark O. Gessner, David Dudgeon, Alexander Prusevich, Pamela Green, Stanley Glidden et al. 2010. "Global Threats to Human Water Security and River Biodiversity." *Nature* 467(7315): 555–61. doi:10.1038/nature09440.

Walther, Gian-Reto. 2010. "Community and Ecosystem Responses to Recent Climate Change." *Philosophical Transactions of the Royal Society of London B: Biological Sciences* 365(1549): 2019–24. doi:10.1098/rstb.2010.0021.

Wang, Daiwei, Tarik C. Gouhier, Bruce A. Menge and Auroop R. Ganguly. 2015. "Intensification and Spatial Homogenization of Coastal Upwelling under Climate Change." *Nature* 518(7539): 390–94. doi:10.1038/nature14235.

Xie, S. P., C. Deser, G. A. Vecchi, J. Ma, H. Y. Teng and A. T. Wittenberg. 2010. "Global Warming Pattern Formation: Sea Surface Temperature and Rainfall." *Journal of Climate*, 966–86. doi:10.1175/2009JCLI3329.1.

Zorita, Eduardo and Hans von Storch. 1999. "The Analog Method as a Simple Statistical Downscaling Technique: Comparison with More Complicated Methods." *Journal of Climate* 12(8): 2474–89. doi:10.1175/1520-0442(1999)012<2474:TAMAAS>2.0.CO;2.

Part III

CASE STUDIES

Chapter Ten

THE NATURE OF ENABLING CONDITIONS OF TRANSBOUNDARY WATER MANAGEMENT: LEARNING FROM THE NEGOTIATION OF THE INDUS AND JORDAN BASIN TREATIES

Enamul Choudhury

Abstract

The complexity of transboundary water management lies in the dynamics of competition and cooperation that arise from the interactions and feedback among variables, processes, actors and institutions that share a basin or aquifer. The nature of complexity is, however, contingent on a variety of contextual characteristics of the interactions as well as the natural characteristics of the basin. Given the complexity of transboundary (hereafter TB) water management and its contingent manifestations, it is important to understand the means of interaction that can equitably explore and distribute the benefits of the basin for the riparians as well as keep its availability sustainable for posterity. Among different frameworks, mediated negotiation has shown resilience to initiate, affect and sustain interaction among riparians even when they remain hostile to each other on other issues. This chapter identifies three enabling conditions that make transboundary negotiation effective in managing complex and prolonged water disputes. In advancing the argument the chapter also makes the point that resiliency of these three enabling conditions rests on operationalizing the values of equity and sustainability in context-specific ways. For empirical illustration of the efficacy of these enabling conditions, the chapter draws upon the case analyses of negotiation that led to the Indus Water Treaty between India and Pakistan and the 1994 peace treaty between Israel and Jordan.

Introduction: The Complexity of Transboundary Water Management

Transboundary water management is complex because of the problems of water allocation in quantities, maintaining water quality and preserving the environment that spans many boundaries—political, social, jurisdictional, physical and ecological. Addressing this complex problem has remained a focal issue in the literature (Wolf 1999; Biswas 1992; Uitto and Duda 2002; Song and Whittington 2004; Dinar et al. 2007; Prescoli and Wolf 2010; Earle et al. 2010; Wolf 2010; Subramanian, Brown and Wolf 2012; Islam and Susskind 2013; Salman 2013; Dellapenna et al. 2013). The specific nature of the problem and, hence, the character of the disputes among the riparians will vary by basin context and the needs and capacities of the riparians. While scientific and engineering solutions are necessary to address the physical, chemical and ecological problems, organizational and political contexts are also integral to finding and sustaining adaptive solutions to disputes as they arise.

The scope of the problem and, hence, the challenge of finding effective ways to address it lies in the fact that there are 276 river basins and over 200 aquifers that cross national boundaries (TWAP 2013). While the majority of the international river basins are shared between 2 countries, there are 13 basins that are shared between 5 and 8 countries, and 5 that are shared between 9 and 11 countries (Watkins 2006). Because these basins cover about 45 percent of the earth's land surface, account for about 80 percent of global river flow and 60 percent of global freshwater supply, and affect about 40 percent of the world's population (Salman 2013, 360), they also serve as focal points of inter-state conflict and cooperation (Giordano et al. 2014, 245; Yoffe et al. 2003). Thus, the consequences of conflict and mismanagement are enormous for public well-being, both local and global, particularly in sustaining the present and future food supply, energy production and the water needs of a growing population that is creating additional demands for water in terms of both quantity, quality and ecological services.

It is estimated that the demand for water is projected to grow by over 40 percent by 2050, and over half of the global population could be living in water-stressed conditions (WWAP 2012). With the growing water footprint of modern nations, it is also projected that water shortages will grow more serious, leading to conditions of widespread conflicts, particularly in water-stressed regions and places where demand for water is outstripping supply (*The Economist* 1999, 52). Even though empirical research has shown that water wars between nations are very few when compared to the evidence of cooperative interaction (Yoffe, Wolf and Giordano 2003; Prescoli and Wolf 2010), nevertheless, water demand amidst water stress and shortages can increase water conflicts and create tensions. While some TB water disputes have been successfully negotiated, others are ongoing, while some have failed. Given the salience of effective governance to manage TB waters, scholars and leaders have focused on the peaceful resolution of disputes as an imperative and a challenge.

Other than the question of demand and supply, additional factors are also making TB waters problems complex. These factors include the values of water governance that serve as normative anchors—that is, values without which managing TB waters cannot be considered effective nor enduring (Hoekstra 2011, 21; Giordano and Wolf 2001; Wolf 1999). One anchor is the value of equity in water allocation and use, while the other is the sustainability of supply and water quality. The argument of equity varies from the application of international laws or conventions to volumetric allocation among riparians, allocations made to the different sectors of the economy and polity (for example, agriculture, energy, urban) or providing access to and making water affordable for the poor, to establishing sustainable practices of usage for securing generational equity. A key focus of equity has been the adequate and affordable provision of safe water to disadvantaged populations and the protection of vulnerable ecosystems.

The value of sustainability argues for addressing, as an example, the long-term effects of climate change and water management practices on the hydrological characteristics of the basin itself as well as its ecology. Protecting both through pollution prevention, resource conservation and efficient water use practices, including relying on virtual water trade, have now become important considerations to conceiving of any effective solution to the TB water problem. Over time, the United Nations, non-profit advocacy groups and community associations have led and upheld both values as critical paths for finding solutions to TB water problems.

Therefore, today, the complexity of water problems is contingent not only on the nature and form of interaction among the stakeholders but also in the ways sustainability and equity manifest themselves within a particular context in the midst of competing demands of multiple stakeholders. Thus, the global understanding of TB interaction today rests on the application of the values of equity and sustainability as normative principles in addressing and resolving TB water issues. Based on the 1997 UN Convention, Articles 5–9 of the Berlin rules specify the principle of equity and its contingent interpretation, while Articles 12–16 establishes the application of the meaning of sustainability and its contingent interpretations (Berlin Rule 2004). Beyond these formal declarations, both the global knowledge community (academics and experts) and the political community (policy makers and community leaders) that spans the public, private, and nongovernmental sectors have been emphasizing both equity and sustainability as guiding values for resolving emerging water problems. Thus, the resolution of complex water problems today goes beyond satisfying sovereign national interests and incorporates the need to address global values.

Given the complexity of sharing TB waters, the question of what explains the interaction of stakeholders and, once they agree to interact, which approach makes their interaction successful are important research and practical questions to address. The aim of this chapter is to first address the contingent nature of interaction and then address the enabling conditions of negotiation that have led to successful and enduring interaction on sharing TB waters. For empirical illustration, the chapter draws upon the events and experience of the negotiation process that resulted in the 1960 Indus Treaty between India and Pakistan, and the peace treaty between Jordan and Israel. The choice of these two cases rests on using the criteria of influential/representative and contrasting/critical cases (Seawright and Gerring 2008; Yin 1989).

An influential case is one that is used as a basis to check the assumptions of a general framework or claim in the existing literature. Here, we are using the Indus and Jordan cases, both widely referred to in the literature to illustrate the efficacy of our proposed enabling conditions of negotiation in the initiation, operation and outcome of stakeholder interaction in a TBW dispute. Their status of being influential cases also rests on the fact that the two treaties have not only sustained until now, but have also addressed new issues that have emerged over time, thus providing empirical evidence of the resilience of negotiated interaction.

A second criterion of case selection is the use of contrasting cases. The two cases chosen convey the contrasting contexts of successful interaction, not only in terms of the time periods involved, cultural differences and differing political contexts, but also in terms of the negotiation techniques used. It is important to reiterate that the aim of using the cases is not to test a theory or generalized model of resolving complex water problems, because there are none. Rather, the aim is to pursue an abductive approach (Fann 1970; Magnani 2001), as opposed to either an inductive or deductive one, in order to explore the conditions for a working framework that can account for the effective ways of addressing and resolving complex water problems.

Based on the two criteria of case selection, the two chosen cases offer an opportunity to inquire into the underlying factors that have led to not only to successful interactions on TB water problems, but if these factors continue to contribute to these interactions. It is these underlying factors of interaction amidst conflicts that the chapter refers to as enabling conditions. These enabling conditions function not as universal or invariant natural laws (for example, laws of gravity), nor represent the multifaceted nature of reality, but function as working concepts that help capture the underlying facts of relatively successful resolutions of complex water problems. We recognize that identification of these enabling conditions is rooted in normative and instrumental reasoning. One simply cannot separate the two when dealing with coupled natural and human systems such as TB water issues. This chapter argues that the interactional conditions become enabling when they are both principled and pragmatic.

There are excellent analyses of these two cases in the literature, both in terms of the chronology of events, the detailed identification of the actors and issues, and the complex interactions among them, as well as the agreements that were reached and the disagreements that arose and still persist. In the Indus case, the leading case analyses include the works of Biswas (1992) Mehta (1988), Alam (2002), Salman and Uperty (2002), Dinar et al. (2007) and Wolf and Newton (2009), while the Jordan case has been narrated by Elmusa (1995), Beaumont (1997), Shamir (1998), Haddadin (2000), Jagerskog (2003) and Abukhater 2013). This chapter draws from these studies—the events, actors, issues and agreements that help illustrate the three enabling conditions. The approach, therefore, is not to provide a complete or comparative case analysis, but to use case illustrations to discern some underlying patterns based on what we currently know about the process of negotiated cooperation involving disputes over TB waters.

The Contingency of Cooperation under Complexity

Scholars and practitioners have already addressed the importance and constraints of cooperation over TB waters, as well as providing useful frameworks and findings for further inquiry. A key framework of analysis points out that conflict and cooperation are not mutually exclusive but operate in a continuum (Wolf, Sadoff and Grey 2005). Further work along this line of reasoning holds that both conflict and cooperation can vary in intensity as well as act in conjunction (Zeitoun and Mirumachi 2008). Another important frame is the focus on the critical function of power and politics (in varied forms) that structures the relationship of riparians and regulates the availability of and access to water for various stakeholders (Frey 1993; Allan and Mirumachi 2010; Warner and Zawarhi 2012). There is also the broad frame of integrated water resource management that seeks to not only efficiently use and preserve available waters for multiple use but also to secure its quality and sustainable supply (Biswas 2004; Saravana, McDonald and Mollinga 2009; Gallego-Ayala 2013).

What these frames of understanding do is not simply to point to the need for and the form of cooperation, but also to the contingency of cooperation itself. Just because, technological capability and political rationality calls for effective cooperation does not mean that it will happen. Cooperation is an outcome of the process of interaction among stakeholders. Hence, it is the means of interaction that leads to cooperation. Just because, a riparian is located downstream does not necessarily mean it will receive less water. A variety of conditions intervene to make interaction of stakeholders effective and sustaining. Thus, the literature on TB water points to a range of contingent possibilities of water management: conflict, cooperation, conflict and cooperation, integration, fragmentation, domination or cooptation. In other words, the literature makes the important point that there are many ways and forms of effective interaction. Although the argument is often made that the uniqueness of each basin prevents generalization of the factors of cooperation (Elhance 2000), nevertheless, there are systematic efforts to identify factors that affect cooperation and conflict across basins. A varied set of such factors have been identified and argued in the literature (Lowi 1993; Frey 1993; Dinar and Dinar 2000 '95–199; Spector 2000; Uttio and Duda 2002; Wolf et al. 2003; Song and Whittington 2004; Mollinga et al. 2007; Zeitoun and Mirumachi 2008, 300; Agnew and Woodhouse 201; Priscoli and Wolf 2009; Giordano et al. 2014; Yamagata, Yang and Galaskiewicz 2013, 251).

The principal factors that tend to create contingency are the following: basin characteristics (geography, hydrology, climate variability); riparian characteristics (securitization of water for territorial integrity or domestic politics, strategic goals, population growth, urbanization, economic development, military power, power difference between riparians, management capacity); locational characteristics (North–South, upstream–downstream, Arid–non-arid); and political–cultural characteristics (democratic–non-democratic; and reputation in terms of peer behavior and network ties).

Given the constellation of factors that make interaction over TB waters uncertain and contingent, the historical fact is that the record of cooperation is more resilient than that which the alarm over water conflicts leads one to expect (Earle et al. 2010; Priscoli and Wolf 2009). Although a variety of interests operating in a situation propel the parties to interact, nevertheless, interaction does not happen automatically. Interaction among stakeholders is contingent on the use of institutional mechanisms that involve value-allocating politics, value-adding processes, and value-realizing practices. There is a variety of such mechanisms—from riparians unilaterally adopting international law, negotiating bilateral or multilateral treaties, to making temporary informal agreements. Among these, bilateral treaties arrived at through negotiation tend to be relied upon more in forging interaction (Earle et al. 2010, 3; Priscoli and Wolf 2009; Jarvis and Wolf 2010; Zawahri, Dinar and Nigatu 2014, 251; Giordano et al. 2014). However, based on the conflict–cooperation framework, the presence of bilateral cooperation on water sharing does not mean the absence of conflict on other water or non-water issues among the parties involved, including conflicting interpretations of the terms of the treaty that has been agreed to. That continuing the interaction possible amidst conflict points to the contingent nature of understanding and solving complex problems.

Approximately 70 percent of the world's transboundary basins are currently governed by treaties. These treaties vary in terms of their scope and coverage as well as the degree of institutionalization (Giardono et al. 2014, 245; Zawahri et al. 2014). Although the presence of a treaty indicates the absence of conflict, in doing so it does not convey the effectiveness or the sustainability of interaction. In other words, treaties by themselves do not create or secure their institutional resilience, which constitutes a key measure of governing TB basins at risk. The question of resiliency of interaction arises because the negotiation process and the form of interaction that results through a treaty are both contingent outcomes (Bercovitch et al. 1991). It follows that to make negotiation effective we need to identify the factors that prevent conflict and promote interaction, as well as to allow for the consideration of constructive conflicts as means to secure enduring interaction.

Although the treaty constitutes the modal medium to address and resolve TB water disputes, there is much need to advance our understanding of the treaty content and the factors influencing treaty design, issues about which our current knowledge needs advancement (Dinar and Dinar 2000, 194; Zawahri, Dinar and Nigatu 2014). To develop our knowledge of these conditions it is instructive to learn from both theory and practice. In terms of theory, the rich scholarship of environmental negotiation provides us with factors that lead to negotiation success. On the other hand, also studying particular negotiation processes that led to successful TB water treaties provides us with the practical ground to specify and verify them. To contribute in advancing the discourse, this chapter identifies and proposes three enabling conditions of negotiated interaction.

Negotiation theory is primarily a process focusing on the behavioral and contextual factors that are brought to bear on understanding and resolving disputes. As a result, it seeks to buffer the asymmetry of power in order for both parties to interact in good faith and with facts grounded in interaction. The difference in negotiation outcome thus is dependent more on the quality of interaction (like trust in the process), bargaining strategy (like technical skills), resource availability (technical and/or financial capacity) and risk minimization (like a joint action or provision) to reach agreement (Coleman and Deutsch 2014; Islam and Susskind 2013; Fisher, Ury and Patton 2011; Jarvis and Wolf 2010; Susskind, Levy and Thomas-Larmer 2000, 17–40, 249–76; Susskind and Babbitt 1992; Bercovitch et al. 1991). A variety of process factors are utilized to affect the aforementioned conditions given the context and actors involved in a negotiation context. These conditions can be classified as falling into three categories: process, design and implementation. The enabling conditions for each of the three categories are shown below:

Process issues: initiating engagement through mechanisms that make disputants recognize their interdependence and engage each other to explore the advantages and disadvantages of the interdependence. The engagement can be enabled through a neutral medium, which can be a third-party mediator, facilitator, process or venue. The meaning of neutrality, here, is operational not substantive. A mediator can and often does have political goals in engaging the process, but as long as the goals are not material issues in the dispute the mediator is operationally neutral. Such neutrality can also be enabled by the position taken by any of the disputing parties or due to the surveillance of the global knowledge and political community. However, the aim of this broadly understood meaning of mediation is to keep disputants engaged and moving to a more concrete understanding of their interdependence and dispute. The enabling conditions, here, are the political willingness to engage, including acceptance of a third party, compromise or acting in good faith (Salman 2013, 407; Choudhury and Islam 2015).

Design issues: sustaining the engagement through designing the interaction process to create the possibility of mutual benefits or share losses. A variety of mechanisms can play an enabling function for parties in a dispute, envisioning the expected benefits to be higher than the cost of non-negotiation. The very fact that the parties are willing to find solutions to the problems they mutually recognize is the beginning process of forging the mutual-gains approach. By mutual gain is meant not a fusion of gains, but the retention of each party's interest as recognized by the other party through negotiated agreements. These mechanisms can range from joint fact-finding, identifying mutual gains, using a single-text to negotiate, or relying on a mediator or arbitrator for equitable distribution of gains and losses. The enabling conditions here also include levelling the playing field in practical ways in relation to the negotiation process, tying the negotiation to the strategic goals of the parties via issue linkages, and creating the opportunity to secure aid for needed policy goals as well as enhancing the political reputation of the parties (Dinar and Dinar 2000, 194; Warner and Zawahri 2012, 215; Choudhury and Islam 2015). The process of securing mutual gain thus includes advancing equity by reducing the power and resource asymmetry in negotiating gains. However, in actual treaties, the design of mutual gains often falls short of what is possible and needed. The possibility of identifying and realizing these potential benefits can be addressed in the third enabling condition.

Implementation issues: securing mutual gains requires operational measures to assure that all parties contribute to the efficacy of the agreement. Once an agreement is reached, whether in the form of a memorandum of understanding or a formal treaty, a variety of mechanisms can enable the agreement to hold and foster cooperation. These enabling mechanisms can range from having an institutional process of mutual monitoring, levying cost of defection, insurance from a third party, or designing an organization to oversee the implementation of the agreement as well as explore cooperation beyond the terms of the agreement. The latter ensures the prevention of future conflict and, hence, the need for outside mediation. Here, the enabling conditions also include data- and information-sharing provision, conflict resolution mechanism, cooperative management of the basin and mechanism for the participation of non-state actors (Giordano et al. 2014, 245; Gerlak et al. 2011, 179; Warner and Zawahri 2012, 215; Choudhury and Islam 2015). In many instance, the capacitation of the implementation process, both in terms of authority and resource, has lagged behind what is needed or what have been neglected to address new issues to interact or new problems to solve. Thus, the potential mutual benefits remain unrealized. This weakness particularly undermines the value of sustainability. However, what is lacking or lagging in treaty design and operation have begun to be addressed by the engagement of local and global non-state actors with the disputing riparians both informally and formally through critique, advocacy and small-scale demonstration projects.

Negotiation of the Indus Water Treaty between India and Pakistan

The dispute over sharing the Indus River was a feature of provincial conflicts in British India and was addressed by the central government though administrative rulings (Salman and Uperty 2002; Dinar et al. 2007, 273). The British rule ended in 1947 with the partition of the subcontinent into the two independent countries of India and Pakistan. Tension arose because of Pakistan's suspicion of India's intentions and its own imperative of securing water for agricultural and energy development. The standstill agreement reached in 1947 between the two countries expired in 1948, and India stemmed the flow of water to Pakistan (Salman 2013; Wolf and Newton 2009, 190). The two sides negotiated an Inter-Dominion agreement to continue the standstill agreement until a permanent agreement was reached. Thus, in the absence of enforceable rule by a legitimate authority, the change in political boundaries and the emergence of uncertainty and suspicion between the riparians made the sharing of the TB waters of Indus into a complex problem.

Enabling Condition I: Active Recognition of Interdependence among Contending Stakeholders

The interdependence of the riparians, whether in terms of water needs, the need for stability, reputation, or any other contextual factors, forms the basis of interaction in sharing TB waters. However, the presence of interdependence will not by itself bring the stakeholders to interact unless they recognize it in an active from. There are many ways such active recognition can arise—from direct contact between parties to third party initiative. In the Indus case, we have a situation in which the initiative of the World Bank to offer the service of its good offices reinforced the willingness of both Pakistan and India to seek such services to engage in negotiation.

The seeking of neutral mediation, particularly by the heads of states, testifies to the necessity of active recognition of interdependence. Similarly, the choice of mediation as the means of conflict resolution also conveys the recognition of interdependence, because the advisement and facilitation process of mediation allows both parties to remain engaged, without the progress in agreements being seen as binding to any side (Salman 2013, 371). Neutrality was sought and accepted by the parties in terms of the mediator being relatively objective in terms of its interest in the dispute, but who had other interests outside of the dispute.

In the Indus case, interdependence thus also extends to the interest of the mediator itself. The willingness of the World Bank to mediate was based on its own need to have a stable climate to make long-term investments in both countries (Salman 2013,

370). In accepting the bank's mediation, the recognition of interdependence from both parties allowed Eugene Black, President of the World Bank, to lay down the "essential principles" of negotiation. The principles were as follows: "That water resources of the Indus Basin should be managed cooperatively and that the problems of the basin should be solved on a functional, not on a political, plane, without relation to past negotiations and past claims" (Wolf and Newton 2009, 192).

Beyond the active recognition of mutual interdependence of the parties, continuing the interaction also rested on the effectiveness of mediation and, hence, on the capability of the mediator in terms of having the requisite political authority, technical expertise and fiscal capacity that the two parties did not possess. The mediator also remained involved with the stakeholders on a long-term basis in order to keep the focus and the pressure on reaching a negotiated resolution. Having a capable mediator engaged on a continuing basis increased familiarity, reduced mutual vulnerability and, hence, buffered the perceived risks of cooperation among the leadership in both India and Pakistan. We find these conditions of effective mediation in the active role played by the bank (Wolf and Newton 2009, 194).

While David Lilienthal, the former head of the Tennessee Valley Authority, nominally began the negotiation process, it was Eugene Black who formally took over and offered his mediation service directly to the political heads of both countries. Throughout the process, Black initiated and maintained direct and personal contacts with the heads of states of India and Pakistan. The negotiation process relied on a working group (composed of engineering teams representing India, Pakistan and the World Bank) to come up with the engineering provisions of a long-range plan. To pave the way, the bank proposed and formed a "Working Party" composed of engineers from both sides to engage each other on the technical measures and means of sharing the Indus waters, which included determining total supplies by catchment and use, particularly for cultivable irrigable areas, and determining the cost of engineering works to meet the water needs of each country (Wolf and Newton 2009, 192). The purpose of the bank's representation in the working group was to mediate technical differences and serve as an impartial adviser to both sides. Finally, interdependence was sustained through the clarification of issues, drafting of proposals, search for areas of agreement between the parties, elaboration of provisional arrangements to circumvent or minimize issues on which the parties remain divided, as well as developing alternate solutions (Salman 2013, 366).

Enabling Condition II: Framing Mutual Interests through Joint Fact-Finding and Creating Mutual Benefits

Agreement to share a competing resource requires the stakeholders to derive benefits or minimize losses from their cooperation. A measure of effective interaction is for the mediator to facilitate the parties themselves to identify and agree on mutual benefits and costs of cooperation, as well as devising the instruments of securing the benefits and minimizing the costs (Salman 2013, 378). In the Indus case, this involved the benefits of securing water for agriculture and energy development though diversion and storage works (Alam 2002, 347).

At the initial stage, the plans submitted by India and Pakistan conveyed a zero-sum approach to water sharing, as each based its claim on its sovereign rights on water or riparian rights under the prevailing international conventions. It took the bank's own proposal to break the impasse. The bank's proposal was based on the principle of negotiating an equitable allocation of the Indus flow and its tributaries based on both historic flow and planned use (Wolf and Newton 2009, 190; Dinar et al. 2007, 276). For example, India agreed to allow the historic supply of water to western rivers in Pakistan until the time Pakistan could build the link canals to secure it from its own sources (Wolf and Newton 2009, 193). Given the constraint of the context and recent events, both countries found the partitioning of the Indus in terms of its western and eastern waters as being equitable on the grounds of securing their sovereignly as newly independent nations, as well as a means to secure international financial support for the engineering works for water storage and energy supply.

In highly charged political contexts, suboptimal solutions like partitioning the works while sidestepping basin-wide consideration and the Kashmir issues, served as the means of creating mutual benefits, and the promise of financial aid ensured those benefits (Wolf and Newton 2009, 194; Salman 2013, 376). The partitioning of the Indus waters between India and Pakistan was seen by both parties as equitable, as it enjoined both countries to cooperate in securing the needed waters for each, as well as securing the bank's financial assistance to build link canals and storage capacity on both sides. For mutual benefits to be perceived as equitable by both sides, Black focused on the water needs of each country and not the favored or long-held positions of each side. This allowed for the circumvention of the zero-sum fears of the stakeholders and shored up confidence in the interactional process of discovering mutual gains.

Furthermore, the bank's facilitation to develop accurate technical knowledge and management capacity to clearly understand how much one party can secure the benefits through unilateral action and how much can be gained through negotiation generated confidence in both sides in framing interests on a long-term and mutual gain basis. The working group composed of engineers from both sides and those working for the bank, agreed to identify benefits in terms of water needs based on the amount of cultivable irrigable land in both countries and the need for engineering works related to water-resource development. Each step of the process was subjected to joint fact-finding verified by the bank's engineering team.

An example of the facilitation of joint fact-finding was the agreement reached by the parties in 1952, whereby India and Pakistan agreed to use data without committing to either its "relevance or materiality," thereby preserving flexibility for both sides as well as preventing unnecessary delays over the disputation of data (Wolf and Newton 2009, 194). Thus, the two countries remained free to formulate their separate plans, which ensured their continuing participation focused on gaining mutual benefit. Although the plans formulated by both countries were based on similar estimates on the total water availability in the Indus Basin, they differed on the amounts to be allocated to each other.

To break the impasse over competing allocation claims, Black offered the bank's own proposal in order to keep the negotiation process alive. While India promptly accepted the proposal to move the negotiation to its final agreement, Pakistan objected on the grounds of unfair consequence in terms of its vulnerability to the water controlled by India. In recognizing Pakistan's concern over its limited water-storage capacity and also that of India's over its storage capacity and the financial cost for creating link canals, the World Bank included both items as criteria to be addressed in a comprehensive plan. This even-handedness on addressing concrete needs that emerged through interaction established equity in the negotiation process. The bank addressed one of India's key concerns: reducing India's cost for the replacement works. The bank further promised financial assistance to build two dams in Pakistan, one in India, as well as to build link canals in both countries. It also provided Pakistan additional financial assistance in foreign exchange to shore up its economy (Salman and Uperty 2002). Thus, clear specification of the costs and benefits, allocation of costs to both countries based on the principle of commensurate benefits and the bank's bearing of the additional costs enabled the framing of immediate and long-term mutual gains for both sides, paving the way for both sides to sign a treaty in September 1960.

Enabling Condition III: Monitor Agreements through a Joint Authority and Build Up Its Capacity to Manage Emergent Problems

To effectively deal with contingencies, the negotiation of an agreement must facilitate the formation of a water authority as another phase of interaction that can realize the proposed mutual benefits through monitoring and upholding the responsibilities of the stakeholders. In operational terms, this involves adequate provision of resources for continuing joint fact-finding as well as devising mechanisms to monitor the terms of the agreement. Capacity enhancement of water authorities to monitor and act on jurisdictional needs and responsibilities is a crucial condition for the efficacy of the interaction without the need for continuing facilitation by an outside mediator. The investments in capacity building also reduce the suspicion among the stakeholders that one party might later take advantage of the vulnerability of the other. To a great extent, the role of the bank in the ten-year transition period facilitated the capacity-building process for both sides, given the constraint that these were newly independent countries lacking such capacity.

The Indus Treaty provided for the formation of the Permanent Indus Commission. The commission was composed of two commissioners, one from each country, and expert in hydrology. The goals were to oversee and promote cooperative arrangements for the implementation of the treaty, the resolution of conflicts over water issues and the development of the Indus Basin (Wolf and Newton 2009, 193; Dinar et al. 2007, 278). The commission was scheduled to meet once a year (with the meeting venue alternating between the two countries) for the purpose of identifying areas of cooperation in implementing the treaty, of resolving any questions that arose in treaty implementation and of submitting an annual report documenting the level of cooperation. A process of referring disagreements to a neutral expert was also established, failing which, a court of arbitration was specified to act as the final resort. Here, the bank also played a facilitative role because of its ability to appoint a neutral expert for any disputes that arose during the ten-year transition period (Salman 2013, 381).

Beyond monitoring treaty agreements, the long-term efficacy of finding mediated solutions rests on the water authority building its adaptive capacity to make adjustments with changing conditions. Such capacity first requires the water authority to develop a common language to exchange technical information as well as to make on-the-ground monitoring in their own, as well as each other's, jurisdictions. The fact that the Indus Commission is still operational—and that it assists both sides to provide updates on the nature of the water problems—testifies to the enabling function of institutional learning in the sustaining the treaty.

The commission has performed its functions well in terms of overseeing the treaty and conflict resolution. For example, it has regularly held its annual meeting, even amidst two wars and many border skirmishes. The commission also has been the conduit for resolving disputes, either through negotiation or arbitration, as has been the case over the Salal Dam (1978), the Wuller Barrage (2004) and Baglihar dam (2010), with the dispute over the Kisanganga (2013) presently under active resolution (Wolf and Newton 2009, 194; Dinar et al. 2007, 278; Kakakhel 2014). However, as of now, the commission has failed to evolve in ways that allows it to cooperate on basin-wide concerns—like integrated watershed development, climate variability and serving ecological needs (Dinar et al. 2007, 278; Miner et al. 2009, 212). A second limitation of the commission has been that the negotiated approach was not replicated for solving subnational water disputes, nor has it created any multiplier effect of negotiating on other outstanding conflicts (Swain 2009). Finally, a limitation that can also be leveled in terms of the commission's incapacity is its lack of movement in addressing the issues of sustainable water supply and usage, both at the basin-wide level as well as within the borders of each country.

Despite these shortcomings, the critics and supporters of the Indus Treaty (both in India and Pakistan), rather than seeking its abrogation favors its expansion to address the changes in water needs and for collaborative projects (Sarfraz 2013 204, 212; Iyer 2005, 3140; Miner et al. 2009, 211). Thus, the interdependence that seemed to have been lost in partitioning the Indus, now appears to be in the process of being reaffirmed through the calls for capacity building of the commission.

Negotiation of the Peace Treaty between Israel and Jordan

The initial effort to resolve the TB water dispute in the Jordan Basin was made in 1951 with the appointment by President Dwight D. Eisenhower of Ambassador Johnston to mediate it. Through vigorous shuttle diplomacy, Johnston sought to bring all sides to agree to a water-allocation plan without undermining the long-term integrity of the basin. After each party (The Arab League countries of Syria, Lebanon, Egypt and Jordan on one side and Israel on the other) rejected the other's plan, in 1955 Johnston negotiated a compromise (known as the unified plan) for both sides to agree on. The unified plan was also backed with the promise of financial aid from the United States for development projects related to implementation of the plan. While the experts on both sides agreed on the water-sharing arrangements of the unified plan, it is on the political implications of sharing waters, particularly over unsettled Palestinian issues and implicit recognition of

Israel by the Arab states (Haddadin 2000), as well as the erosion of Arab trust in Johnston (Salman 2013, 395) that led to the failure of the unified plan becoming a treaty.

Despite its political failure, on technical grounds, the Johnston plan remained the basis for the riparians to legitimate their claims on the TB waters of the Jordan Basin as well as served as their template in negotiation (Elmusa 1995, 69; Haddadin 2000, 277). Thus, while the unified plan failed to become a formal treaty, it nevertheless succeeded in making the riparians acknowledge their mutual interdependence in sharing of TB water as well as recognizing the water needs of each other for a long time to come. The plan also clearly established the condition that interaction on sharing TB water is contingent on the formal recognition and negotiation of political interdependence among the riparians (Jagerskog 2003). The reason the long-term technical agreement led the way for a swift political agreement is because of the complex interdependence of political issues with technical issues, as well as the technical/professional mindset of experts on both sides, who informally engaged and recognized each other's water needs.

Enabling Condition I: Active Recognition of Interdependence among Contending Stakeholders

Even though the Johnston plan failed to secure an agreement, the recognition of interdependence on a basin-wide basis and involving all the riparians established not only the principle of interaction but also the possibility of cooperation for sharing the TB waters in the Jordan Valley. We find this basis of cooperation later manifested in the building of the Maqarin Dam on the Yarmuk River—from Jordanian officials seeking Israel's permission in 1975, followed by the 1980 diplomacy of US envoy, Phillip Habib, to the mediation of Ambassador Richard Armitage, to brokering a deal in 1989. By 1990 an agreement began to take shape only to be stalled, first by the 1991 Gulf War, and then by the 1991 Jordan–Israel peace talks in Madrid (Wolf and Newton 2009, 201).

However, the context of interaction was different in 1990s than it was in 1950s. A changing geopolitical context altered the interrelations among the riparians and, hence, changed the focus and scope of interaction. With the change, a new pattern of recognizing active interdependence emerged among the water experts of Jordan and Israel in the form of informal interaction, commonly known as the "picnic table talks" (Haddadin 2000; Shamir 1998, 275), which kept the recognition alive. In 1979 a makeshift bridge was built on the Yarmuk River for Israeli water officials to hold bilateral talks on sharing TB waters with their Jordanian counterparts. The fact that informal interaction continued for 15 years despite the absence of formal recognition of Israel by Jordan, the ebb and flow of political tensions in the region, and the mutual misgivings that existed on both sides, point to the enabling function of interdependence to bring and keep both sides together.

In 1991, the United States initiated a new cooperation venture through the Madrid Conference. The initiative established a two-track approach to negotiate a comprehensive peace—one key element of which was sharing TB waters. One track was nominally multilateral and included Israel and Jordan—with Jordan formally representing the Palestinian Authority. The second track included Israel and Jordan in order to address their bilateral issues and interests (Haddadin 2000). The separation of the Palestinian and Jordanian tracks also signaled the willingness of all parties to interact on the basis of tangible interests rather than negotiate their political differences. In particular, this enabled Israel and Jordan to negotiate mutual interests. This situational condition of separating Palestinian issues from the consideration of Jordanian needs, while allowing the latter to be addressed in terms of negotiation between the Palestinian Authority and Israel affected equity on issues to be addressed at a practical level.

The recognition of concrete interdependence was facilitated by holding informal side meetings of experts on water, energy and the environment for exploring and crafting a common agenda for cooperation (Haddadin 2000, 277). These pre-negotiation meetings were very effective for reaching agreements on agenda items, thus paving the way for specific negotiations on sharing waters and joint projects for water-resource development (Haddadin 2000, 277). In the pre-negotiation meeting, US diplomats often played the mediational role as "gavel holder" (Wolf and Newton 2009, 202) to bring the parties together and facilitate their interactions (Haddadin 2000, 278).

The enabling role of third-party negotiation continued with President Bill Clinton holding talks with the political leadership of the disputing parties at the highest level in the White House. Such talks not only provided an implicit recognition of Palestine, but in 1994 provided a basis for Jordan to sign the peace treaty with Israel, knowing that the Palestinian interests were being addressed separately (Haddadin 2000, 278). Thus, the mediation forged a politically pragmatic basis of identifying mutual interdependence between Israel and Jordan, especially regarding their need for and use of water resources. Contextual factors were also at play to bring the two sides together. Arab states by this time became disunited, allowing interactions to focus on concrete issues rather than on the questions of identity or power. An example of such a concrete issue was Jordan's need for water to accommodate the population that had migrated from Iraq due to the 1991 Gulf War (Haddadin 2000, 277), and the warming relation between the United States and Jordan due to the realignment arising from the emergent unipolar world (Haddadin 2000, 280; Wolf and Newton 2009, 202).

Enabling Condition II: Framing Mutual Interests through Joint Fact-Finding and Creating Mutual Benefits

Once interdependency was based on concrete issues and recognized at the highest levels of government, it paved the path for the search for mutual benefits to make the interaction effective. For example, on September 26, 1994, Israel forwarded a draft peace treaty for Jordan's consideration. It took only two days for Jordan to make its counter proposal and begin the negotiation on substantive issues. The proposals dealt with water issues comprehensively (involving water storage, water quality and protection, and ground water) but left the actual amounts of water sharing unspecified (Haddadin 2000, 279). The eventual agreement, signed in October 1994, provided Israel with a fixed amount of water based on summer and winter flows (guaranteeing Israel the supply), and provided

Jordan with the residual amount (which can be more or less depending on the variability of the climate (Elmusa 1995, 69) and Jordan tapping other sources.

The negotiation centered on creating mutual gains for both parties. Each side sought and secured a different outcome from the treaty. The treaty for both sides to be at peace, and water sharing served as only one item (Shamir 1998, 279). Thus, the scope of mutual benefits extended to issues beyond water allocation to include cooperation on data exchange, building storage facilities and protecting water sources from pollution (Haddadin 2000, 281; Elmusa `995). Thus, joint fact-finding was extended from the initial phase of interacting on technical issues to the formation of the treaty. The interactions based on finding mutual gains also extended beyond the terms of the agreement. For example, in securing the winter flow for Israel, the treaty paved the way for Israel to withdraw its long-held objection to the joint construction of the Maqarin Dam (renamed as Al-Wehda Dam) by Jordan and Syria in 2004 (Wolf and Newton 2009, 201).

With the peace treaty, Jordan increased its overall water supply by 25 percent. It also received nearly all of the East Bank waters and secured the means to pursue joint management and development of TB waters. This arrangement helped Jordan overcome its water shortages, especially in the household sector, and such gains allowed Jordan to secure public support for the PT (Elmusa 1995, 64). On the other hand it benefitted Israel to secure peace with a second Arab state without significantly affecting its total water supply (Elmusa 1995, 64). It also allowed Israel to gain a stable environment to pursue its long-term economic development plans (Manna 2006, 59).

Another aspect of mutual benefit could be found in the agreement made over direct, indirect and future water allocations. Other than direct grants of water, the peace treaty sets up water-exchange arrangements and future water developments. For example, it allows Israel to extract water for a limited period (subject to renegotiation) from a groundwater source (Wadi Araba) that lies under Israeli occupation, while at the same time granting its sovereignty to Jordan. It also allows for Jordan to receive additional waters from a "yet-to-be identified source" and allocates water to both parties from the lower parts of the Jordan and Yarmuk Rivers. It also provides for water-storage facilities for Jordan to store flood waters from the Jordan and Yarmuk rivers (Elmusa 1995). Such an arrangement prevents either side pursuing unilateral actions for its own gains at the cost of the other party or coaxing the other party to accept unequitable water allocations.

Despite the naming of the treaty as permanent, it remains complex and ambiguous, thereby making it possible for each side to defend different perspectives on how the treaty benefits them. It follows that the ambiguity in the treaty's provisions allows for interaction of stakeholders to develop while at the same time satisfying their immediate needs (Fischhendler 2008). Thus, in the treaty some water issues were resolved with clear allocation while other arrangements remained vague for future cooperation and development. For example, Jordan's allocation from unidentified Israeli or joint sources provided the impetus to cooperate on developing such sources (i.e., desalination plants or building storage for flood waters), rather than relying exclusively on present sources. This made the idea of mutual gains forward looking, even though it kept the status quo unchanged for the present (Haddadin 2000, 278). For instance, the ambiguity of interlinkages of water systems to secure present and future supply provides for the entry of one country's interest into the territory of another. This recognition lays the pragmatic ground for mutual trust to form and be demonstrated through commitments made to be honored (Shamir 1998, 280).

Based on such constructive ambiguity, one can legitimately argue the treaty to be inequitable and a direct reflection of the asymmetrical power relations between the parties involved. In fact, this is one conclusion made on the terms of the treaty (Elmusa 1995, 69; Beaumont 1997, 415; Agnew and Woodhouse 2011; Abukhater 2013). A key basis of the assessment rests on Jordan receiving less water than what was specified in the Johnston plan, especially when Jordan argued its needs to be based on the Johnston plan allocation.

However, there is another way to claim the treaty to be based on equity. The treaty uses the term "rightful allocation" rather than "water rights" to convey the principle of equity (Shamir 1998, 277; Salman 2013, 397) as the basis of interaction through mutual gains. The use of this terminology also corresponds to the evolution of international water laws although, as Elmusa (1995) perceptively noted, "neither the law nor the criteria for translating its principles into quantities are explicitly mentioned" (65). The assessment of equity can also be based on creating a foundation of mutual gains that includes the interest of future generations and the ecology while also honoring the precedent of the earlier negotiated agreement (Wolf and Newton 2009, 201).

Finally, interaction on the basis of mutual benefits is made possible not only on pragmatic and political grounds of equity, but also because of its fidelity to the value of sustainability as a measure of effective water management. The peace treaty provides for the protection of water quality and the environment (an integral part of the treaty), which also includes monitoring and minimization of waste discharge and of pollution prevention on both sides. However, the means of securing these goals are not clearly specified, leaving predominantly environmental concerns without effective representation (Shamir 1998, 280). Thus, the ambiguity in the treaty language can also be considered as enabling in terms of ways of furthering mutual gains, once the outstanding issues are resolved and mutual confidence is built. Thus, the nature and prospect of mutual gains moved beyond water sharing to other areas that were absent in the Indus case.

Enabling Condition III: Monitoring Agreements through a Joint Authority and Building Their Capacity to Manage Emergent Problems

The peace treaty provides for the formation of the Joint Water Committee (JWC) and two subcommittees (one for the northern sources and the other for southern sources) to oversee the implementation of the treaty provision and explore areas of joint management (Shamir 1998, 277). The projects include development of energy and water sources, protecting the natural environment, joint tourism development and the development of the Jordan Rift Valley (Shamir 1998, 277; Elmusa 1995). The JWC consists of three members from each side, but is not an autonomous expert body, as it remains under the direct supervision of the highest level of the respective governments (Shamir 1998, 276).

Despite the political control over the JWC, its formation and use reflects the recognition of institutional capacity and flexibility to address water issues as they may arise in the process of implementing the treaty, as well as issues of cooperation that may arise in the future (Shamir 1998, 276). For example, the task of creating additional waters to meet shortages was left to the JWC to formulate (Shamir 1998, 280), and in doing so the implementation process, by design, embedded cooperative action without mandating it. Another example of flexibility lies in its ability to address new issues such as drought as an aspect of dealing with climate vulnerability. The officials of JWC negotiated a temporary arrangement, thereby modifying treaty allocations to reflect water availability. As Olivia Odom and Arron Wolf (2011) note, "Even though water availability restricted the parties from fulfilling Treaty obligations, the institutional arrangement created by the Treaty enabled the parties to peacefully cooperate in time of severe drought" (703).

The design of the JWC also accounts for conflict and conflict resolution. In fact, it invites constructive conflict to make cooperation effective. For example, the supply of 50 MCM/y of drinkable water by Israel to Jordan was still not realized—even after two years beyond the deadline—because of the non-specification of the source in the treaty. The dispute was based on Jordan holding it to be Israel's responsibility to supply the amount from its sources, while Israel held it to be provided from sources on both sides, with Jordan bearing the cost of provision. The dispute was resolved through political intervention, with the Israeli water minister, Ariel Sharon, agreeing to supply an additional 25 MCM/y to Jordan from the Israeli system, although it was not stipulated in the 1994 treaty (Manna 2006, 62; Shamir 1998, 279).

There is no formal third-party dispute resolution-mechanism specified in the implementation of the treaty provisions. However, Article 29 of the treaty provides for conflict resolution (Shamir 1998). It states: "Disputes arising out of the application or implementation of this Treaty shall be resolved by negotiations. Any such disputes which cannot be settled by negotiations shall be resolved by conciliation or submitted to arbitration." The non-specification of the mechanisms to effect such conciliation or arbitration requires that both parties to the dispute need to negotiate the modality.

In reality, many critical conflicts that arose were resolved, not by the JWC per se, but by high-level political officials, including the heads of state (Zawahri 2008, 470–71). In these high-level conflict-resolution cases, the JWC served as the conduit of technical information, agreement and follow up monitoring. The JWC thus brought two otherwise conflicting sides to interact and continue their interaction in spite of their conflicts and misgivings. The very continuity of the treaty conveys the resilience of the institutional mechanisms to sustain the interaction that led to cooperation. The resilience rests on institutional designs that maintain flexibility to realize mutual benefits on a continuous basis as well as resolve disputes that arise in the process of interaction.

Conclusion

The complexity of TB water management lies in the dynamics of interconnections and feedback that operates across variables, processes, actors and institutions, as well as the operationalization of the principles of equity and sustainability, both as the goals and

criteria of assessing the effectiveness of interaction. Based on the contingency of complexity, the enabling conditions of interaction find their resilience, not in their pragmatic
sensitivity and institutional capacities to bring conflicting sides to interact, but also in incorporating and sustaining the values of equity and sustainability in the interactional process.

Using the negotiation of the Indus treaty between Pakistan and India and the peace
treaty between Israel and Jordan as paradigmatic cases of resolving complex TB water
problems, this chapter identified three broad conditions that led to negotiated cooperation that endures to this day. These three conditions are attributed as enabling because
they find general support in the conflict-resolution literature as well as in their ability to
account for the durability of the treaties in specific basins at risk. The three enabling
conditions and their manifestation in the two cases examined in this chapter are
summarized below:

- Enabling Condition I: Recognition of interdependency as the basis of cooperation.
  Such recognition can arise from an active and sustained mediation by a third party
  (in the case of India and Pakistan) or indirectly through informal meetings among
  officials; bilateral negotiations between the sides coupled with indirect facilitation by a
  third party (in the case of Jordan and Israel).
- Enabling Condition II: The pursuit and identification of mutual benefits as the basis
  of continuous interaction and cooperation. What renders the mutual benefit process
  cooperative rather than competitive is the acknowledgment by all sides of equity and
  sustainability as the guiding values to seek, create and secure in practice. In the case of
  the Indus Treaty such mutual benefits included securing the water needs of each country, and using this supply for national development. The agreement was both leveraged
  by and formed the basis of funding from the World Bank. In the case of the Israel–
  Jordan peace treaty, mutual benefits included not only sharing the existing waters but
  also developing new sources of water, separately or jointly. Here, too, the agreement
  created the conditions for economic aid and economic development in both countries.
- Enabling Condition III: The creation of institutional mechanisms to monitor the terms
  of the agreement and develop its capacity to advance interaction in new areas as they
  arise. In the Indus case, in addition to forming a permanent joint authority (Permanent
  Indus Commission) representing the water experts from both sides, the World Bank as
  a third party also was involved in the ten-year transition period in implementing the
  treaty. The Indus Commission has successfully resolved disputes or channeled disputes
  through the mechanisms stipulated in the treaty itself. In the Israel–Jordan peace treaty
  the JWC was also composed of experts from both sides, but it enjoys less autonomy in
  dispute resolution. On the other hand it is more forward-looking in terms of working
  on joint projects.

The enabling function of these conditions does not rest on a fixed set of techniques
or measures, but on the context-specific conditions—involving both actors and
situations—that become active in a negotiation process. As both cases demonstrate,
the negotiation process takes a long time and is never complete. Thus, there are other
factors or sufficient conditions that affect the relative efficacy of each of the enabling

conditions in a negotiation process. Given the complexity and contextual nature of TB water issues, these sufficient conditions cannot be pre-specified, as they are contingent on the issues and stakeholders involved in a dispute. For example, the issues of environmental sustainability and economic development are now pressing upon the Indus Treaty to be reevaluated, and issues of geopolitical change in the region that made the peace treaty possible.

Given the open-ended nature of contextual factors, sustaining the enabling condition provides stability to institutional processes to addressing and incorporating the contextual factors on an incremental basis. This capacity largely rests on the functioning and evolution of the joint commissions. In general, research findings provide encouraging signs of the evolution of institutional capacities. For example, treaties have been found to become more comprehensive over time, both in terms of the issues they address and the tools they use to manage those issues cooperatively. In addition to addressing the traditional issues of water allocation, irrigation and hydropower generation, environmental issues are also gaining greater attention (Giordano et al. 2014, 245). The growing capacitation of institutions, however, needs to remain flexible to address the contingent needs of a basin as they arise—for instance issues such as adapting to climate change, relying more on virtual waters and using conservation technologies more effectively to gain water efficiency and improve water quality, what Sadoff and Grey (2002) so aptly referred to as deriving benefits from water.

References

Abukhater, Ahmed. 2013. Water as a Catalyst for Peace: *Transboundary Water Management and Conflict Resolution*. London and New York: Routledge.

Agnew, Clive and Phillip Woodhouse. 2011. *Water Resources and Development*. New York: Routledge.

Alam, Undala. 2002. "Questioning the Water Wars Rationale: A Case Study of the Indus Waters Treaty." *Geographical Journal* 168(4): 354–64.

Allan, Tony J. A. and Naho Mirumachi. 2010. "Why Negotiate? Asymmetric Endowments, Asymmetric Power and the Invisible Nexus of Water, Trade and Power that Brings Apparent Water Security." In: *Transboundary Water Management: Principles and Practice*, edited by Anton Earle, Anders Jagerskog and Joakim Ojendal, 13–27. Washington DC: Earthscan.

Bercovitch, Jacob, Theodore J. Anagnoson and Donnette L. Wille. 1991. "Some Conceptual Issues and Empirical Trends in the Study of Successful Mediation in International Relations." *Journal of Peace Research* 28(1): 7–17.

Bercovitch, Jacob. 1996. *Resolving International Conflicts: The Theory and Practice Mediation*. Boulder: Lynne Rienner Publishers, 28.

Berlin Rules on Water Resources. 2004. http://internationalwaterlaw.org/documents/intldocs/ILA_Berlin_Rules-2004.pdf (accessed 02/15/15).

Beaumont, Peter. 1997. "Dividing the Waters of the River Jordan: An Analysis of the 1994 Israel–Jordan Peace Treaty." *International Journal of Water Resources Development* 13(3): 415–24.

Biswas, Asit K. 2004. "Integrated Water Resources Management: A Reassessment." *Water International* 29(2): 248–56.

Biswas, Asit K. 1992. "Indus Water Treaty: The Negotiating Process." *Water International* 17: 201–09.

Choudhury, Enamul and Shafiqul Islam. 2015. "Nature of Transboundary Water Conflicts: Issues of Complexity and the Enabling Conditions for Negotiated Cooperation." *Journal of Contemporary Water Research and Education* 155(1): 43–52.

Coleman, Peter T., Morton Deutsch, Eric C. Marcus (eds). 2014. *The Handbook of Conflict Resolution: Theory and Practice, 3rd edition.* San Francisco: Jossey-Bass.

Dellapenna, Joseph, Joyeeta Gupta, Wenjing Li and Falk Schmidt. 2013. "Thinking about the Future of Global Water Governance." *Ecology and Society* 18(3): 28.

Dinar, Ariel, Shlomi Dinar, Stephen McCaffrey, and Daene McKinney. 2007. *Bridges over Water: Understanding Transboundary Water Conflict, Negotiation and Cooperation.* Hackensack, NJ: World Scientific Publishing Co.

Dinar, Shlomi and Ariel Dinar. 2000. "Negotiating in International Watercourses: Diplomacy, Conflict and Cooperation." *International Negotiation* 5(2): 193–200.

Dinar, Shlomi. 2000. "Negotiations and International Relations: A Framework for Hydropolitics." *International Negotiation* 5(2): 375–40.

Earle, Anton, Anders Jägerskog and Joakim Öjendal. 2010. "Introduction: Setting the Scene for Transboundary Water Management Approaches." In: *Transboundary Water Management: Principles and Practice,* edited by Anton Earle, Anders Jägerskog and Joakim Öjendal, 1–13. Washington DC: Earthscan.

The Economist. 1999. "Preventing Conflict in the Next Century." In: *The World in 2000.* London: Economist Publications, 51–52.

Elhance, Arun P. 2000. "Hydropolitics: Grounds for Despair, Reasons for Hope." *International Negotiation* 5(2): 201–22.

Elmusa, Sharif. 1995. "The Jordanian-Israel Water Agreement: A Model or an Exception." *Journal of Palestine Studies* 24(3): 63–73.

Fann, K. T. 1970. *Peirce's Theory of Abduction.* The Hague: Martinius Nijhoff.

Frey, Frederick W. 1993. "The Political Context of Conflict and Cooperation over International River Basins." *Water International* 18(1): 54–68.

Fisher, Roger, William Ury and Bruce. Patton. 2011. *Getting to Yes: Negotiating Agreement Without Giving In.* Revised edition. New York: Penguin Books.

Fischhendler, Itay 2008. "Ambiguity in Transboundary Environmental Dispute Resolution: The Israeli–Jordanian Water Agreement." *Journal of Peace Research* 45(1): 91–110.

Gallego-Ayala, Jordi. 2013. "Trends in Integrated Water Resources Management Research: A Literature Review." *Water Policy* 15: 628–47.

Gerlak, Andrea K., Jonathan Lautze and Mark. Giordano. 2011. "Water Resources Data and Information Exchange in Transboundary Water Treaties." *International Environmental Agreements: Politics, Law and Economics* 11(2): 179–99.

Giordano, Mark, and Aaron Wolf. 2001. "Incorporating Equity into International Water Agreements." *Social Justice Research* 14(4): 349–66.

Giordano, Mark, and Aaron Wolf. 2003. "Sharing Waters; Post Rio IWM." *Natural Resources Forum* 27: 163–71.

Giordano, Mark, Alena Drieschova, James Duncan, James A. Sayama, Yoshiko, De Stefano, Lucia, and Aaron T. Wolf. 2014. "A Review of the Evolution and State of Transboundary Freshwater Treaties." *International Environmental Agreements: Politics, Law and Economics* 14(3): 245–64.

Haddadin, Munther. 2000. "Negotiated Resolution of the Jordan-Israel Water Conflict." *International Negotiation* 5(2): 263–88.

Hoekstra, Arjen Y. 2011. "The Global Dimensions of Water Governance: Why the River Basin Approach Is No Longer Sufficient and Why Cooperative Action at Global level is Needed." *Water* 3: 21–46.

Islam, Shafiul and Lawrence Susskind. 2013. *Water Diplomacy: A Negotiated Approach to Managing Complex Water Networks.* New York: Routledge.

Iyer, Ravi R. 2005. "Indus Treaty: A Different View." *Economic and Political Weekly* 40(29): 3140–44.

Jarvis, Todd and Aaron T. Wolf. 2010. *Managing Water Negotiations and Conflicts in Concept and in Practice,* edited by Anton Earle, Anders Jagerskog and Joakim Ojendal, 125–42. Washington DC: Earthscan.

Jagerskog, Anders. 2003. "Why States Cooperate over Shared Waters: The Water Negotiations in the Jordan River Basin." PhD dissertation, Department of Water and Environmental Studies, Linkoping University.

Kakakhel, Shafqat. 2014. "The Indus Waters Treaty: Negotiation, Implementation, New Challenges, and Future Prospects." *Indus Waters* 9(2).

Kliot, Nurit, Deborah Shmueli and Uri Shamir. 2001. "Institutions for Management of Transboundary Water Resources: Their Nature, Characteristics and Shortcomings." *Water Policy* 3: 229–55.

Lowi, Mariam. 1993. *Water and Power: The Politics of a Scarce Resource in the Jordan River Basin.* Cambridge: Cambridge University Press.

Magnani, Lorenzo. 2001. *Abduction, Reason, and Science: Processes of Discovery and Explanation.* New York: Kluwer Academic Plenum Publishers.

Manna, Maya. 2006. "Water and the Treaty of Peace between Israel and Jordan." Roger Williams University Center for Macro Projects and Diplomacy Working Paper Series 10(Spring): 58–64.

Mehta, Jodhbir S. 1988. "The Indus Water Treaty: A Case Study in the Resolution of an International River Basin Conflict." *Natural Resources Forum* 12(1): 69–77.

Miner, Mary, Gauri Patankar, Shama Gamkar and David Eaton. 2009. "Water Sharing between India and Pakistan: A Critical Evaluation of the Indus Water Treaty." *Water International* 34(2): 204–16.

Mollinga, Peter P., Ruth Meinzen-Dick and Douglas J. Merrey. 2007. "Politics, Plurality and Problemsheds: A Strategic Action Approach for Agricultural Water Resources Management Reform." *Development Policy Review* 25(6): 699–719.

Odom, Olivia and Aaron T. Wolf. 2011. "Institutional Resilience and Climate Variability in International Treaties: The Jordan River Basin as 'Proof-of-Concept.'" *Hydrological Science Journal* 56(4): 703–10.

Prescoli, Jerome D. and Aaron T. Wolf. 2010. *Managing and Transforming Water Conflicts,* New York: Cambridge University Press.

Sadoff, Claudia W. and David Grey. 2002. "Beyond the River: The Benefits of Cooperation on International Rivers." *Water Policy* 4: 389–403.

Salman, Salman M. A. 2013. "Mediation of International Water Disputes: The Indus, the Jordan, and the Nile Basins Interventions." In: *International Law and Freshwater: The Multiple Challenges,* edited by L. Boisson de Chazournes, C. Leb, M. Tignino. Northampton, MA: Edward Elgar.

Salman, Salman M. A. and Kishor Uprety. 2002. *Conflict and Cooperation on South Asia's International Rivers: A Legal Perspective.* Washington, DC: The World Bank.

Saravanan, V. Saravanan, Geoffrey T. McDonald and Peter P. Mollinga. 2009. "Critical Review of Integrated Water Resource Management: Moving beyond Polarised Discourse." *Natural Resources Forum* 33: 76–86.

Sarfraz, Hamid. 2013. Revisiting the 1960 Indus Water Treaty. *Water International* 38(2): 204–16.

Seawright, Jason and Gerring, John. 2008. "Case Selection Techniques in Case Study Research: A Menu of Qualitative and Quantitative Options." *Political Research Quarterly* 61(2): 294–338.

Shamir, Uri. 1998. "Water Agreements between Israel and Its Neighbors." In: *Transformations of Middle Eastern Natural Environments: Legacies and Lessons,* edited by J. Albert, M. Bernhardson and R. Kenna, Number 103, Bulletin Series, 274–96. Yale School of Forestry and Environmental Studies.

Song, Jennifer and Dale Whittington. 2004. "Why Have Some Countries on International Rivers Been Successful in Negotiating Treaties? A Global Perspective." *Water Resources Research* 40.

Spector, Bertram I. 2000. "Motivating Water Diplomacy: Finding the Situational Incentives to Negotiate." *International Negotiation* 5(2): 223–36.

Subramanian, Ashok, B. Brown and Aaron Wolf. 2012. *Reaching across the Waters: Facing the Risks of Cooperation in International Waters.* Washington, DC: World Bank.

Susskind, Lawrence, Paul F. Levey and Jennifer Thomas-Larmer. 2000. *Negotiating Environmental Agreements.* Washington, DC: Island Press.

Susskind, Lawrence and Ellen Babbitt. 1992. "Overcoming Obstacles to Effective Mediation of International Disputes." In: *Mediation in International Relations: Multiple Approaches to Conflict Management,* edited by J. Bercovitch and J. Rubin. New York: Saint Martin's Press.

Swain, Ashok. 2009. "The Indus II and Siachen Peace Park: Pushing the India-Pakistan Peace Process Forward." *The Round Table* 98(404): 569–82.

Transboundary Waters Assessment Programme (TWAP). 2013. http://twap-rivers.org (accessed on February 18, 2015).

Watkins, Kevin. 2006. "Beyond Scarcity: Power, Poverty and the Global Water Crisis." In: *UNDP 2006 Human Development Report*, edited by Bruce Ross-Larson, Meta de Coquereaumont and Christopher Trott. New York: Palgrave Macmillan.

UN Convention on the Law of the Non-Navigational Uses of International Watercourses. 1997. http://legal.un.org/ilc/texts/instruments/english/conventions/8_3_1997.pdf (accessed on February 15, 2015).

Uitto, Juha I. and Alfred M. Duda. 2002. "Management of Transboundary Water Resources: Lessons from International Cooperation for Conflict Prevention." *The Geographical Journal* 168(4): 365–78.

Warner, Jeroen and Neda Zawahri. 2012. "Hegemony and Asymmetry: Multiple-chessboard Games on Transboundary Rivers." *International Environmental Agreements: Politics, Law and Economics* 12(3): 215–29.

Wolf, Aaron and Joshua Newton. 2009. Case Studies of Transboundary Dispute Resolution, appendix C. In: *Managing and Transforming Water Conflicts*, edited by A. Wolf and J. Delli Priscoli, Cambridge: Cambridge University Press.

Wolf, Aaron, T. 1999a. "'Water Wars' and Water Reality: Conflict and Cooperation Along International Waterways." *Environmental Change, Adaptation, and Security* 65): 251–65.

———. 1999b. "Criteria for Equitable Allocations: The Heart of International Water Conflict." *Natural Resources Forum* 23: 3–30.

———. 2007. "Shared Waters: Conflict and Cooperation." *Annual Review of Environment and Resources* 32(3): 241–69.

———. 2010. "Possible Futures for Transboundary Water Resources." *World Politics Review*, January 20. http://www.worldpoliticsreview.com/articles/4961/possible-futures-for-transboundary-water-resources.

WWAP. 2012. United Nations World Water Development Report 4. UNESCO, UN-Water, WWAP. March.

Yamagata, Yoshiki, Jue Yang and Joseph Galaskiewicz. 2013. "A Contingency Theory of Policy Innovation: How Different Theories Explain the Ratification of the UNFCCC and Kyoto Protocol." *International Environmental Agreements: Politics, Law and Economics* 13(3): 251–70.

Yin, Robert K. 1989. *Case Study Research: Designs and Methods*. Newbury Park, CA: Sage.

Yoffe, Shira B., Aaron T. Wolf and Mark. Giordano. 2003. "Conflict and Cooperation over International Freshwater Resources: Indicators of Basins at Risk." *Journal of the American Water Resources Association* 39(5): 1109–26.

Zawahri, Neda A. 2008. "Designing River Commissions to Implement Treaties and Manage Water Disputes: The Story of the Joint Water Committee and Permanent Indus Commission." *Water International* 33(4): 464–74.

———. 2009. "Third Party Mediation of International River Disputes: Lessons from the Indus River." *International Negotiation* 14: 281–310.

Zawahri, Neda A., Ariel Dinar and Getachew Nigatu. 2014. 'Governing International Freshwater Resources: An Analysis of Treaty Design." *International Environmental Agreements: Politics, Law and Economics* 16(3): 307–31.

Zeitoun, Mark and Naho Miramuchi. 2008. "Transboundary Water Interaction I: Reconsidering Conflict and Cooperation." *International Environmental Agreements* 8: 297–316.

Zeitoun, Mark, Naho Mirumachi and Jeroen Warner. 2011. "Transboundary Water Interaction II: The Influence of 'Soft' Power." *International Environmental Agreements* 11: 159–78.

Chapter Eleven

MEDIATION IN THE ISRAELI–PALESTINIAN WATER CONFLICT: A PRACTITIONER'S VIEW

Patrick Huntjens

Abstract

Water, food, and energy challenges—study after study makes clear—are among the most significant contributors to violent domestic and international conflict in the world today. Access to clean and sufficient water is critical, not only for human health, the environment and economic development, but also for establishing stability and sustaining peace. Since 1991 in the Middle East Peace Process, water has been one of six key issues discussed, along with the borders between Israel and Palestine, refugees, illegal settlements, Jerusalem and security demands. The myriad dimensions of the conflict are heightened by inequality in access to water. To resolve the conflict—any conflict—negotiation, mediation and conciliation are needed. To be successful, though, such processes must be rooted in in-depth contextual understanding. This chapter analyzes two case studies in the Israeli–Palestinian water conflict, focusing on how to use interventions to build trust among stakeholders and how to determine whether policy learning has occurred and, if so, whether it has had any effect. The author of this chapter was involved in both case studies as a third-party mediator/facilitator. The chapter finds that third-party and multitrack diplomacy are critical to maintaining dialogue under uncertain political conditions, particularly when formal negotiations have come to a halt, as they have between Israel and the Palestinian Authority. Especially in these times, it is important that a peaceful diplomatic solution to vital contested issues is still possible.

Introduction

Water-related conflicts and the need to resolve them are included on political agendas worldwide. Numerous studies make it clear that water, food and energy challenges are primary contributors to international and domestic conflict (Brock 2011; Gleick et al. 2014). The necessity of putting water diplomacy within the appropriate policy framework has thus become more apparent to politicians.

Water disputes rarely take place in isolation, however, and are typically part of an already-complex and likely violent conflict—such as that between Israel and Palestine.

Could water become the way to peace, trust and cooperation? Cooperation on water by the Israelis and Palestinians is a golden opportunity. Finding a different modality of negotiation first, however, is essential. Until then, parties will continue the zero-sum game, sticking to their existing rights and principles rather than searching for mutual gains.

Access to clean and sufficient water is critical in the Middle East, not only for human health, the environment and economic development, but also for establishing stability and sustaining peace. Since 1991, water has been one of six key regional issues—the others being the finite borders between Israel and Palestine, return of refugees, illegal West Bank settlements, the status of Jerusalem and Israel's security guarantee demands. The conflict has multiple dimensions but is heightened by the injustice of inequality in access to water. According to Amnesty International (2009), Palestinian consumption in the Occupied Palestinian Territory is on average about 70 liters a day per person—well below the 100 liters recommended by the World Health Organization (WHO)—whereas Israeli consumption averages about 200 liters (Reed and Reed, 2011). At the same time, the two parties depend on shared water resources. The bodies of water essential to both are so interconnected that simple division is impossible (see Figure 11.1). Roughly two-thirds of their fresh water resources are shared, and thus some agreement for joint management is essential (Brooks, Trottier and Doliner 2013).

Sharing waters should be approached with an eye to sharing opportunities, and the aim of water diplomacy should be to identify and strengthen such benefits. Water diplomacy can be successful when parties with conflicting interests recognize that failure to collaborate is likely to lead to a worse outcome for all (Huntjens, and Van Schaik 2014).

This analysis looks at two case-study trajectories of mediation and facilitation in the Israeli–Palestinian water conflict, focusing on how participatory and diplomatic interventions can be applied to build trust among stakeholders and to reach agreement and sustainable implementation, and how we know that policy learning has occurred and whether it has had any effect. The author was deeply involved in both case studies, as an independent and external third party mediator/facilitator during the key events described in this chapter.

The first study is the development of the NGO-based Integrated and Transboundary Master Plan for the Lower Jordan River Basin (Master Plan LJRB). The second study is the operationalization of the Water Annex (GI Water Annex), an element of the 2003 Geneva Accord proposing a detailed solution for a final peace agreement. The Track II agreements reached in these case studies were each shaped within the parameters of a two-state solution, as a transboundary master plan and a model peace treaty, respectively. As such, the agreements may contribute to options for future peace negotiations.

Adaptive Water Management and Water Mediation

Cooperation over shared water resources is complex and uncertain and, in the Israeli–Palestinian conflict, is also highly sensitive to securitization politics. An issue becomes "securitized" when it is deemed an essential component of national security (Trottier 2008; Zeitoun 2007). Water issues are complex because of their intricate coupling with multiple issues within the natural and societal domains (Islam and Susskind 2013). Water

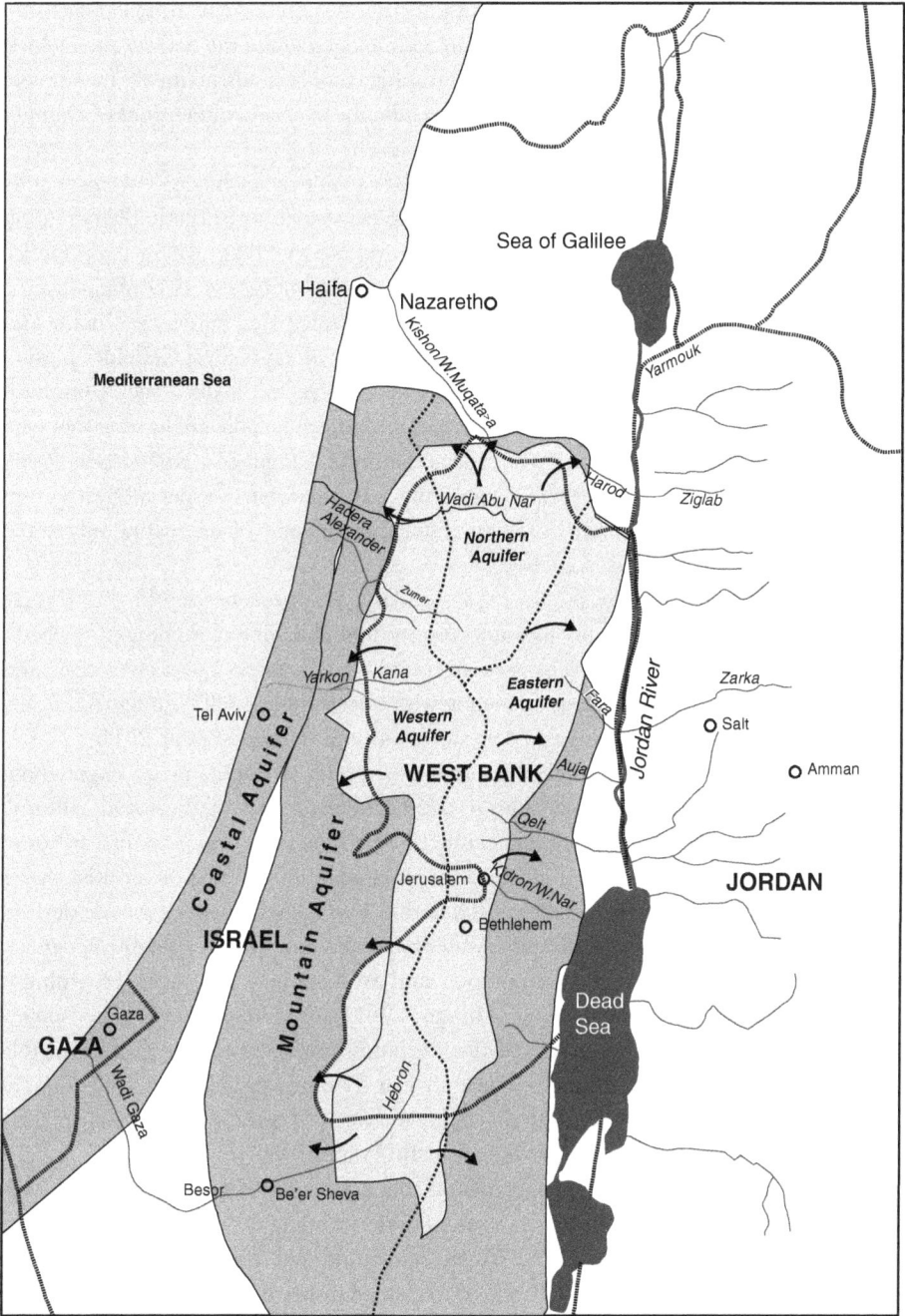

Figure 11.1 Shared waters of Israel and the Occupied Palestine Territories.
Source: EcoPeace Middle East.

negotiations must also take issues related to uncertainty, nonlinearity and feedback into account. Uncertainties related to water resources management in the case studies include unpredictable developments (climatic, demographic, economic or political), incomplete knowledge, ambiguity and conflicting views. Because the pace and dimensions of changes are accelerating, such uncertainties are on the rise.

Uncertainty is also increased by the tendency for closer policy linkages across sectors (Richardson 2000). Overcrowding increases complexity, such as in the implementation of the EU Water Framework Directive (Kastens and Newig 2008), where the presence of too many participants made constructive work nearly impossible. A consequence of overcrowding might be that processes do not have the anticipated influence (Koontz and Newig 2014). Improved understanding is needed on how long-term stakeholder participation can encourage achievement of sustainable development goals. Policy communities and networks can become linked in a messy and unpredictable chain of actors who do not know each other well and do not speak the same "language" (Richardson 2000). That they may bring quite different policy frames to the table is especially important (Schön and Rein 1994; Fligstein 1997). Such stakeholders may be a network only in the very loosest of senses (Richardson 2000).

Cooperation over water resources requires adaptive approaches sensitive to diverse viewpoints and values, ambiguity and uncertainty, and changing and competing needs. Societal and iterative learning processes are required to address these challenges and should be based on a commitment to dealing with uncertainties, deliberating technically sound alternatives and reframing problems and solutions (Huntjens et al. 2011).

In practice, water-related conflict resolution is mostly the outcome of negotiation, mediation and conciliation rooted in an in-depth understanding of the social, cultural, environmental, economic and political contexts. Transformation from conflict to cooperation is often only possible through constant renegotiation, building on iterative cycles of learning, between many stakeholders at different levels. Useful insights can be derived from the theory and practice of adaptive water management, which explicitly recognizes that the systems to be managed are complex and evolving and that societal learning is an important source of policy change (Huntjens 2011). Within conflict-oriented policy theory, on the other hand, the nature of the mechanism or agent of change and the role of knowledge in that process often remains unclear (Bennett and Howlett 1992; Castles 1990). Conventional analysis tends to downplay the role of power asymmetry in creating and maintaining conflicts over water (Zeitoun and Warner 2006).

The goal of adaptive water management is to increase the adaptive capacity of both society and the water system based on a sound understanding of what determines a system's resilience and vulnerability (Pahl-Wostl, Sendzimir et al. 2007). The challenge is to increase the ability of the system to respond to change rather than to react to its impacts.

Much has been written on adaptive water management and its potential to deal with complexity and uncertainty (Adger et al. 2005; Folke et al. 2005; Pahl-Wostl 2007; Pahl-Wostl, Sendzimir et al. 2007; Huntjens et al. 2011). Less is known about how to operationalize this knowledge to practice in a situation of contested knowledge, ambiguity or conflicting views on the seriousness of a problem, its causes and potential solutions. This chapter supports adaptive water management. The case studies are analyzed to determine how participatory and diplomatic interventions can be applied successfully to build trust among stakeholders and to reach agreement and sustainable implementation and, secondly, how we know that policy learning has occurred and whether it has had any effect.

Methodology

The analysis focuses on four stages of mediation and facilitation: stakeholder assessment and engagement, joint fact-finding, facilitating multiparty problem-solving, and developing forms of agreement that take the need for adaptive management into account.

Case Study 1. NGO-Based Integrated and Transboundary Master Plan for the Lower Jordan River Basin

The Lower Jordan River originates at the Sea of Galilee and meanders along 200 kilometers through the Jordan Valley to the Dead Sea. Its basin, though small in size, is well known for its remarkable geographic features, ancient civilizations, and religious relevance. The basin is shared by five riparians: Lebanon, Syria, Jordan, Israel and Palestine, which together define the current political landscape of the area. Most of the water is drained for agriculture and drinking. The river is seriously polluted from discharge of saline water, untreated wastewater, and other contaminants.

The goal of the cross-border master plan is to integrate Israeli, Palestinian and Jordanian plans to produce a healthy ecosystem, distribute water fairly and provide open public access to the river (EcoPeace, Royal HaskoningDHV 2015). A joint vision for 2050 shows the benefits of a cooperative, confident and peaceful region with a healthy economy and strong development perspectives for those living in the basin. The objective was to stimulate Track I stakeholders. The Israeli portion was completed in cooperation with the governmental Rehabilitation of the Jordan River project, which was in turn coordinated with Jordanian authorities and the Joint Water Committee.

EcoPeace Middle East, together with the Stockholm International Water Institute (SIWI) and the Global Nature Fund (GNF), awarded this contract to the Dutch-based international consultancy firm Royal HaskoningDHV to prepare a master plan to encourage Israelis, Palestinians and Jordanians jointly to rehabilitate the river and restore its ecology and hydrological functions. Important elements include creation of free access, good security conditions and a healthy economic basis. The Hague Institute for Global Justice was contracted by Royal HaskoningDHV to facilitate the participatory planning process by organizing a series of stakeholder consultation meetings in Jordan, Israel and the West Bank from 2012 to 2014.

Case Study 2. Water Annex of the Geneva Accord

The Geneva Initiative (GI) project, titled "Vision on Water Within the Permanent Status Agreement," began in 2013. The goal was to contribute to solving the water problem at the bilateral and regional levels. The project was conducted jointly by the two GI partners, PPC–GI (Palestinian Peace Coalition) and HLEP (Geneva Initiative Israel, H.L.

Education for Peace), with the assistance and cooperation of The Hague Institute for Global Justice on meetings that took place in The Hague.

During the seminars, the key goal was a supplementary chapter to address outstanding issues not included in the GI Water Annex, focusing on operational issues and implementation arrangements. Four major steps were taken to reach a Track II agreement (an addendum to the annex) in November 2014. The first was a stakeholder assessment in which several private and not-for-attribution interviews were conducted to establish insights into potential areas of conflict and common interest. Second, a joint fact-finding process was conducted to minimize the risk of politicized discussions. Third was facilitation multiparty problem-solving. Last, the areas of agreement were assembled and agreed upon.

A Framework for Analyzing Different Levels of Policy Learning

Policy learning is a concept in the public administration field (Hall 1988; Bennet and Howlett 1992; Sanderson 2002; Leicester 2007; Grin and Loeber 2007; Sabatier 1988; Sabatier and Jenkins-Smith 1993). This chapter defines it as a "deliberate attempt to adjust the goals or techniques of policy in the light of the consequences of past policy and new information so as to better attain the ultimate objects of governance" (Hall 1988). It involves a socially conditioned discursive or argumentative process of developing cognitive schemes or frames that question policy goals and assumptions (Sanderson 2002).

Policy changes have been explained in terms of learning by Sabatier and Jenkins-Smith (1999) through their Advocacy Coalition Framework (ACF). One limitation in the ACF is that advocacy coalitions take their identities from core beliefs (Weible et al. 2009). This conservatism leads Jenkins-Smith and Sabatier (1993) to propose that collective learning appears, not from change within policy coalitions, but as a result of the changing influence of policy coalitions on the whole. Movement is argued to be stimulated by shocks and trends exogenous to the system—wider political change, legislative reform or stressors such as climate change. The ACF does not explicitly account for, or is ambiguous about, the roles of ideas and self-interest (see Kübler 1999; Compston and Madsen 2001).

As Argyris and Schön (1996) have shown, changing values is far more difficult than changing practices. Individuals tend to avoid challenging established values for one of three reasons:

- individual risk aversion that leads actors to avoid direct interpersonal confrontations and public discussion of sensitive issues that might expose them to future negative repercussions;
- a desire to protect others by avoiding testing assumptions when it might evoke negative feelings and by keeping others from exposure to blame; or
- a wish to control the situation by keeping one's own view private and avoiding any public questioning that might refute it.

Policies change in various ways. Some are new and innovative, others merely incremental refinements (Hogwood and Peters 1983; Polsby 1984). Policy learning may differ in intensity (Pahl-Wostl, Craps et al. 2007). In distinguishing among and classifying learning processes, definitions are a useful starting point (Hargrove 2002):

Table 11.1 Loop learning framework.

Type	Indicators
Single loop learning	1. Small changes are made to specific practices or behaviors, based on what has or has not worked in the past. Things are done better without necessarily examining or challenging underlying beliefs and assumptions (Kahane 2004). Goals, values, plans and rules are operationalized rather than questioned (Argyris and Schön 1974) 2. Goals, values, frameworks and, to a significant extent, strategies are taken for granted. The emphasis is on techniques and making techniques more efficient (Usher and Bryant 1989, 87).
Double loop learning	1. Modifications (as the result of learning) are occurring, or have occurred, in personnel, programs, and legal and organizational structures that incorporate new information (including policy feedback) and causal understandings that yield more intellectually perceptive processes, a wider range of capabilities, and more effective policy (Brown 2000, 3). 2. Actor networks are changed by including new stakeholders, supporting reflection on assumptions, and showing new possibilities. The social network of stakeholders is used for learning and dealing with change (Folke et al. 2005; Geels, Elzen, and Green 2004). 3. Uncertainties are identified as a first step to find solutions (Brugnach et al. 2008), and then taken into account in policymaking (Huntjens et al. 2010).
Triple loop learning	1. Horizons of possibility are expanded (Hargrove 2002, 118). 2. A paradigm shift takes place that alters our way of thinking and behavior (Hargrove 2002,119, Pahl-Wostl 2006). 3. A major structural change takes place in the regulatory framework for water resources management (WRM) planning.

- **Single loop learning (SLL)** is a refinement of established actions to improve performance without changing guiding assumptions or taking alternative actions into account.
- **Double loop learning (DLL)** is a change in frame of reference and guiding assumptions.
- **Triple loop learning (TLL)** is a transformation of context to change factors that determine the frame of reference. It refers to transitions of the entire regime in which values and norms are shaped and stabilized by structural context.

The concept of loop learning has been operationalized into an analytical framework, summarized in Table 11.1 (Huntjens 2011; Huntjens et al. 2011).

Analysis

The analysis focuses on four stages of the cooperation and mediation process, with illustrations from each case: stakeholder assessment and engagement, joint fact-finding, facilitating multiparty problem-solving, and finding forms of agreement. In addition, each case is evaluated in terms of level of policy learning being observed, based on the indicators described in Table 11.1.

Stakeholder Assessment and Engagement

Stakeholder participation is effective in improving policy outcomes for two reasons: it increases stakeholder ownership, and stakeholders often have access to information enabling them to devise solutions better than, or complementary to, those delivered from the top down. Perhaps the most important aspect of their participation is that it can align government objectives with those of local people, mobile public support and reduce objections. This gives local stakeholders incentives to manage water resources well and can empower them by giving them influence over outcomes during implementation (Huntjens et al. 2011). Building relationships between governments is therefore not enough. The role for nonstate actors—such as water users, nongovernmental organizations (NGOs), and networks of scientists and universities—is important in treaty implementation because such a role can add an important dimension to trust-building efforts (Susskind and Islam 2012).

Stakeholder networks are important to the success of adaptive management and can provide critical on-the-ground feedback, especially as governments experiment with new technologies or ways of managing water supply and pricing. In addition, strong stakeholder interest in promoting alternate outcomes can push governments to keep searching for joint gain solutions (Susskind and Islam 2012). In the Israeli–Palestinian conflict, NGOs such as EcoPeace Middle East and the Geneva Initiative convened experts and advocates from both sides, which proved decidedly helpful to mediators.

GI Water Annex: Stakeholder Engagement

The Geneva Initiative project first conducted an assessment and review with stakeholders, whose input was integrated into project implementation. The following activities were organized:

(1) **Unilateral meetings** included two on the Palestinian side that focused on defining and determining current water status, key problems and mechanisms for networking and cooperation. Some thirty individuals attended, including representatives from the water, wastewater and environment sectors, as well as interested parliamentarians and politicians. HLEP conducted three meetings with relevant senior stakeholders in Israel with a direct connection to the issue of water use in Israel and the impact of an agreement on the water issue within the framework of a peace treaty. The meetings included representatives from the Union of Local Authorities in Israel, Israel's Ministry of Foreign Affairs and parliamentarians.
(2) **Field visits** by the Palestinian and Israeli GI team to the Jordan Valley
(3) **Local meetings** included four for the Palestinian team of experts. These were attended by the GI water core team, representatives of the Palestinian Water Authority, the Palestinian Negotiation Support Unit, the Ministry of Planning,

the Ministry of Agriculture, the environment authority, the Association of Local Authorities in Palestine, Al Quds University and others. Another four were conducted for the Israeli team of experts. These were attended by the core members of the Israeli team, various experts participating at some of the meetings.

(4) **Joint meetings** were held for the Palestinian and Israeli teams of experts, which enabled both teams to reach a joint proposal for the water addendum. They were attended by eight participants from both sides at each of the four meetings.

(5) **Two seminars with international experts** were organized by The Hague Institute for Global Justice, in collaboration with GI, bringing together both teams of experts, Jordanian experts, representatives of GI with international experts (mainly from the Netherlands). The second session also hosted representatives of the Blue Peace global water initiatives and of the Swiss Ministry of Foreign Affairs.

An additional stakeholder analysis undertaken by The Hague Institute for Global Justice involved private and not-for-attribution interviews with delegation members by a neutral entity, which yielded a matrix delineating areas of possible overlap and of likely conflict. It was used to structure an agenda, ground rules, timetable and fact-finding procedures. Setting up a methodology for peer-reviewed expert contributions was agreed to for the fact-finding component. Topics were selected based on the common ground identified during the first seminar. For the points of contention, it was agreed to identify ways (such as joint research proposals) to explore facts and uncertainties on the ground.

Master Plan LJRBL: Stakeholder Engagement

In identifying stakeholders in the Master Plan LJRB, a distinction was made between public, private and voluntary organizations. Table 11.2 presents an overview of representation of water-user groups in the private and voluntary sectors in Palestine, Israel and Jordan. In the public sector, a wide variety of governmental stakeholders and academic institutions dealing with land, water, environmental and economic issues in the study area were identified: at least 30 in Israel, 15 in Palestine and 17 in Jordan. For a detailed overview of the stakeholder analysis, we refer to the March 2014 baseline report.

One of the key challenges for the Master Plan LJRB is that excluding or including stakeholders for consultations can have far-reaching consequences, not only on the mediation, but also for the constituency of the plans. Hence, special attention was given to potential stakeholders, who are expected to have a positive influence through compromises based on common needs of stakeholders with perceived antagonist's interests, such within the irrigated agriculture sector being one of the core LJRB economic activities (EcoPeace, Royal HaskoningDHV 2015). Furthermore, due consideration was given to the fact that stakeholders' interest in the consultation issue can change during the consultation process, and a stakeholder can become more supportive or less supportive of the initiative. Balancing between economic and environmental interests of various stakeholders is a sensitive process. Identification and selection of stakeholders was therefore considered a critical step, one that influences constituencies.

Phase 1: Assessment Phase	**First consultation workshops** Focusing on project objectives, common understanding on the baseline situation, and assessment of ideas on how to solve current problems	**2 National Workshops**
Phase 1: Interventions	**Second consultation workshops** Presentation of long list of interventions; evaluation of alternative interventions; moving towards short list	**2 National Workshops**
Phase 2: National Masterplan	**Third consultation workshops** Presentation and discussion of proposed NGO National Masterplan	**2 National Workshops**
Phase 2: Integration	**Fourth consultation workshops** Final Workshop focusing on Integrated Transboundary NGO Master Plan	**1 Regional Workshop**

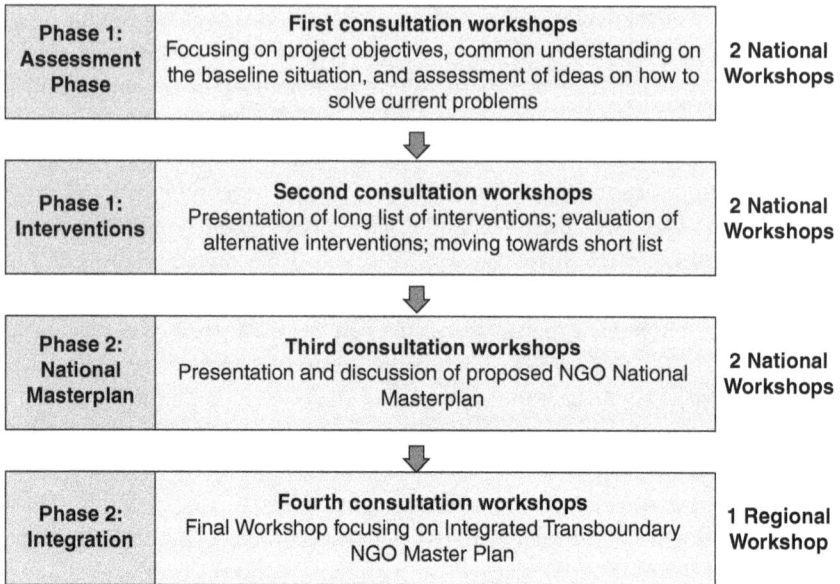

Figure 11.2   Stakeholder consultation scheme for the Master Plan LJRB.
Source: Developed by P. Huntjens.

The stakeholder consultation process for developing the Master Plan LJRB was built on the assumption that all stakeholders have relevant experience, knowledge and information to inform and improve the quality of the planning process as well as any actions that may result. Hence, the participatory planning process brought relevant stakeholders, or those who have a stake in a given issue or decision, into dialogue with one another. The key objective was to enhance trust between actors, to share information and to generate solutions and relevant good practices that have a constituency among stakeholders.

A particular challenge regarding stakeholder engagement was general suspicion about regional cooperation. Resolving this challenge requires that deep-rooted disputes over land and water issues often can be overcome, and people and their governments can be convinced that transboundary cooperation provides new opportunities, not only for restoring the natural and ecological values of the Jordan Valley, but also for the social and economic benefit of all. Substantial efforts in that direction included communication of a transparent planning framework based on clear assumptions—specifically, a two-state solution and the rehabilitation of the Jordan River (for the latter, see Safier 2011; Sagive 2013). That the planning process would not entail a formally endorsed Track I governmental agreement was also often important to clarify. Instead, a joint vision for 2050 was framed as the final product to show the benefits of a cooperative, confident and peaceful region with a healthy economy and strong development perspectives.

Another approach to dealing with suspicion was stakeholder engagement on a national level. This to some extent mitigated suspicion and prejudices early on, allowing a focus

Table 11.2 Water user groups in Palestine (bold), Israel (italic), and Jordan (normal).

Interest Category	Private Sector	Voluntary Sector
Agricultural water users	**– Chamber of Commerce and Industry (private investors in irrigated agriculture)** *Farmers' Federation of Israel* – Private investors in irrigated agriculture – Irbid and Balqa Chambers of Commerce and Industry	**Palestinian Farmers Union** **General Union of Palestinian Peasants and Cooperatives** **Bedouin shepherds** Community groups managing irrigation systems Water Users Association Jordan Farmers Union Union for Farmer Women in Jordan
Residential water users		**Union of Palestinian Women Committees** **Consumers Association** *Council of Women's Organizations in Israel* General Federation of Jordanian Women Arab Women Organization of Jordan Wise Water Women Initiative
Industrial water users	**Israeli-Palestinian Chamber of Commerce** *Israeli-Palestinian Chamber of Commerce* Irbid and Balqa Chambers of Commerce and industry	
Sustainable development in Jordan Valley	**Applied Research Institute of Jerusalem** **Alternative Tourism Group**	**EcoPeace Middle East Network** **Jordan Valley Solidarity Network** **Palestinian Environmental NGO network** *EcoPeace Middle East Network* *Society of Protection of Nature in Israel* *Adam Teva v'Din* *Zalul*
	Highland Water Forum	EcoPeace Middle East Network Jordan Environmental society Jordanian Hashemite Fund for Human Development Jordan Youth Innovation Forum

on identifying problems and possible solutions. Only in the final stage were national plans integrated into a single transboundary master plan.

Joint Fact-Finding

Unlike purely political negotiations, water mediation must also take account of scientific forecasts and analyses. If each side produces its own technical analyses, any resulting confrontations usually mean that scientific input is devalued. The way to avoid this, and to ensure that scientific input does not become a casualty of political disagreement, is to structure an appropriate joint fact-finding process.

GI Water Annex: Joint Fact-Finding

The Israeli, Palestinian and Jordanian water teams included relevant subject matter experts and Geneva Initiative members with relevant experience. Other local and international experts were also included and consulted.

In preparation for the water seminars, a methodology for joint fact-finding was developed, based on peer-reviewed expert contributions that included the following steps:

(1) Topics selected based on the agreed joint agenda.
(2) Draft expert contribution prepared by at least two experts.
(3) Contributions peer reviewed by an external reviewer.
(4) Five draft contributions sent to GI delegations.
(5) Expert contributions presented.
(6) Second review by GI delegations and wider expert group.

Master Plan LJRB: Joint Fact-Finding

In the Master Plan LJRB, joint fact-finding was undertaken by a multidisciplinary team, involving local and international experts on spatial planning, water management, ecology, agriculture, economics, sociology and water governance. The fact-finding itself was shaped by an extensive baseline report of the transboundary Lower Jordan Valley prepared by the consultant in March 2014. This process included a series of workshops with key stakeholders and expert input to discuss the data collection and based on an iterative process of collection, processing, deliberation and validation. Only after approval of the client, the results of this process were included in the baseline report. To avoid costly delays, baseline time planning was disentangled from the planning process. This proved a wise decision given that finalization and approval took almost a year longer than foreseen. This delay was mainly caused by disagreement on data inputs, including current water demands, population statistics and related growth projections, current state of the economy and related socioeconomic data. The demographic projections for the Jordanian part, for example, were adjusted from 600,000 (based on national trends) to 750,000 in the year 2050. The reason for the adjustment is that the Jordan Basin does not reflect national trends. Instead, the higher rate is based in part on the Syrian refugee situation.

The baseline report describes the Lower Jordan River Basin as a whole in terms of its geography, land use and infrastructure, water resources and supply, environment and ecology, cultural heritage and climate-change impacts. It also describes the socioeconomic situation, circumstances and future scenarios, agriculture, tourism, industries and water requirements. It also provides an overview of the major stakeholders and their organizations in the basin. Finally, it describes basin governance, geopolitical context and international legal and cooperation agreements.

The baseline report concludes with the major challenges for the people in the basin and their governments. In short, the challenge is to embark on more efficient and rational water management practices with a stronger emphasis on environmental values and the need to urgently address some of the persistent sanitary problems.

Fact-finding for the master plan includes several unique features. First, it provides a plan for the entire LJRB. Second, it provides for the first time an integrated baseline assessment in which the water balance of the LJRB is being modeled, showing the current water demands and supplies and related projections. This allows for a better understanding and evaluation of the impacts of proposed measures and, thus, how to achieve a sustainable water balance for the entire system. Third, it considers the resources as a multifunctional system for the good of ecology and economy. Fourth, unlike a traditional planning process, which includes several scenarios and for each scenario several strategies, the Master Plan LJRB deliberately includes only one scenario. Its reason is that the client intends to develop, not a governmental master plan, but instead a future joint vision, which shows the benefits of cooperation.

Facilitating Multiparty Problem-Solving

The parties must then explore ways to meet the most important interests on all sides simultaneously. Science-intensive policy disputes, such as transboundary water negotiations, offer opportunities for "value creation" and problem-solving.

Multiparty problem-solving takes many forms and can occur throughout the adaptive process. However, more meaningful participatory processes involve social learning and stakeholder information exchange linked to management decisions (Stringer et al. 2006). Social learning means learning together to manage together (Craps 2003; Pahl-Wostl 2002; Craps et al. 2007). An important hypothesis in the literature is that stakeholder collaboration enhances social learning (Boonstra 2004; Hisschemöller 2005; Muro and Jeffrey 2008; Stringer et al. 2006). It helps build trust, develop a common view on the issues at stake, resolve conflicts and arrive at joint solutions that are technically sound and actually implemented in practice. It helps all stakeholders achieve better results than they could otherwise (Craps 2003). Successful mechanisms include interactive forums such as workshops (Walters et al. 2000), focus groups and group model building (Huntjens et al. 2014) and scenario development (Peterson et al. 2003; Pahl-Wostl 2002; Van den Belt 2004). In each of these, social learning is facilitated and information flows are multidirectional.

GI Water Annex: Multiparty Problem-Solving

During the water seminars, the goal was to develop an addendum to address outstanding issues, with a focus on operational issues and implementation arrangements. The meetings, held in May and November of 2014, constituted a platform for consultation between experts from the region and international experts drafting the addendum. The problem-solving was based on the methodology for peer-reviewed expert contributions and followed by interactive plenary sessions. Facilitation was based on ground rules, including those of Chatham House. Delegations and experts were encouraged to explore more ways to create more value and generate a broader vision on sharing benefits. The latter is described in this chapter as increasing the pie (i.e., mutual gains) by means of a multi-functional usage approach. Last, the areas of agreement—principles, definitions, the need for cross-border wastewater management—were finalized. Discussions revolved around points of contention, key obstacles, and options (see Table 11.3).

Master Plan LJRB: Multiparty Problem-Solving

The Master Plan LJRB was developed in close cooperation with a wide variety of relevant stakeholders in the basin during a series of workshops. Two events are highlighted in the participatory planning process because they represent several relevant methods for multiparty problem-solving:

(1)   March 17, 2014. A **group model building** (GMB) event was used to jointly define problems, identify possible solutions and measures and determine priorities.
(2)   June 26, 2014. **Focus group meetings** and **multicriteria analysis** addressed a set of interventions proposed by the project team and evaluated by the stakeholders.

During the first meeting, GMB was used to facilitate deep involvement of a group in building a model of a particular management system for the Jordanian Master Plan LJRB. GMB is increasingly attracting attention in the fields of complex decision making, public policymaking, and implementation (Vennix 1999; Vennix et al. 2000; Zagonel and Rohrbaugh 2007; Cockerill et al. 2006, 2007). In particular, it is recognized as a useful method for stakeholder participation to develop and improve decision support models and to integrate existing information with local and stakeholder knowledge (Stave 2002;

Table 11.3 Zone of possible agreement.

1. Determining mainstream definitions used by international leading institutions and organizations on international water courses, transboundary water courses, and comparable factors
2. Developing criteria for shared and rightful redistribution of water
3. Increasing demand management and awareness-raising about the options for demand management
4. Initiating mutually beneficial projects—such as pilot projects on data sharing, joint research, or education—to build trust
5. Strengthening joint fact-finding as essential to building trust
6. Recognizing the urgent need for cross-border wastewater management
7. Operationalizing water pricing and creating incentives for cost recovery and the polluter pays principle
8. Further operationalizing the Future Joint Water Commission
9. Strengthening Palestine Water Authority institutions and capacity to harmonize donor input and arrive at equal standing with Israeli counterparts
10. Addressing long-term solutions versus urgent short-term solutions, how to avoid irreversible environmental problems, taking into account the necessity of humanitarian needs
11. Determining a joint research agenda to integrate issues and stimulate cooperation within academia

Den Exter 2004; Hare et al. 2006). The objective is to improve group understanding about that system and its problems, and possible solutions to better management decisions (Vennix 1999; Hare 2003). Both the model and the understanding among stakeholders are important products of the process (Huntjens et al. 2014).

The variety of techniques used for GMB relates to its many purposes and levels of participation (den Exter 2004). In the Master Plan LJRB, the goals in using GMB included encouraging societal learning and communication and improving the chances of success by (a) improving inputs from expert knowledge, (b) improving the uptake and commitment of participants to the process, and (c) improving the democratic accountability of the process. From a scientific perspective, GMB has been used to involve multiple experts to increase the validity and robustness of an existing conceptual or computational model on ecosystem or water management functioning (such as Costanza et al. 1990; Costanza and Ruth 1998).

Because they were unfamiliar with the method, participants began hesitantly, but soon developed momentum. A round-robin method helped people overcome reservations about saying something in the group. Feedback evaluation forms indicated that the participants valued the integrative aspect of the method and the possibility of integrating their own ideas in a structured way.

During the second meeting, interventions were evaluated using multicriteria analysis in focus groups. Increasingly used in participatory research for adaptive water management (Hirsch et al. 2010), focus groups are broadly defined as meetings to obtain public understanding on a distinct area of interest in a permissive environment (Morgan 1997). During the Masterplan LJRB meeting in June 2014, 4 groups were established to evaluate 20 interventions. Facilitation took place in four parallel sessions, with a goal of creating an optimal and appropriate input, and with ample opportunity for discussion. Participants evaluated the interventions using a set of 11 predefined criteria. In a relaxed atmosphere, each group (six to eight people) shared ideas and perceptions on each criterion. In a smaller group, participants usually feel that they have more influence, and persuading reticent participants to contribute is easier. One obstacle encountered was that not all participants were accustomed to reflecting abstractly on the proposed interventions, which included reuse of wastewater, post-harvesting support, extension services improvement, pollution control, ecological restoration, tourism development and institutional strengthening projects. Each intervention took at least an hour to be properly evaluated (see Table 11.4).

Developing Forms of Agreement That Take Adaptive Management into Account

Treaties and institutional arrangements on water resources cannot remain static. Factors such as water requirements, use patterns and efficiency of management change with time, as do water management paradigms, practices and processes (Varis, Biswas, and Tortajada 2008, xi). Formulating dynamic treaties is not an easy task, but it is a critical one.

Because transboundary water management involves so much complexity and uncertainty, final agreements are quite different from purely political ones (Islam and Susskind 2013). They call almost always for some form of collaborative adaptive management and therefore need to include joint monitoring elements as well as dispute-handling clauses.

Table 11.4 Criteria for participatory evaluation of interventions.

1. Good economic status
2. Agricultural water availability
3. Industrial water availability (including food processing)
4. Socioeconomic impacts
5. Good human status
6. Access to drinking water and sanitation
7. Vector-borne diseases and other health impacts
8. Good environmental status
9. Habitat disturbance
10. Water quality and quantity impacts
11. Planning and implementation
12. Technical considerations
13. Compatibility to other plans
14. Investment costs
15. Political considerations

GI Water Annex: Forms of Agreement

At the beginning of the Geneva Initiative project, the two teams raised several issues not fully addressed in the 2009 Water Annex that had become apparent in the intervening years. These were refined, and solutions developed to address them. The project concluded with the draft of an agreed model, considered an addendum to the original document. Both the addendum itself and the experience generated during the project may help significantly in future negotiations on a permanent status agreement. The addendum addressed a number of key issues, which included the following:

• principles for rightful redivision of the shared resources,
• management of particular water resources as a single administrative unit,
• cooperation on monitoring water flows and quantities,
• economic principles for sustainable and efficient use and management of shared water resources, and
• mechanisms for regional cooperation.

Additional recommended steps were also developed that could not be implemented within the framework of the project:

• a doctrine for pricing for water,
• a model for management of transboundary sewage,
• ways to quantify reallocation of water resources,
• ways to ensure that current activities in the field will not preclude implementation of Geneva Initiative proposals, and
• steps to be taken during implementation that will unfold fully only after a permanent agreement is signed.

The project also created a body of local and international experts ready to advocate for a negotiated solution and to make this solution credible in international as well as domestic eyes. Last, it facilitated outreach to key sectors in Palestine and Israel and to the international community—including specialized experts, politicians and civil society leaders who will draw on the project's success to advocate for the feasibility of a final-status agreement.

Master Plan LJRB: Forms of Agreement

The final master plan presents an investment planning scheme of $4.4 billion toward the long-term vision for 2050. Specifically, it outlines the possibility that the region may gain a clean and healthy environment and sufficient flows in the Jordan River to sustain healthy ecosystems. At the same time, the river will act as natural water conveyor and source for water supply in the Jordan Basin. Water will be shared equitably among the three riparian countries, and the basin will be freely accessible for all nationalities within an appropriate security framework. Regional stability will encourage local, private and foreign investment. In short, the reforms will generate an investment climate and a regulatory business environment that promotes sustainable development. In other words, the aim is for the proposed interventions to be used as advocacy tools with national stake-holders, international financiers and other international actors to increase the political will to adopt the proposed interventions, whether in full or in part.

The master plan includes several recommendations for Lower Jordan Basin governance, such as institutional strengthening for responsible authorities, advocacy and local community empowerment, social responsibility and international cooperation and security. One governance objective is to establish robust and flexible institutions and policy processes that continue to work satisfactorily when confronted with social and physical challenges, which is fundamental to the concept of adaptive management. Building trust and reciprocity are important elements for sustainability (Huntjens et al. 2012). The long-term interventions proposed are all based on the assumption of a two-state solution by 2050. Integrated and transboundary management is possible only when the parties have confidence in trilateral collaboration and the willingness to find solutions. Short-term interventions, however, can largely be implemented without a two-state-solution or final water agreements. Such "no-regret" actions—interventions that will not negatively influence the water balance of the other party—include protection of groundwater infiltration zones, reduction and management of water demands, reduction of water losses, and unaccounted-for-water, treatment and removal of pollutions sources, particularly wastewater, salt water, polluted runoff from fish ponds and integrated solid and hazardous wastewater, and full treatment and reuse of domestic wastewater for agricultural purposes.

Maximizing the economic and environmental development perspectives in the Lower Jordan Basin requires that transboundary cooperation be strengthened, particularly in Jordan, Israel and Palestine. This may include preparing for a joint Lower Jordan Basin Organization, updating the Jordanian–Israeli water agreements taking into account the increased water stress in Jordan and a new water division based on proportional distribution of water, jointly restoring the Lower Jordan River and developing joint economic initiatives. In the long run, the challenge might be to work toward an integrated Jordan Basin Commission for all riparian countries. In the interim, a Lower Jordan Basin Commission involving Jordan, Palestine and Israel should be advanced.

Assessment of Policy Learning

Table 11.5 presents a summary assessment of the observed policy learning in each case study, by using the indicators as presented in Table 11.1.

Table 11.5 Assessment of observed policy learning.

Full name of Track II agreement	NGO-based Integrated and Transboundary Master Plan for the Lower Jordan River Basin	Addendum to Water Annex of the Geneva Accord
Current status	Finalized and published in May 2015	Finalized in February 2015, publication on hold
Objectives	To integrate Israeli, Palestinian, and Jordanian plans and thus produce a healthy ecosystem, distribute water fairly and provide open public access to the river	To develop a supplementary paper to address outstanding issues not included in the GI Water Annex, with a focus on operational issues and implementation arrangements
Entirely new management measures proposed?	Yes, maximum five	Yes, maximum five
Entirely new physical interventions proposed?	Yes, maximum ten	Yes, maximum ten
Uncertainties being recognized and addressed?	Yes, master plan builds on different socioeconomic and climate scenarios, but on a single political scenario (a two-state solution)	Limited, Geneva Accord builds on a single political scenario (a two-state solution), but options for interim period have been discussed, such as to avoid irreversible environmental damage and human health risks due to untreated cross-border wastewater
Changes in the actor network?	Mostly stakeholders from water, spatial planning, agriculture, wastewater and environment sectors, but also representatives from public health, tourism, cultural heritage, and economic sectors	Dominated by stakeholders from water, wastewater, and environment sectors
Changes in the regulatory framework?	Recommendations for new Joint Water Committee, Jordan Basin Commission, and other governance arrangements	Recommendations for new Joint Water Committee

Table 11.5 Continued.

Full name of Track II agreement	NGO-based Integrated and Transboundary Master Plan for the Lower Jordan River Basin	Addendum to Water Annex of the Geneva Accord
New information entered into the process and adopted?	Yes, via bottom-up participatory process and expert judgement, including widespread adoption in large set of interventions	Yes, via expert judgment, but with selective adoption
New norms and values?	Regional vision for LJRB showing benefits for all parties (including future riparian states), and multifunctional usage approach	Precautionary principle, polluter pays principle, and ongoing transition from zero-sum to mutual gains approach
Dominant type of policy learning	Double loop learning, some instances of triple loop learning	Single and double loop learning

Discussion: From Zero-Sum to Mutual Gains

The Geneva Initiative agreement in November 2014 sheds light on possible solutions to meet current and future water needs of both peoples on the basis of fairness. A 1999 report found that most international negotiations over water during the past century proceeded on the basis of each side's recognizing the needs of the other sides rather than on a priori principles or rights (Wolf 1999). Both Israeli and Palestinian delegations realize that they depend on each other for their water supply. An approach recognizing that the same water can be used for multiple purposes is more realistic and provides more benefits for all. In the Water Annex addendum, this approach was consolidated in Article 2.2, stating that rightful redivision of the shared water resources will depend on sharing the benefits derived from the use rather than the allocation of water. This illustrates an important transition away from zero-sum thinking—that water can be used only once and by one party only (Islam and Susskind 2013)—in that the 2009 Water Annex focused on allocation.

In the Master Plan LJRB, a similar line of thinking was framed as multifunctional usage, in which water is used over and over again, an approach considered crucial to bypassing the contentious issue of rights to transboundary waters. The major pitfall to quantitative allocation is that changes are perceived as a threat to national security and mean a deadlock in negotiations.

In conclusion, an essential game-changer for both case studies was an ongoing paradigm shift among key stakeholders from quantitative allocation toward multifunctional usage. Examples in the Master Plan LJRB include treating and reusing domestic water supply for agriculture, already applied widely in Israel but not yet in Palestine. To build trust and show that cooperation pays off, the Master Plan identified ample opportunities, such as along the borders of the West Bank in Jenin–Magen Sha'un, the Tulkarim and Qalqilya-Netantya region and the Wadi Al-Nar–Kidron Valley.

Another noteworthy example concerns the rehabilitation of the Lower Jordan River itself by creating a multi-functional river system (Safier 2011; Sagive 2013) in which the Jordan is used for ecological purposes, for tourism and economic development, for domestic water supply and for agricultural water supply, altogether multiplying the value of each cubic meter of water.

Policy Learning

In both case studies evidence of loop learning is clearly visible (see Table 11.5). While the Masterplan LJRB process is dominated by double loop learning, with some instances of triple loop learning, the GI process is dominated by single and double loop learning. By comparing both case studies, two distinctive elements appear to be important for supporting higher levels of policy learning, as illustrated by the Masterplan LJRB process: (a) broad and horizontal stakeholder participation, and (b) new information entering into the process via a bottom-up and participatory process in combination with expert judgement. Huntjens et al. (2012) have introduced policy learning—through exploring uncertainties, deliberating alternatives and reframing problems and solutions—as an essential element for improving water cooperation, especially in a situation of changing environments, such as political change, climate change, financial crises and natural disasters. Policy learning approaches generally hold that actors can learn from their experiences, and that they can modify their present actions on the basis of their interpretation of how previous actions have fared (Bennet and Howlett 1992).

Building Trust

Resolving the general suspicion among stakeholders about regional cooperation required substantial efforts in the Masterplan LJRB process, including communication of a transparent planning framework based on clear assumptions. In the GI process it was agreed to initiate mutually beneficial projects—such as pilot projects on data-sharing, joint research, or education—to build trust. Actors in both case studies confirmed that joint fact-finding is a core element of building trust.

In addition, successful mechanisms that helped to build trust included interactive forums such as workshops (Walters et al. 2000), focus groups and group model building (Huntjens et al. 2014a). In each of these, mutual learning was facilitated, and information flows were multidirectional. These mechanisms support trust building by developing a common view on the issues at stake, resolving conflicts and arriving at joint solutions that are technically sound and actually can be implemented in practice. It helps all stakeholders achieve better results than they could otherwise (Craps 2003).

More work is needed on how trust is built, starting with areas that Huntjens et al. (2012) suggest, such as: early communication of uncertainties, joint participative knowledge production, open access to, and shared information sources, transparency about the decision-making process and sharing of responsibilities (e.g. on regional knowledge transfer). Transparency and trust-building are closely related (Abrams et al. 2003), and special attention is given to the role of leaders who are able to provide key functions for adaptive governance, such as "building trust, making sense, managing conflict, linking actors, initiating partnership among actor groups, compiling and generating knowledge, and mobilizing broad support for change" (Folke et al. 2005).

Multilevel Governance

Stakeholder participation allowed development of the Master Plan LJRB to be fine-tuned to the local context. In practice, local stakeholders (such as irrigation farmers and well owners) already manage most of the water resources. In this sense, participation can be seen as much as participation by government agencies in local governance arrangements as the reverse. Public and stakeholder participation is an essential link across local, national and regional planning levels. Different interest groups should therefore be able to participate in the planning process at multiple levels.

The aim of the proposed master plan interventions is their use as advocacy tools for national stakeholders, international financiers and other international actors to increase the political will for adoption of the proposed interventions.

The interventions described in Annex 1 (2015) include a suggested institutional setting for each. Financing has yet to be secured and will require additional preparation and design, including elaboration of the proposed institutional and governance aspects, depending on funding stipulations. National authorities are anticipated to play the primary role in implementation. The Israeli portion of the plan, for example, was developed in cooperation with the government's Rehabilitation of the Jordan River project, Jordanian authorities and the Joint Water Committee. This coordination is an important contribution to acceptance of the NGO vision.

Municipalities also have an important role in that they represent the population in the basin and provide water, wastewater collection and solid-waste management services. The subsidiary principle is again relevant here.

Finally, EcoPeace Middle East is expected to play a key role as one of the major NGOs active in the Lower Jordan River Basin, particularly with regard to organizing grassroots environmental protection activities and to engaging and organizing local stakeholders. As to multilevel governance, the Master Plan LJRB shows that multi-stakeholder participation can have not only a significant influence in shaping projects, but can also feed information to policymakers for future policies. Feedback during consultation meetings shaped not only local interventions, but also broader management. Nevertheless, achieving such feedback across institutional levels seems to require an agent or institution, acceptable to all groups, who can ensure that such dialogue does take place (see also Stringer et al. 2006). In the Master Plan LJRB, this role was assumed by EcoPeace Middle East, and in the Water Annex case, by the Geneva Initiative. Without these players, it is unlikely that these projects would have had the successes they did.

Conclusion

The initiation, format, content and outcome of water mediations may vary, as the case studies make clear. Whereas the Master Plan LJRB was characterized by broad stakeholder participation with professional backing, and fine-tuning bottom-up learning with high-level Track II strategies and long-term visions, the GI Water Annex was more modest in terms of stakeholder participation and relied primarily on expert judgment. In both, the process was initiated by Track II parties and guided by a firm conviction that the Israeli–Palestinian conflict can be resolved only by a two state-solution, and that successful cooperation over shared water resources offers the potential to build mutual trust and understanding more generally. Even if water scarcity is not the primary driver of the conflict, a solution to the shared water problem is critical to a peaceful resolution.

However, reaching agreement and ensuring cooperation remain complex and uncertain. The multiple dimensions of the larger conflict complicate joint fact-finding and multiparty problem-solving. Successful mechanisms therefore include interactive techniques such as multi-stakeholder workshops, focus groups, and group model building. In each, policy learning is facilitated, and information flows between stakeholders are multidirectional.

To conclude, cooperation over shared water resources requires adaptive approaches. Societal and iterative learning processes based on commitment to dealing with uncertainties, deliberating alternatives and reframing problems and solutions are critical. Water-related conflict resolution is thus mostly the outcome of processes of negotiation, mediation and conciliation rooted in in-depth contextual understanding. Transformation of water conflict into cooperation is often only possible through constant negotiation.

So far, joint efforts for dealing with water have been hampered by the lack of regional stability. Lack of dialogue between states and civil society organizations has led in turn to a lack of shared knowledge. This chapter underscores the importance of third-party and multitrack diplomacy when political conditions are uncertain and formal negotiations have come to a halt. The Master Plan LJRB and Geneva Initiative accord show that agreement on vital issues is (still) possible.

Acknowledgments

The author thanks the Geneva Initiative, The Netherlands Ministry of Foreign Affairs, and The Hague Institute for Global Justice for their support of the operationalization of the GI Water Annex. The author is also grateful to EcoPeace Middle East, the Global Nature Fund, and Stockholm International Water Institute as clients, and Royal HaskoningDHV as implementing agency, for their support of the NGO-based Integrated and Transboundary Master Plan for the Lower Jordan River Basin, financed by the European Commission. Last, the author thanks two anonymous reviewers and Shafiqul Islam, Ram Aviram, Abdelrahman Tamimi, Jeroen Kool and Rens de Man for reviewing earlier versions of this chapter.

References

Adger, W. Neil, Nigel W. Arnell, and Emma L. Tompkins. 2005. "Successful Adaptation to Climate Change across Scales." *Global Environmental Change* 15(2): 77–86.

Amnesty International. 2009. *Troubled Waters: Palestinians Denied Fair Access to Water.* London: Amnesty International.

Argyris, Chris. 1999. *On Organizational Learning*, 2nd edn. Malden: Blackwell.

Argyris, Chris and Donald A. Schön. 1996. *Organizational Learning II: Theory, Method and Practice.* Reading, MA: Addison-Wesley.

———. 1974. *Theory in Practice: Increasing Professional Effectiveness*. San Francisco: Jossey-Bass.

Bennett, Colin J. and Michael Howlett. 1992. "The Lessons of Learning: Reconciling Theories of Policy Learning and Policy Change." *Policy Sciences* 25(3): 275–94.

Boonstra, Jaap J. 2004. "Introduction." In: *Dynamics of Organizational Change and Learning*. Chichester: John Wiley.

Brock, Hannah. 2011. "Competition over Resources: Drivers of Insecurity and the Global South." Oxford: Oxford Research Group.

Brooks, David B., Julie Trottier and Laura Doliner. 2013. "Changing the Nature of Transboundary Water Agreements: The Israeli–Palestinian case." *Water International* 38(6): 671–86. doi:10.1080/02508060.2013.810038.

Brown, H. Douglas. 2000. *Principles of Language Teaching and Learning*, 4th edn. White Plains, NY: Longman.

Brugnach, Marcela, Art Dewulf, Claudia Pahl-Wostl and Tharsi Taillieu. 2008. "Towards a Relational Concept of Uncertainty: About Knowing too Little, Knowing too Differently, and Accepting Not to Know." *Ecology and Society* 13(2): 30.

Cockerill, Kristan, Howard D. Passell and Vince C. Tidwell. 2006. "Cooperative Modeling: Building Bridges Between Science and the Public." *Journal of the American Water Resources Association* 42(2): 457–71.

Cockerill, Kristan, Vince C. Tidwell, Howard D. Passell and Leonard A. Malczynski. 2007. "Cooperative Modeling Lessons for Environmental Management." *Environmental Practice* 9(1): 28–41.

Compston, Hugh and Per Kongshoj Madsen. 2001. "Conceptual Innovation and Public Policy: Unemployment and Paid Leave Schemes in Denmark." *Journal of European Social Policy* 11(2): 117–32.

Costanza, Robert and Matthias Ruth. 1998. "Using Dynamic Modeling to Scope Environmental Problems and Build Consensus." *Environmental Management* 22(2): 183–95.

Costanza, Robert, Fred H. Sklar and Mary L. White. 1990. "Modeling Coastal Landscape Dynamics." *BioScience* 40(2): 91–107.

Craps, Marc, ed.. 2003. "Social Learning in River Basin Management." HarmoniCOP WP2 Reference Document. Leuven: Centre for Organizational and Personnel Psychology. http://www.harmonicop.uni-osnabrueck.de/_files/_down/SocialLearning.pdf.

EcoPeace, Royal HaskoningDHV. 2015. "Transboundary NGO Master Plan for the Lower Jordan River Basin." Amman: EcoPeace Middle East. http://www.foeme.org

den Exter, Kristin A. 2004. "Integrating Environmental Science and Management: The Role of System Dynamics Modelling." PhD thesis, Southern Cross University, Australia.

Fligstein, Neil. 1997. "Social Skill and Institutional Theory." *American Behavioral Scientist* 40(4): 397–405.

Folke, Carl, Thomas Hahn, Per Olsson and Jon Norberg. 2005. "Adaptive Governance of Social-Ecological Systems." *Annual Review of Environmental Resources* 30: 441–73.

Geels, Frank W., Boelie Elzen and Ken Green. 2004. Introduction. In: *Systems Innovation and the Transition to Sustainability: Theory, Evidence and Policy*, 1–16. Cheltenham: Edward Elgar.

Gleick, Peter H., Newsha Ajami, Juliet Christian-Smith, Heather Cooley, Kristina Donnelly, Julian Fulton, Mai-Lan Ha, Matthew Heberger, Eli Moore, Jason Morrison, Stewart Orr, Peter Schulte and Veena Srinivasan. 2014. *The World's Water, vol. 8: The Biennial Report on Freshwater Resources*. Washington, DC: Island Press.

Grin, John and Anne Loeber. 2007. "Theories of Policy Learning: Agency, Structure and Change." In: *Handbook of Public Policy Analysis. Theory, Politics, and Methods*, ed. Frank Fischer, Gerald J. Miller and Mara S. Sidney, 201–19. Boca Raton, FL: CRC Press.

Hall, Peter A. 1993. "Policy Paradigms, Social Learning, and the State." *Comparative Politics* 25(3): 275–96.

Hare, Matt P. 2003. "A Guide to Group Model Building: How to Help Stakeholders Participate in Building and Discussing Models in Order to Improve Understanding of Resource Management." Osnabrück: Seecon Deutschland GmbH.

Hare, Matt P., Olivier Barreteau, M. Bruce Beck, Erik Mostert, J. David Tàbara and Rebecca Letcher. 2006. "Methods for Stakeholder Participation in Water Management." In: *Sustainable Management of Water Resources: An Integrated Approach*, edited by Carlo Giupponi, Anthony J. Jakeman, Derek Karssenberg and Matt P. Hare, 177–225. Chichester: Edward Elgar.

Hargrove, Robert. 2002. *Masterful Coaching*, rev. edn. San Francisco: John Wiley.

Hirsch, Darya, Geraldine Abrami, Raffaele Giordano, Stefan Liersch, Nilufar Matin and Maja Schlüter. 2010. "Participatory Research for Adaptive Water Management in a Transition Country: A Case Study from Uzbekistan." *Ecology and Society* 15(3): 23. http://www.ecologyandsociety.org/vol15/iss3/art23.

Hisschemöller, Matthjis. 2005. "Participation as Knowledge Production and the Limits of Democracy." In: *Democratization of Expertise? Exploring Novel Forms of Scientific Advice in Political Decision-Making*, edited by Peter Weingart and Sabine Maasen, 189–208. Dordrecht: Springer.

Hogwood, Brian W. and B. Guy Peters. 1983. *Policy Dynamics*. Brighton: Wheatsheaf Books.

Huntjens, Patrick. 2011. *Water Management and Water Governance in a Changing Climate: Experiences and Insights on Climate Change Adaptation from Europe, Africa, Asia and Australia*. Delft: Eburon Academic.

Huntjens, Patrick, Claudia Pahl-Wostl, and John Grin. 2010. "Climate Change Adaptation in European River Basins." *Regional Environmental Change* 10(4): 263–84.

Huntjens, Patrick, Claudia Pahl-Wostl, Benoit Rihoux, Maja Schlüter, Zsuzsanna Flachner, Susana Neto, Romana Koskova, Chris Dickens and Isah Nabide Kiti. 2011. "Adaptive Water Management and Policy Learning in a Changing Climate: A Formal Comparative Analysis of Eight Water Management Regimes in Europe, Asia, and Africa." *Environmental Policy and Governance* 21(3): 145–63.

Huntjens, Patrick, Louis Lebel, Claudia Pahl-Wostl, Jeff Camkin, Roland Schulze, and Nicole Kranz. 2012. "Institutional Design Propositions for the Governance of Adaptation to Climate Change in the Water Sector." *Global Environmental Change* 22(1): 67–81.

Huntjens, Patrick, Bouke Ottow, and Ralph Lasage. 2014a. "Participation in Climate Adaptation in the Lower Vam Co River Basin in Vietnam." In: *Action Research for Climate Change Adaptation: Developing and Applying Knowledge for Governance*, edited by Arwin van Buuren, Jasper Eshuis and Mathija van Vliet. Abingdon: Routledge.

Huntjens, Patrick and Henk van Schaik. 2014b. Afterword. In: *Water Security and Peace Conference Proceedings*. The Hague: Water Diplomacy Consortium.

Islam, Shafiqul and Lawrence Susskind. 2013. *Water Diplomacy: A Negotiated Approach to Managing Complex Water Networks*. New York: Routledge.

Kahane, Adam. 2004. *Solving Tough Problems: An Open Way of Talking, Listening, and Creating New Realities*. San Francisco: Berrett-Koehler.

Kastens, Britta and Jens Newig. 2008. "Will Participation Foster the Successful Implementation of the Water Framework Directive? The Case of Agricultural Groundwater Protection in Northwest Germany." *Local Environment* 13(1): 27–41.

Koontz, Tomas M. and Jens Newig. 2014. "Cross-Level Information and Influence in Mandated Participatory Planning: Alternative Pathways to Sustainable Water Management in Germany's Implementation of the EU Water Framework Directive." *Land Use Policy* 38(May): 594–604.

Kübler, Daniel. 1999. "Ideas as Catalytic Elements for Policy Change: Advocacy Coalitions and Drug Policy in Switzerland." In: *Public Policy and Political Ideas*, edited by Dietmar Braun and Andreas Busch, 116–35. Cheltenham: Edward Elgar.

Leicester, Graham. 2007. "Policy Learning: Can Government Discover the Treasure Within?" *European Journal of Education* 42(2): 173–84.

Morgan, David L. 1997. *Focus Groups as Qualitative Research*, 2nd edn. London: Sage.

Muro, Melanie and Paul Jeffrey. 2008. "A Critical Review of the Theory and Application of Social Learning in Participatory Natural Resources Management." *Journal of Environmental Planning and Management* 51(3): 325–44.

Pahl-Wostl, Claudia. 2002. "Participative and Stakeholder-Based Policy Design, Evaluation and Modeling Processes." *Integrated Assessment* 3(1): 3–14.

———. 2006. "The Importance of Social Learning in Restoring the Multifunctionality of Rivers and Floodplains." *Ecology and Society* 11(5): 10.

———. 2007. "Transition Towards Adaptive Management of Water Facing Climate and Global Change." *Water Resources Management* 21(1): 49–62.

Pahl-Wostl, Claudia, Marc Craps, Art Dewulf, Erik Mostert, David Tabara and Tharsi Taillieu. 2007. "Social Learning and Water Resources Management." *Ecology and Society* 12(2): 5. http://www.ecologyandsociety.org/vol12/iss2/art5.

Pahl-Wostl, Claudia, Jan Sendzimir, Paul Jeffrey, Jeroene Aerts, Ger Berkamp and Katharine Cross. 2007. "Managing Change Toward Adaptive Water Management Through Social Learning." *Ecology and Society* 12(2): 30. http://www.ecologyandsociety.org/vol12/iss2/art30

Peterson, Garry S., Graeme S. Cumming and Stephen R. Carpenter. 2003. "Scenario Planning: A Tool for Conservation in an Uncertain World." *Conservation Biology* 17(2): 358–66.

Polsby, Nelson W. 1984. *Political Innovation in America: The Politics of Policy Initiation*. New Haven: Yale University Press.

Reed, Brian and Bob Reed. 2011. "Minimum Water Quantity Needed for Domestic Use in Emergencies," Technical Note no. 9, Geneva: WHO, http://www.who.int/water_sanitation_health/publications/2011/tn9_how_much_water_en.pdf

Richardson, Jeremy. 2000. "Government, Interest Groups and Policy Change." *Political Studies* 48(5): 1006–25.

Sabatier, Paul A. 1988. "An Advocacy Coalition Framework of Policy Change and the Role of Policy-Oriented Learning Therein." *Policy Sciences* 21(2/3, Fall): 129–68.

Sabatier, Paul A. and Hank C. Jenkins-Smith. 1999. "The Advocacy Coalition Framework: An Assessment." In: *Theories of the Policy Process*, ed. Paul A. Sabatier. Boulder, CO: Westview Press.

———, eds. 1993. *Policy Change and Learning: An Advocacy Coalition Approach*. Boulder, CO: Westview Press.

Safier, Gilad. 2011. "Roadmap for the Rehabilitation of the Lower Jordan River." Amman: EcoPeace/Friends of the Earth Middle East.

Sagive, Michal, Mohammed Obidallah, Hana Al-Asad'd et al. 2013. "Cross-Border 'Priority Initiatives' of the Good Water Neighbors Project." Amman: Friends of the Earth Middle East. http://foeme.org/uploads/13841891381~%5E$%5E~Community_Based_Problem_Solving_on_Water_Issues_2013.pdf

Salathiel, Lauren. 2012. "'Take Me Over the Jordan': Concept Document to Rehabilitate, Promote Prosperity, and Help Bring Peace to the Lower Jordan River Valley." Friends of the Earth Middle East, October 21, 2012. https://foeme.wordpress.com/2012/10/21/take-me-over-the-jordan-the-first-step-in-a-regional-river-plan.

Sanderson, Ian. 2002. "Evaluation, Policy Learning and Evidence-Based Policy Making." *Public Administration* 80(1): 1–22.

Schön, Donald A. and Martin Rein. 1994. *Frame Reflection. Toward a Resolution of Intractable Policy Controversies*. New York: Harper Collins.

Stave, Krystyna A. 2002. "Using System Dynamics to Improve Public Participation in Environmental Decisions." *System Dynamics Review* 18(2): 139–67.

Stringer, Lindsay C., Andrew J. Dougill, Evan Fraser, Klaus Hubacek, Christina Prell and Mark S. Reed. 2006. "Unpacking 'Participation' in the Adaptive Management of Social-Ecological Systems: A Critical Review." *Ecology and Society* 11(2): 39. http://www.ecologyandsociety.org/vol11/iss2/art39.

Susskind, Lawrence and Shafiqul Islam. 2012. "Water Diplomacy: Creating Value and Building Trust in Transboundary Water Negotiations." *Science & Diplomacy* 1(3). http://www.sciencediplomacy.org/perspective/2012/water-diplomacy.

Trottier, Julie. 2008. "Water Crises: Political Construction or Physical Reality?" *Contemporary Politics* 14(2): 197–214.

Usher, Robin and Ian Bryant. 1989. *Adult Education as Theory, Practice and Research*. London: Routledge.

van den Belt, Marjan. 2004. *Mediated Modeling: A System Dynamics Approach to Environmental Consensus Building*. Washington, DC: Island Press.

Varis, Olli, Cecilia Tortajada and Askit K. Biswas. 2008. Preface. In: *Management of Transboundary Rivers and Lakes*. Berlin: Springer-Verlag.

Vennix, Jac A. M. 1999. "Group Model-Building: Tackling Messy Problems." *System Dynamics Review* 15(4): 379–401.

Vennix, Jac A. M., Cécile M. Thijssen, and Etienne A. Rouwette. 2000. "Group Model Building: A Decision Room Approach." *Simulation and Gaming* 31(3): 359–79.

Walters, Carl, Josh Korman, Lawrence E. Stevens and Barry Gold. 2000. "Ecosystem Modeling for Evaluation of Adaptive Management Policies in the Grand Canyon." *Conservation Ecology* 4(2): 1. http://www.consecol.org/vol4/iss2/art1/.

Weible, Christopher M., Paul A. Sabatier and Kelly McQueen. 2009. "Themes and Variations: Taking Stock of the Advocacy Coalition Framework." *Policy Studies Journal* 37(1): 121–40.

Wolf, Aaron T. 1999. "Criteria for Equitable Allocations: The Heart of International Water Conflict." *Natural Resources Forum* 23(1): 3–30.

Zagonel, Aldo A. and John Rohrbaugh. 2007. "Using Group Model Building to Inform Public Policy Making and Implementation." In: *Complex Decision Making: Theory and Practice*, edited by H. Qudrat-Ullah, J. M. Spector and P. Davidson, 113–38. New York: Springer.

Zeitoun, Mark. 2007. "The Conflict vs. Cooperation Paradox: Fighting Over or Sharing of Palestinian-Israeli Groundwater." *Water International* 32(1): 105–20. http://www.tandf.co.uk/journals/pdf/papers/rwin_2007_Zeitoun.pdf.

Zeitoun, Mark and Jeroen Warner, "Hydro-hegemony: A Framework for Analysis of Transboundary Water Conflicts." *Water Policy* 8(5): 435–60.

Chapter Twelve

RISK DISTRIBUTION AND THE ADOPTION OF FLEXIBILITY: DESALINATION EXPANSION IN QATAR

Abdulla AlMisnad, Richard de Neufville
and Margaret Garcia

Abstract

Future needs for urban water supply are often highly uncertain, especially in cities with rapid, variable population growth. Technological innovations and volatility in energy prices aggravate the risks of investments in water infrastructure systems. In such uncertain conditions, Flexible Design can significantly increase expected performance compared to conventional designs, as Chapter 6 in this book demonstrates. Flexible Design achieves this result by embedding Real Options in the design and using them to adjust the system performance over time according to need. The problem is: How should we provide the contractual incentives to insure that the designers and suppliers of the system deliver appropriate flexible designs? This chapter shows, through a detailed case study of desalination projects for Qatar, that the form of the contract between the water authority and the suppliers greatly influences the overall design and economic performance. This chapter applies a waterfall analysis to trace how contracts allocate benefits and differentially motivate designers to create or avoid flexible designs. This result means that, contrary to current practice, joint consideration of physical design and contractual arrangements can improve the design of water supply systems. Additionally, understanding the impact of contract choice on the willingness of a public–private partnership (PPP) to implement a flexible design offers insights for water infrastructure projects with other types of multi-stakeholder relationships.

Introduction

This chapter argues that, to achieve the expected outcomes, designers and negotiators need to jointly consider both the physical and contractual arrangements of a project. The reasoning is:

(1)   The future outcomes of any project are uncertain because of inescapable changes in needs, supplies, costs, and technologies.

(2)   Contractual arrangements inevitably allocate these outcomes differentially between water suppliers and end users—largely, but not only, because their costs and benefits occur unevenly over time.

(3)   Designers will shape the project in response to the risks associated with the contracts, which may not be best from the overall societal perspective.

(4)   Therefore, to get the best project designs, we need to consider jointly the physical and contractual arrangements for a project.

Putting the case colloquially, we should not think of maximizing the size of the pie and sharing it among stakeholders as being two distinct issues. Our rules for sharing the pie will influence the size of the pie. While the chapter demonstrates this phenomenon through a case study, the general principle applies widely to water infrastructure projects where multiple parties share the risks and benefits.

Conventional planning of infrastructure focuses on identifying and creating the best project to meet assumed needs in an assumed environment. Such reliance on assumed projections is unfortunate. Experience demonstrates, again and again, that actual future needs differ from earlier predictions, and that the operating environment changes from original anticipations.

The difference between original predictions and eventual reality is notably true for water resource planning. Indeed, populations migrate, land uses shift from agricultural to urban, and technology changes so that patterns of water consumption shift. Moreover, important primary costs change (such as those associated with energy) and new technologies arise, shifting the relative attractiveness of alternative designs. Political, legal and climatological disturbances also affect the availability of water to specific regions. In short, the reality is that we cannot accurately predict the future of the world for which we design.

The reality of change and uncertainty is impelling water resource planners to focus increasingly on designs that are flexible, that enable smooth adaptation to actual needs and circumstances. Chapter 6 in this book presents the case in detail (Wong-Turlington et al. 2015). Flexible designs differ from conventionally developed plans in two ways: (a) to the extent possible they defer commitments until the evolution of demand and supply demonstrate needs; and (b) they embed the capability for easy transition to future needs into the initial phases of the project. That is, flexible design creates Real Options in the projects that system managers can exercise when and if needed. Flexible designs thus enable the system to avoid downside risks and take advantage of upside opportunities.

The shift toward Flexible Design implies a substantial change in the way public authorities or clients commission and pay for the facilities that will serve them. Water authorities and contractors increasingly recognize that they are involved in a multi-stage process that both parties have to think through carefully—instead of having a single-shot relationship around a single project. The logic of flexible design implies an implicit long-term continuing relationship between the authorities who commission a project and the contractors who deliver it. This is because a flexible design means that any initial project is likely to be a first step in a system delivered over time; the initial project is likely to change capacity or capability in response to the reality of future needs and conditions. Although those who commission the project can change contractors over time, and the contractors themselves may drop out, the default arrangement is a long-term relationship.

The reality of change and uncertainty in needs and the environment raises an important issue: How can we most effectively arrange the long-term dynamics between the public authorities and their contractors? What kind of contract is most suitable? The outcomes of any development are uncertain and, at best, we can only know them as a distribution of eventual benefits and costs. This raises crucial questions: How shall authorities and contractors share these risks? What kind of contract should govern the relationship between those who commission the infrastructure and their contractors? Here are the essential issues:

- The contract defines the rules of the game that the contractors must follow;
- The contractors will "game" the system to maximize their benefits, minimize their risks;
- The design best for the contractors is not likely to maximize the expected value of the system as a whole.

Therefore, Flexible Design calls for a careful consideration of the ways that contracts promote or impede the delivery of infrastructure that maximizes value.

This chapter explores the connection between contractual arrangements and design. It presents an analysis of the impact of contract arrangements and risk preferences on the adoption of flexibility in public–private partnerships. It uses a case study of public authority contracting with private designers/operators of desalination facilities in Qatar to demonstrate the process of analysis and illustrate how authorities might think about and choose between their alternative arrangements.

The relationship between the public authority and its contractor is important to consider, as many large water projects involve such pairs of interests. Moreover, understanding this relationship and the impact of the choice of contract on the ability and willingness of both parties to implement a flexible design offers insights for water infrastructure projects with other types of multi-stakeholder relationships. For example, to implement a project in a transboundary watershed, such as a reservoir on a shared river, all stakeholders with jurisdiction or ownership over the resource need to reach an agreement allowing a project to proceed. They must also then maintain this relationship to allow for future changes. Similar to the contract, the structure of the agreement affects the distribution of risks and benefits across stakeholders and, therefore, the attractiveness of flexible options to each party. Thus, the case study offers insights for multi-stakeholder water projects and helps inform future investigation into the impact of agreement choice on the incentives for Flexible Design.

Before proceeding to the case study, the presentation begins with a brief discussion of public–private partnerships, which are typical in the desalination industry. After a brief presentation of the situation in Qatar, a flexibility analysis of desalination expansion documents the long-term value of alternative flexible strategies, both over a conventional inflexible design, and relative to each other. A waterfall analysis then determines how the water authority and the contractor would share the benefits of flexibility according to several contracts. Finally, an analysis of these interactions indicates how contractors would respond to contracts and what this means for the overall value of the system. The case thus illustrates the way different contracts can shape the design and affect the overall expected performance of the system. The concluding discussion stresses the importance of considering the effect of contract choice on the overall system performance.

Public–Private Partnerships and Their Risks

Public–private partnerships (PPP) is a generic name for long-term contractual relationships between public authorities that contract with private companies for the provision of a service that some infrastructure enables. The PPP contract specifies how the public and private entities will split the project risks and benefits.

PPP contracts come in many forms. This case study of desalination looks at the Build-Own-Operate-Transfer (BOOT) contracts, which are increasingly common in the industry. For a more extensive discussion of this and other forms, see AlMisnad (2014) and World Bank (2013).

In a BOOT contract, the role of the private firm is to finance, design, build and operate the infrastructure. The public agency acts as a client who pays for the provision of a service, in this case water. The agency compensates the private company on the basis of performance, not of cost. It buys water directly and may pay for the availability of the capacity to produce water (Weber and Alfen 2010).

The water purchase agreement is a key element of a BOOT contract. It specifies how the water authority will pay for water production or sales—that is, how the public and private stakeholders will share the risks associated with uncertainty in the future demand for water. To cover the construction cost of the plant, the contract typically involves a capacity payment known as a take-or-pay clause. In general, it guarantees that the public agency will pay the private firm some minimum amount (Voutchkov 2012; Wolfs and Woodroffe 2002).

Take-or-pay clauses are features of BOOT contracts for desalination projects worldwide, in both developing and developed nations. Some examples are:

- The San Diego County Water Authority agreed to purchase a minimum of 48,000 acre-feet per year (~170,000 m³/day) of water at a set price from the desalination plant in Carlsbad, California (Poseidon Water 2014).
- The Singapore Tuas Desalination Plant has a take-or-pay clause (Chowdhury 2011).
- Abu Dhabi's Al Taweelah water and power plant is under a BOOT contract that has both capacity and variable payment elements (Wade 1999).

There are several ways to structure take-or-pay clauses for flexible plants. These base client payments on different notions of capacity. The most obvious approach considers installed capacity. But this limits the ability of private firms to benefit from flexibility. It might thus be better for the water authority to base payment on the ultimate capacity

Table 12.1 Summary of BOOT contract provisions.

Approach	Payment to Private Firm Based on
Installed capacity	Difference between installed capacity and actual water demand, in each period
Fixed Capacity	Ultimate capacity
Capacity savings	Water sales, installed capacity, and difference between installed and ultimate capacity. This variation encourages contractors to reduce investment costs

of the plant. Alternatively, it might want to pay the private firm based on capacity saved (see Table 12.1). The case study examines the implications of each of these possibilities.

Water infrastructure projects face many risks, including water shortage, expansion and renegotiation risks. Water shortages are perhaps the biggest risk associated with water supply, particularly for public agencies. Ensuring the availability of water for the general public is the main purpose behind the construction of desalination plants. Changes in either supply or demand can lead to shortages. Expansion risks refer to the uncertainty regarding the price and duration of expanding the capacity of the desalination plants in order to meet increases in water demand. There are two kinds of capacity expansions: planned capacity expansions according to the flexible design rule and unplanned ones that occur due to unexpected short-term mismatches between available capacity and water demand. Under both of these expansions, there is a risk that both financing and construction may not be readily available, leading either to an increase in expansion costs or in duration. Finally, there is the risk of contract renegotiation after a project is in operation. Renegotiations typically occur due to changes in the socioeconomic conditions and are quite common in water infrastructure projects (Guasch et al. 2007; Cruz and Marques 2013). Guasch (2004) found that roughly 80 percent of the 160 projects they surveyed involved renegotiations, and the common causes include lack of perceived fairness and low profits. In this context, flexible contracts that can adapt to changing conditions reduce risk.

Overall, Flexible Design decreases the risks and increases the benefits by enabling adaptation to minimize downside exposure and leverage upside opportunities. In doing so, flexible designs shift risks and benefits between the public agency and the private company. This affects the willingness of participants to adopt flexible designs.

Case Study: Desalination in Qatar

The case study considers desalination projects in Qatar. This is triply appropriate because:

(1)   Desalination plants are highly modular and thus eminently suitable for flexible designs that would deploy capacity and capability over time;

(2)   Qatar depends almost exclusively on desalination for its water supply; and

(3)   Contractors typically operate desalination plants, and thus inherently have a close relationship with the water resources agencies.

Background

Qatar is a water-scarce nation. It is located on the Arabian Peninsula, which is predominately arid desert. Qatar has no permanent surface water bodies, a low annual average rainfall of 80 mm, and limited fresh groundwater resources. Yet it is rapidly developing, and has a high demand for water that greatly exceeds its natural renewable supply (Darwish and Mohtar 2007). Desalination from the Persian Gulf entirely supplies Qatar's municipal water demand, which accounts for almost 40 percent of its water use (Aquastat 2008).

Desalination processes use energy to remove salt and other impurities from water. They are well suited to regions that are both water poor and energy rich. Much of the world's desalination infrastructure is thus located in the six nations that comprise the Gulf Cooperation Council (GCC) on the Persian Gulf side of the Arabian Peninsula.

The Multi-Stage Flash (MSF) thermal process has dominated desalination capacity in the GCC region. As of 2013, roughly 85 percent of all global MSF capacity was located in the GCC (Global Water Intelligence 2013a). This technology is robust and reliable. However, it is energy-intensive and thus has become increasingly expensive in terms of cost per unit of water produced (Al-Rashed and Sherif 2000; Dawoud 2005; 2030 Water Resources Group 2009).

Advances in membrane technology since 2000 have led to a trend-breaking boom in reverse osmosis (RO) desalination. As of 2013, RO made up roughly three quarters of the capacity of all desalination plants and 60 percent of seawater desalination plants (Global Water Intelligence 2013b).

RO designs are generally modular. This enables easy capacity expansion and thus promotes flexibility. So long as the design has made space available and sufficiently prepared the site, operators can simply add modular units to an existing plant to expand capacity. Modular RO plants usually consist of self-contained process units of up to 20,000 m^3/day, that take advantage of economies of scale associated with the ready availability of off-the-shelf pumps and heat exchangers (Sakhrani 2015). Designers typically arrange these units in trains up to a capacity of 200,000 m^3/day. To date, this size has been the upper end for a single RO plant, as larger units result in operational complexities (see Figure 12.1). To achieve greater capacity, designers construct new plants. It is common practice to design large plants with intakes that can accommodate future expansion, sometimes by as much as 100 percent (Voutchkov 2012). In sum, RO desalination is well suited for flexible design.

Desalination projects, in Qatar and elsewhere, are typically delivered as PPP. Public agencies dominate the overall municipal water sector. In 2007, they made up over 90 percent of the total market in terms of the number of people served (Marin 2009). Because desalination plants largely provide municipal water, the public sector typically is responsible for awarding projects. The major suppliers of large-scale saltwater RO plants include Veolia and Degrémont (France), Acciona (Spain), IDE (Israel), and Hyflux (Singapore). These large

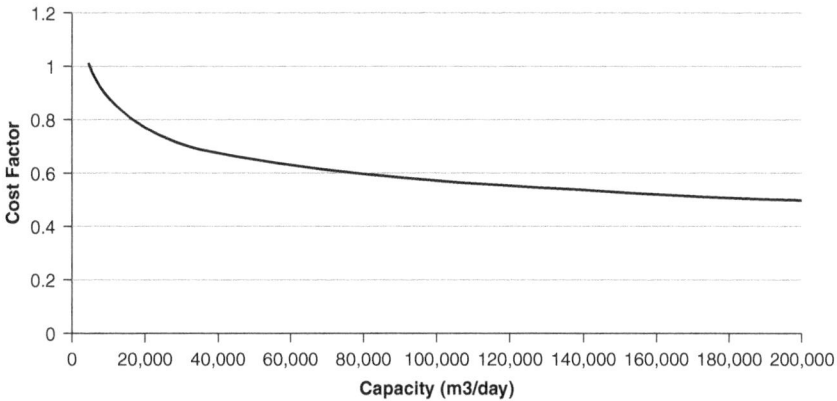

Figure 12.1 Economies of scale for RO desalination plants (Voutchkov 2010).

firms are representative of the typical private partner in a RO desalination PPP. In terms of worldwide capacity in this market, these companies account for 38 percent of the total and 50 percent of the large (>50,000 m³/day) plants. These conglomerates offer complete technical packages that include engineering design, construction and operations.

Uncertainties

Many uncertainties influence the performance of desalination plants. The major uncertainties concern population growth, rates of consumption, energy costs and trend-breaking technological advances. For Qatar, the greatest uncertainty concerns the underlying demand for municipal water, which depends on the size and need of the population. Qatar's population and its pattern of growth have fluctuated tremendously. Its annual growth rate varied between 2 percent and 18 percent from 2000 to 2010 (see Figure 12.2). Much of this instability is associated with changes in the number of temporary migrant construction workers, who have constituted roughly a quarter of the total population (Qatar Information Exchange 2014). Although Qatar tightly controls its borders and regulates work visas, and theoretically should be able to manage its population growth, it has not done so. The variability of Qatar's population is both large and highly unpredictable.

The uncertainty surrounding Qatar's population and its demand for water is likely to continue. Changes in the national GDP drive investments in national infrastructure that in turn impel increases in migrant population. In the decade after 2000, Qatar's GDP ballooned because of the increase in energy prices and the expansion of its liquid natural gas (LNG) sector (Ibrahim and Harrigan 2012). As of 2015 Qatar's future is particularly uncertain. It committed to much construction to prepare for hosting the FIFA 2022 World Cup, which could draw enough spectators to increase total population by 10 percent. However, the collapse of world energy prices must affect Qatari GDP. This continuing unpredictability of GDP and population growth implies large uncertainties in Qatar's future demand for drinking water.

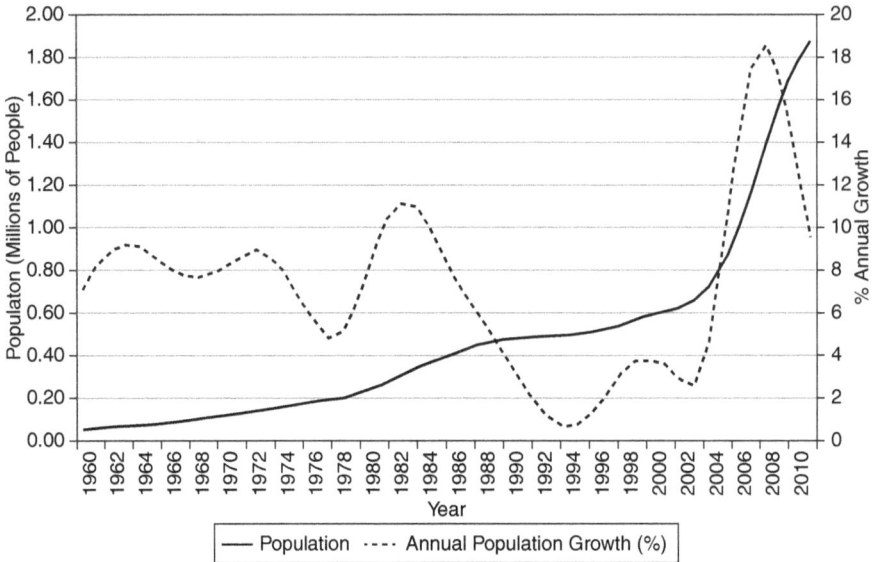

Figure 12.2 Historical data on population growth in Qatar (World Bank 2014).

In this context, Qatar has struggled to create enough desalination capacity to keep up with demand (Qatar General Secretariat for Development 2011). Since 2000 it has had to fast-track the construction of desalination projects in order to meet unexpected increases in demand. Between 2000 and 2013, at least five major desalination projects came online (see Figure 12.3).

Flexibility Analysis

The flexibility analysis calculates the value of alternative designs in the context of uncertainty. Based on the best available data, it simulates the performance of the design over the range of possible scenarios. Chapter 6 in this book describes the approach in detail (Wong-Turlington et al. 2015).

Model Details

A simulation model of the system is the heart of the analysis. In this case it concerns the ability of desalination plants to meet future water demands. It performs a discounted cash-flow analysis to determine the net present value (NPV) of the desalination facilities for any specified production of potable water under the prevailing conditions. Monte Carlo simulation samples the growth rate to determine population and subsequently the demand for each year. Existing supplies are first used to meet demand. The new desalination plant is then called upon to supply the remaining demand. The model computes costs based on plant capacity and the quantity of water produced, and revenues according to the demand served and the contract in place. Under Flexible Design alternatives, decision rules built into the model trigger plant expansion. AlMisnad (2014) provides full details on the Excel® spreadsheet model and its parameters.

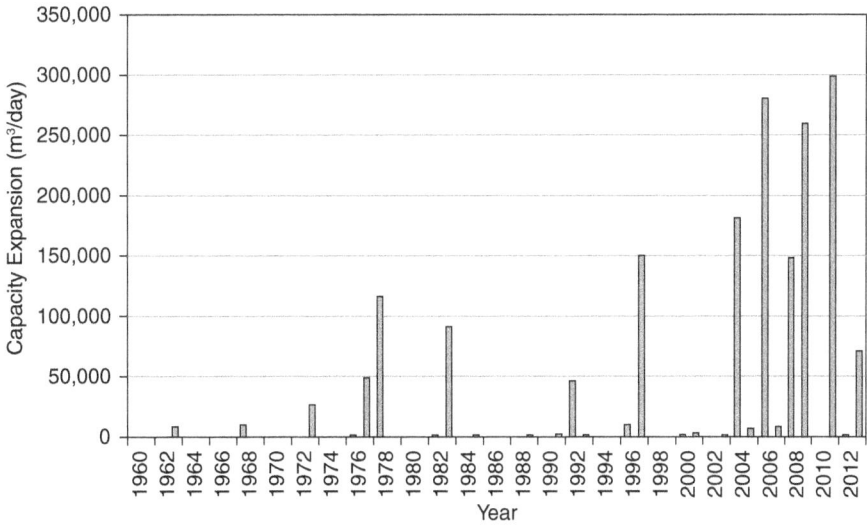

Figure 12.3 Timing of desalination projects in Qatar (Global Water Intelligence 2014).

The case study examined designs for a stand-alone, reverse osmosis plant incorporating dedicated seawater intake, pretreatment, and a disposal system. The facility ties into the existing water-distribution system and electric grid and therefore does not require any additional distribution pipelines or electricity-generation capacity. Given fast-track practices in Qatar, the analysis assumes that plant construction takes 12 months, which is slightly faster than the average plant construction time of 12 to 24 months (Voutchkov 2012). According to the Qatari water authority, Kahramaa (MEED Insight 2012), the plant site allows for an ultimate desalination capacity of 400,000 m³/day.

The analysis used cost information for a 200,000 m³/day RO plant in Qatar (Global Water Intelligence 2014; Voutchkov 2012). It assessed capital costs at US$430 million for the entire 400,000 m³/day capacity plant. Operating costs have fixed and variable components. Fixed costs include labor, maintenance, environmental monitoring and other overhead costs. These depend on installed capacity regardless of utilization. Variable costs depend on the volume of water produced. These include the cost of chemicals, membranes, brine disposal and—most importantly—of energy. The model assumed energy costs at an average RO plant efficiency of 3.1 kw/m³ of water produced, and an average energy tariff of US$0.06 /kwh.

The analysis evaluated alternative designs over 20 years. It used real values for costs, and thus did not factor in inflation. In accordance with local practice, it fixed the discount rate at 10 percent and assumed a zero-percent tax rate.

Inflexible Design Base Case

The base case design follows the conventional practice of building a single plant to meet the forecast level of demand over the planning period—400,000 m³/day in this case. It is inflexible in that it leaves no room for change in capacity. The model is run without Monte Carlo sampling and instead uses the expected value of population in each year. To take advantage of economies of scale as discussed by Voutchkov (2012), the plan is to build two identical trains of 200,000 m³/day each before the first year of operation. Based on predicted demand, the estimated the NPV of this design was a cost US$910 million, split between capital costs (US$433 million) and operating costs (US$477 million).

Uncertainty Consideration

Since we know that the future will vary from any single forecast, a realistic analysis must consider uncertainty. This analysis focused on variability in demand, which depends on the uncertainty in total population. It reflects the fact that the population of Qatar has two distinct components: native Qataris and migratory workers. The analysis modeled the uncertain growth of these populations differently. It assumed that annual growth in the Qatari population (about 750,000) followed a normal distribution centered on 2 percent. It modeled changes in the migratory population (about 900,000 construction workers and other expats) using a uniform distribution of annual growth rates ranging from -3 to 12 percent, which is appropriate because of the tight relationship with the number of construction projects and the assumption that these can start or end in any year.

The flexibility analysis created scenarios of possible levels of water demand over the 20 years of the evaluation. It did this by first simulating the total population of Qatar by year, as the sum of the two possible levels of population of Qataris and non-Qatari workers, and then multiplying the population by the per capita water use, which is assumed constant. Using Monte Carlo simulation (facilitated here by the @Risk® add-in to Excel®) the simulation can generate thousands of scenarios of the independent growth rates of the component populations and thus thousands of future water demand projections.

Figure 12.4 contrasts the deterministic forecast of water demand with some of the simulations of the uncertain demand. As shown, the uncertain scenarios can be completely under or over the deterministic projection, indicated by the dashed line. Moreover, the uncertain demand could exceed the capacity of the plant, indicated by the dotted line. Conversely, it could also happen that the demand never requires the full capacity of the plant—as has occurred throughout Australia (Ferguson 2014).

Uncertainty in demand, of course, leads to uncertainty in the estimate of operating costs. Realistically, these are not a fixed amount but a distribution. In this case, 95 percent of the distribution ranged from US$398 to 553 million (a difference of US$155 million). In short, the appraisal shows with uncertainty the potential for enormous consequences for the client and the supplier. While supply-side uncertainties dominate some systems, the reliance on desalination in Qatar makes the focus on demand and cost uncertainties appropriate.

Flexible Designs

Flexible designs start with an initial configuration and delay some significant commitments of capacity or capability until the need for them seems assured. They thus reduce

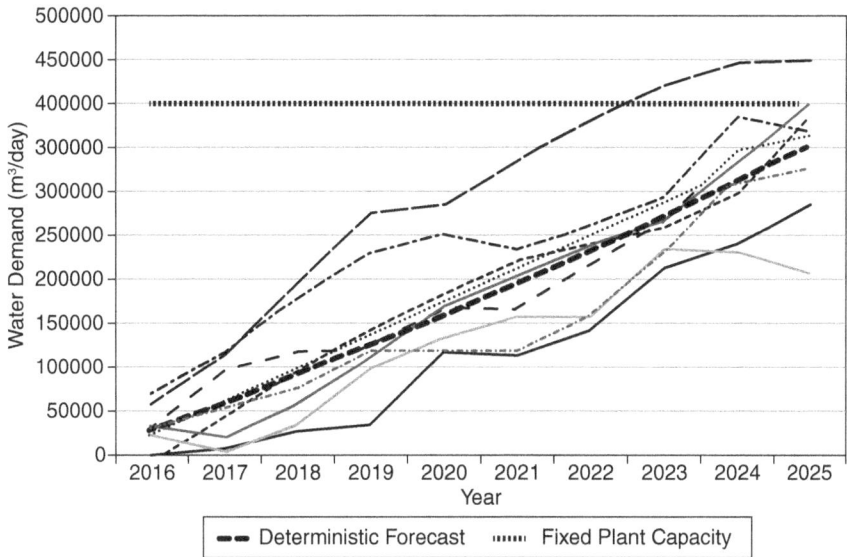

Figure 12.4 Some possible evolutions of water demand.

the present-value cost of these later commitments (or even eliminate them, if the need does not materialize). Flexible designs can also increase the present-value projects by taking advantage of new opportunities—for example by investing in more efficient new technologies previously unavailable. Overall, flexible designs offer the possibility of triple win-win solutions: they lower future costs and increase possible benefits, while reducing initial investments. They provide more value on less capital, that is, higher rates of return (De Neufville and Scholtes 2011).

This case study examined four possible flexible designs. Each describes a different way of expanding capacity relative to unfolding water demand. Given that in this case the major source of uncertainty is water demand, the appropriate flexibility concerns when and how the providers will expand capacity. Specifically, these designs differ in three dimensions: the initial installed desalination capacity, the size of the expansion tranche, and the number of years ahead considered when deciding to expand capacity. These designs do not represent the total menu of potential designs. They illustrate a spectrum of increasingly flexible designs and thereby indicate how greater flexibility influences project value. The four designs are (see Table 12.2):

(1) **Phased**: This design reflects a relatively conservative approach to flexibility. It builds half of the ultimate plant capacity (200,000 m³/d) in year one, and the second half when and if the need arises based on a two-year forecast of water demand.
(2) **Flexible I**: This design initially installs 160,000 m³/day capacity with adequate pre-investments to support expansion up to one desalination train of 200,000 m³/day capacity. Expansions occur in tranches of 80,000 m³/day based on a two-year forecast.

Table 12.2 Summary of design cases, from least to most flexible.

Design	Initial Capacity (m³/day)	Expansion Tranche (m³/day)	Forecast Year
Inflexible (base case)	400,000 (100%)	None	None
Phased	200,000 (50%)	200,000 (50%)	2
Flexible I	160,000 (40%)	80,000 (20%)	2
Flexible II	160,000 (40%)	Multiples of 20,000	2
Flexible III	80,000 (20%)	Multiples of 20,000	1

(3) **Flexible II**: This is similar to Flexible I except that expansion occurs in tranches of 20,000 m³/day as required to closely match a two-year forecast of water demand.

(4) **Flexible III:** This design starts with a small plant of 80,000 m³/day with adequate pre-investments to support expansions of 20,000 m³/day. Expansion occurs to match one-year forecasts. It is the most flexible expansion approach considered in this analysis.

The analysis of the flexible designs augmented the deterministic analysis by incorporating the means to (a) identify when a demand scenario warranted expansion, and (b) add this cost and capacity of the system. Using the same Excel® spreadsheet augmented by @Risk® to perform the simulation analysis, the model incorporated IF statements that test available capacity against projected demand in each year. When these tests indicate that capacity would be insufficient, they trigger the implementation of the additional capacity at the appropriate time and cost. In detail, the analysis assumes that all expansions can occur within four months, in line with the discussions of potential expansion timelines (Darwish and Najem 2005). The De Neufville and Scholtes (2011) text describes the procedure in further detail.

Figure 12.5 shows how, increasingly, flexible designs improve the value of the project. More flexible designs reduce the present costs of both the capital investments and total costs. The effects are significant: the most flexible design reduces expected capital costs by over US$100 million. Of course, there is a range of uncertainty about the amount of the improvement, since different scenarios lead to different outcomes, as the ranges on the data in Figure 12.5 indicate. Figure 12.6 makes the advantage of flexible designs explicit. It shows that, in the case of Qatar, flexible designs could decrease the expected value of total costs anywhere from 12 to 22 percent.

This analysis shows that flexible strategies can significantly improve the long-term expected value of a project when compared to a conventional, inflexible, design. This is another specific instance of the general phenomenon (De Neufville and Scholtes 2011; Wong-Turlington 2015). Viewed overall, flexible designs are desirable.

The question is: Can the negotiation between partners in the PPP process lead to the implementation of the most beneficial flexible designs? Pragmatically, we must consider

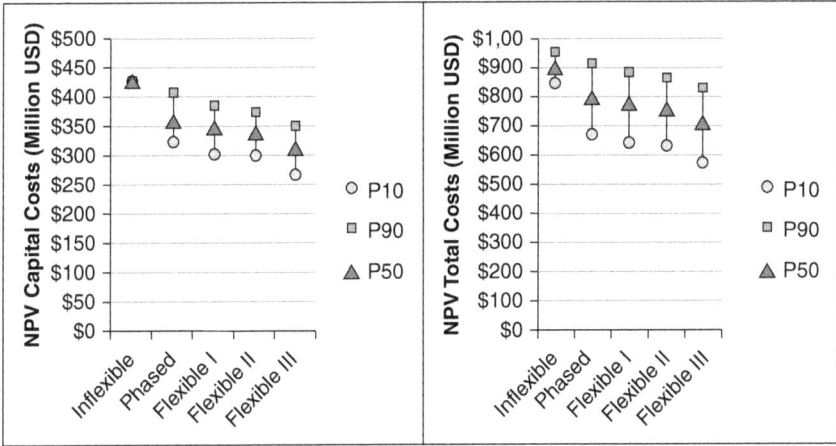

Figure 12.5 Present value of capital and total costs of alternative designs.

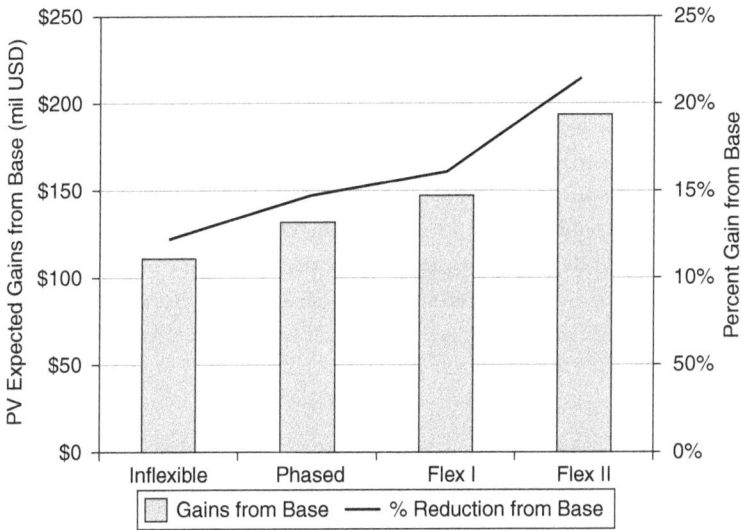

Figure 12.6 Project gains associated with increasingly flexible designs.

how the PPP contract distributes the benefits of flexible design. If stakeholders do not sufficiently share these benefits, they may block designs that are good overall.

Waterfall Analysis and Results

A waterfall analysis determines how a contract distributes the value of a project to its stakeholders. In general, some stakeholders have priority and get paid before others. The metaphor reflects the idea that money cascades down to several recipients. This analysis is useful in determining how stakeholders will respond to various contracts.

The waterfall analysis calculated how each type of contract listed in Table 12.1 distributes the total gains to the desalination project from the implementation of flexibility to the project participants. The public agency gains value by reducing its sum of payments to the private firm. The private firm gains value by increasing its NPV. The aim of this analysis is to show the financial incentives for each participant to pursue flexibility under each contract scenario, pointing to the fact that the choice of contract can shape the adoption of flexibility.

The *Installed Capacity* contract discourages private firms from flexible designs. Its payment structure channels most of the flexibility gains towards the public sector. Conversely, greater flexibility more severely reduces both the revenues and the expected internal rate of return (IRR) of private firms, as Figure 12.7 shows. This contract thus discourages private firms from offering flexible designs.

Conversely, the *Fixed Capacity* contract channels most of the benefits of flexible designs to contractors. It thus encourages them to maximize design flexibility, as Figure 12.8 indicates. The greater the flexibility, the more a contractor would earn and the higher the contractor's NPV. However, this contract gives the public agency little incentive to encourage flexible design.

The *Capacity Savings* contract enables both the private firm and the public agency to benefit from the capacity savings due to flexible designs. The private firm gets paid both for capacity installed *and* for capacity saved compared to the fixed capacity. It is a hybrid between the two previous approaches. As Figure 12.9 indicates, the arrangement allows both the private and public sectors to share the increasing project gains from flexibility. The details of the split depend, of course, on the details of the contract.

Table 12.3 summarizes how each contract type shapes the preference for the participant towards the adoption of flexible designs that inherently improve the overall value of a project.

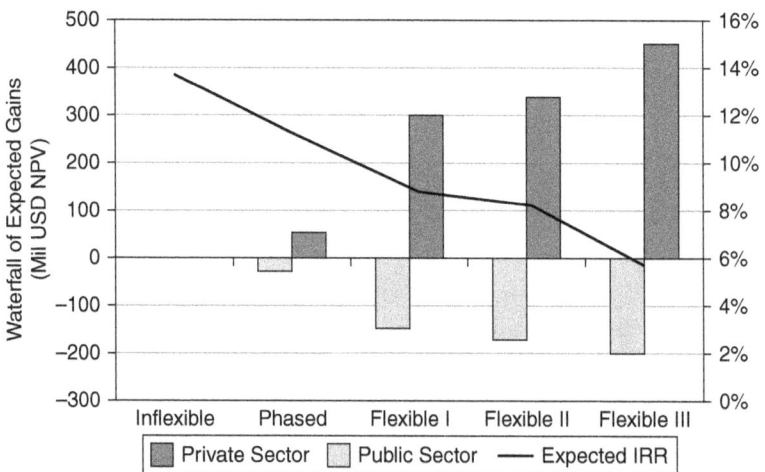

Figure 12.7 Waterfall for an Installed Capacity contract.

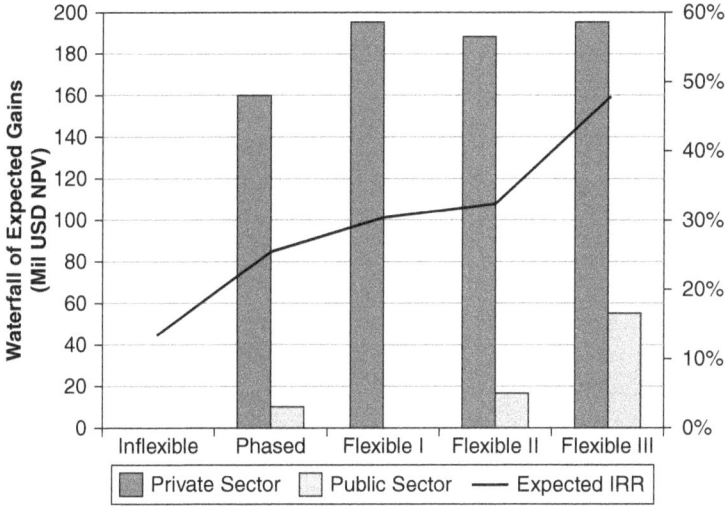

Figure 12.8 Waterfall for a Fixed Capacity contract.

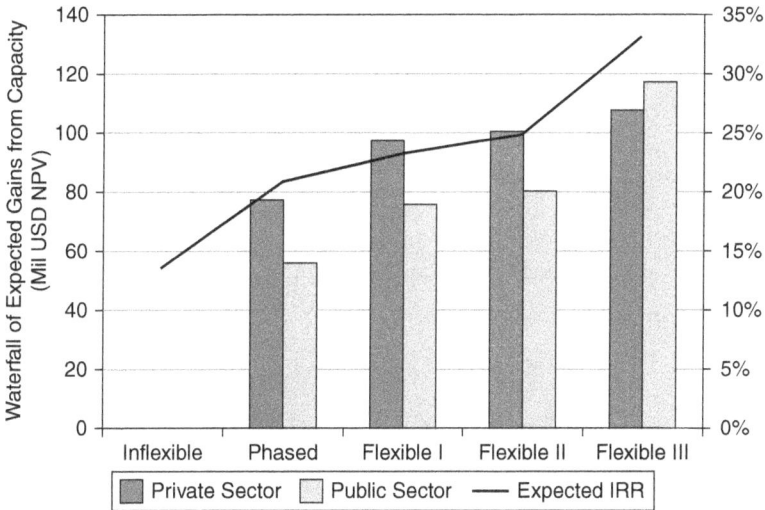

Figure 12.9 Waterfall for a Capacity Savings contract.

It demonstrates the potential influence of contract arrangements on whether the public and private stakeholders with will be able to agree to achieve the benefits of flexible design.

Risk Analysis and Results for Risk-Neutral Participants

Building on the discussion so far, this section analyzes the nature of the interactions between the public agencies and the private contractors, insofar as these determine the

Table 12.3 Summary of contract incentives to adopt flexible design.

Provisions	Public	Private
Installed Capacity	Benefits	Incentive Absent
Fixed Capacity	Incentive Absent	Benefits
Capacity savings	Shared benefits	

extent to which they will agree to adopt flexible designs that increase the overall project value. The question is: To what extent will they jointly be able to agree on designs that maximize project value?

The analysis traces out the consequence for the adoption of design flexibility of the two-step interaction or game between the public agency and the contracting company. In accordance with common practice, the analysis assumes that the public authority proposes a specific contract for a project and requests proposals from private suppliers. The participants make two key decisions during the bidding process: first the public agency chooses the terms of the contractual arrangements that it issues as part of the tender. Then, the private firm responds by proposing a design that presumably best meets their interests in the circumstances. The level of flexibility that ultimately results from this interaction thus depends on the contract-design pair that results in the greatest gains to the public agency.

The analysis considers two contract scenarios: Installed Capacity and Capacity Savings. The Fixed Capacity contract is not considered because the public agency ultimately selects the contracts and the Fixed Capacity contract does little to increase the value of the project (see Table 12.3 and Figure 12.8).

As a first pass, this analysis assumes that both the public agency and private company are *risk-neutral*. This presumes that the public agency is:

(1) Comfortable with potential shortfalls from the desalination facility, because it can use its water-storage capacity or other sources to compensate for any water shortages, and is

(2) Not concerned with any fairness issues associated with the private firm making high returns.

Complementarily, this assumes that the private firm is a diversified corporation that can manage the expansion risks associated with the project.

In this situation, what happens if the public agency offers an Installed Capacity contract? This arrangement might lead to its highest possible gains for the public agency (about US$350 million) if it obtained the most flexible design, as Figure 12.7 indicates. Under this contract, however, the private firm maximizes its value with an inflexible design and would respond accordingly, as the gray rectangle in Figure 12.10 indicates. The overall result is that there would be no flexibility, no overall improvement to the value of the project and no gains to either participant.

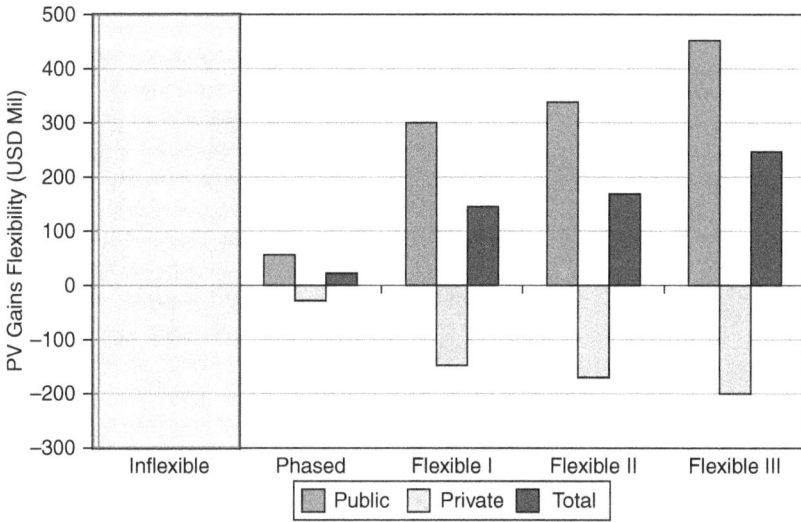

Figure 12.10 Interaction outcomes from an Installed Capacity contract.

What happens if the public agency offers a Capacity Savings contract? The public agency could potentially gain much less than under the alternative contract (only US$120 million) if the contractor proposes the most flexible design. However, this contract incentivizes the private firm to propose designs with significant flexibility, as the gray rectangle in Figure 12.11 shows. The result in this case is that the project is much more valuable, and both the public agency and the private company can expect significant gains.

The conclusion is that although the public agency might have expected to maximize its value by pursuing the Installed Capacity contract, it actually would realize the greatest gains by offering a Capacity Savings contract. This results from the interaction between the preferences of the two participants. As in much of life, to get something one has to give something and share!

Risk Analysis and Results for Risk-Averse Participants

Many public agencies and private companies are not risk-neutral; they are sensitive to risk. Most often they are risk-averse, seeking to pay a premium to avoid downside risks. (Some potential participants may be risk-positive, acting as gamblers, as it were.) How might such preferences affect the choice of design?

First consider the situation in which the *public agency is risk-averse, and the private company is risk-neutral*. For example, the water authority may be concerned about water shortages because of a limited confidence in national water storage. It may thus wish to avoid capacity-expansion rules that are too flexible and might risk water shortages. It may thus modify the contract to impose penalties in case the contractor fails to provide sufficient

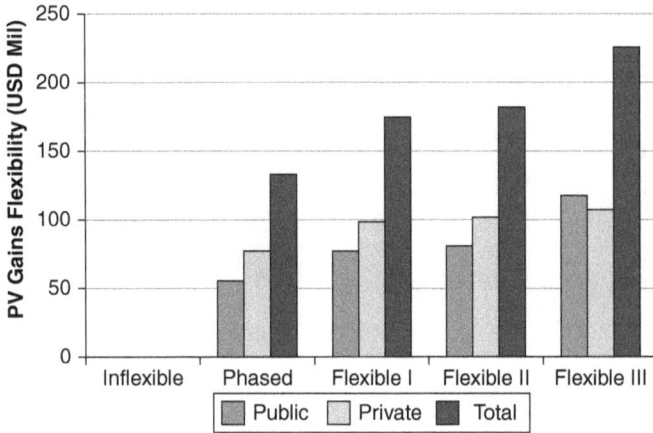

Figure 12.11 Interaction outcomes under a Capacity Savings contract.

capacity to avoid water shortages that might occur because of unexpected surges in demand.

The Capacity Savings contract still leads to the better overall results. However, the result is somewhat different. The potential imposition of fines incentivizes the private firm to be more cautious, to create greater capacity earlier than it would under the most flexible design, as the gray box in Figure 12.12 indicates. The overall result is smaller, but still significant, gains to the project and to both the public agency and the private company.

Finally, consider the situation in which both project participants are risk-averse. This case assumes that while the private firm can manage the expansion risks associated with flexibility, it will only do so if it feels that the returns are high enough to justify these risks. These conditions may lead it to pursue a less flexible design. The gray box in Figure 12.13 suggests the possible result. For example, the company might respond to the tender by proposing a phased design. Such a minimal amount of flexibility would still lead to gains over an inflexible, single-step design and would lead to mutual gains for both the public agency and the private company.

Reflections on Case Study

The case study of desalination for Qatar provides a basis for reflections about the:

(1) Value of flexibility in design;
(2) Influence of contracts on the design process, including the extent of flexibility incorporated; and the
(3) Importance of considering uncertainties in negotiating contracts in general.

The analysis first provides another example of how flexible designs could, given the inevitable uncertainty surrounding future needs, significantly increase the overall value of a

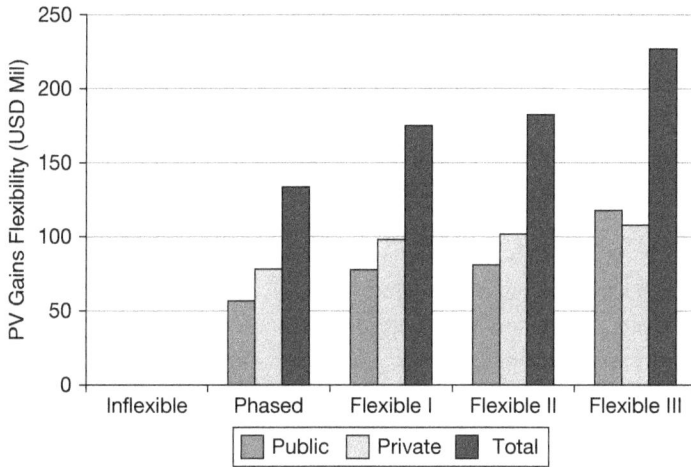

Figure 12.12 Interaction outcomes under a Capacity Savings contract with penalties.

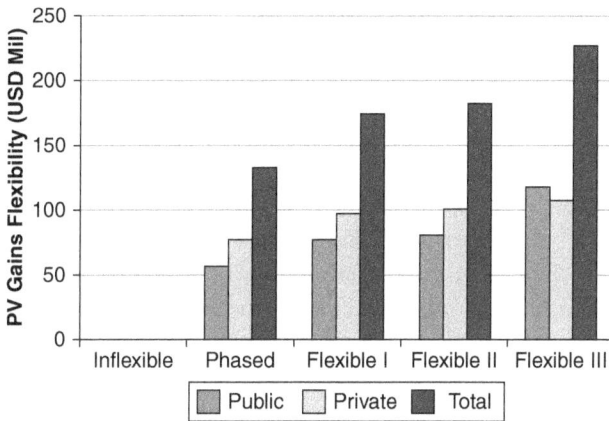

Figure 12.13 Interaction outcomes under a Capacity Savings with penalties for risk-averse participants.

water project. There is clear advantage to embedding the capability to adjust the project easily, through changes in capacity or more generally of function, to adjust to needs and contextual conditions as they evolve.

The case then demonstrates, by example, the problem of getting the partners in the design process to set up and implement a flexible design. A commitment to flexibility implies an ongoing commitment to manage the project to provide appropriate capacity and capability over time in the face of uncertainty. Both the providers of the service and the clients face considerable risk that they inevitably manage through their designs and choice of contractual arrangements. The case then demonstrates that the choice of contractual arrangements can result in significant differences in the level of flexibility and, thus, of resulting project value, which is ultimately achieved. This means that we need to recognize the need to consider contractual and design arrangements simultaneously.

The analysis also suggests a way of understanding some pervasive, persistent problems in the PPP procurement of water resources (and other) projects. Uncertainty has a strong effect on the distribution of benefits and costs to the stakeholders. Contractual arrangements both shape and are shaped by this distribution. When developing a contract, public agencies select an arrangement that increases gains and decreases risks for the agency. Private contractors then respond to the contract by proposing a design that maximizes their value. Stakeholders risk-preferences also shape how they respond to this combination of uncertainty and contract conditions. Risk-neutral stakeholders maximize their expected value while risk-averse stakeholders balance maximizing expected value with minimizing downside risks. By highlighting these effects, the analysis illustrates the challenge of encouraging designs that maximize overall benefits. Indeed, contracts between public water authorities and private contractors are notoriously problematic as evidenced by their frequent renegotiation (Guasch et al. 2007; Cruz and Marques 2013). Insufficient consideration of uncertainty is one driver of these renegotiations. This analysis highlights the need for flexible contracts that can deal with inevitable surprises and that fairly share gains and cover expenses, not just for expected outcomes, but across a distribution of plausible conditions (Gil 2007).

Conclusions

The adoption of Flexible Design can add considerable value to infrastructure projects. The ability and willingness of project participants to implement Flexible Design, however, depends on the specific contractual arrangements. Through a case study on desalination in Qatar, this analysis shows that different contracts result in different distributions of the gains and the risks from the adoption of flexibility between the project participants. Furthermore, the participants' degrees of risk neutrality or risk aversion influence their willingness to take on the risks of a flexible design. While contractual arrangements only distribute the gains from flexibility between the project participants without adding value because of the ways in which they can lead to decisions to adopt different levels of flexibility, the choice of contract structure can influence the ultimate gains to the project from flexibility.

This analysis does not identify an optimal contract structure that allows project participants to maximize the value of their projects. It points to the fact that given their role in the adoption of flexibility and the associated gains to the project, project participants should evaluate the effects that different contractual arrangements might have on their decisions to adopt flexibility in the particular context of their project. To this point, Sakhrani (2015) used an experimental exploration of the design of desalination plants to demonstrate the value of negotiated collaborative design as a way to bridge some of these issues. Given the need to balance the incentives and rewards, the more desirable contracts may not have simple formulas. Instead, they will likely be hybrids that combine several beneficial features.

Various contextual considerations can affect the applicability of the results beyond the specific context of a small, rich developing nation located in the Arabian Peninsula. An important consideration is the nature of the uncertainty that a particular project faces. For desalination projects, we can expect flexibility to add value to the overall project so long as the uncertainties surrounding the demand for desalinated water are high. This uncertainty does not need to arise from changes in population, however, and can be the result of uncertainties associated with changes to construction and operating costs, other water supplies, environmental conditions and technology. Flexibility has the potential to add value in the face of all these uncertainties.

A second important consideration is the ability of the project participants to manage the uncertainties. For example, the availability of the necessary infrastructure, both physical and institutional, to support flexibility is particularly important. The use of flexibility allows project proponents to modify capacity or function in response to changing input costs, water demand, technological developments and changing environmental conditions. To do so, however, the appropriate institutional framework that allows for the ability to monitor changing conditions and feedback information into the decision-making process needs to be in place. While private firms may be in a better position to monitor some of these uncertainties, public agencies also play a role. The ability of governments to implement institutional reforms, including the implementation of water-demand reduction measures, has the potential to increase the gains from flexibility already considered for this case.

These considerations are especially true for infrastructure projects involving multiple stakeholders, such as projects on a shared river. While the Qatar case study demonstrates the value of explicit consideration of how PPP contracts distribute the risks and benefits of flexibility, the application of these ideas to other types of multi-stakeholder agreements introduces new challenges. In cases where multiple states or countries share the gains and risks of a project, the analysis must also consider the impact of the potentially changing institutional arrangements on the ability of various parties to follow through on their part of the agreement, as well as the different risk preferences of each party. However, understanding the impact of contract choice on the willingness of public and private parties to pursue flexible designs can inform future work on the impact of agreement structure with other types of multi-stakeholder relationships.

References

2030 Water Resources Group. 2009. *Charting Our Water Future: Economic Frameworks to Inform Decision Making*. Accessed February 22, 2016. http://www.mckinsey.com/client_service/sustainability/latest_thinking/charting_our_water_future

Al-Rashed, Muhammad and Mohsen M. Sherif. 2000. "Water Resources in the GCC Countries: An Overview." *Water Resources Management*, 14(1): 59–75.

AlMisnad, Abdulla. 2014. "Desalination under Uncertainty: Understanding the Role of Contractual Arrangements on the Adoption of Flexibility," S.M. Thesis, MIT Technology and Policy Program, Cambridge, MA.

Chowdhury, Abu Naser and Robert L. K. Tiong. 2011. "Financing and Case Study of Singapore PPP Desalination Project." *China Engineering Consultants Inc.*, 90.

Cruz, Carlos Oliveira and Rui Cunha Marques. 2013. "Contractual Flexibility" and "Renegotiation." In: *Infrastructure Public-Private Partnerships*, 53–81 and 113–50. Springer: Berlin. .http://dx.doi.org/10.1007/978-3-642-36910-0_3

Darwish, Mohammed A. and Najem Al-Najem. 2005. "The Water Problem in Kuwait." *Desalination*, 177(1): 167–77.

Darwish, Mohamed A. and Rabi Mohtar. 2013. "Qatar Water Challenges." *Desalination and Water Treatment*, 51(1–3): 75–86.

Dawoud, Mohamed A. 2005. "The Role of Desalination in Augmentation of Water Supply in GCC Countries." *Desalination* 186(1): 187–198.

De Neufville, Richard and Stefan Scholtes. 2011. *Flexibility in Engineering Design*. Cambridge, MA: MIT Press.

FAO. Food and Agriculture Organization of the United Nations. 2008. "Qatar Profile 2008." In *AQUASTAT—FAO's Information System on Water and Agriculture.* Accessed February 22, 2016. http://www.fao.org/nr/water/aquastat/countries_regions/QAT/index.stm

Ferguson, John. 2014. "Billions in Desalination Costs for Not One Drop of Water." *The Australian.* Accessed October 18, 2014. http://www.theaustralian.com.au/national-affairs/state-politics/billions-in-desalination-costs-for-not-a-drop-of-water/story-e6frgczx-1227094416376

Gil, Nuno. 2007. "On the Value of Project Safeguards: Embedding Real Options in Complex Products and Systems." *Research Policy* 36(7): 980–99. doi:10.1016/j.respol.2007.03.004.

Global Water Intelligence. 2013a. "Plant Analysis." *DesalData.* Accessed December 19, 2013. http://www.desaldata.com/ projects/analysis

———. 2013b. "SWRO Capacity Doubled in Same Footprint." *DesalData.* Accessed October 25, 2013. .http://www.desaldata.com/news/35199

———. 2014. "Qatar–Desal Data." *DesalData.* Accessed January 5, 2014. http://desaldata.com/countries/63

Guasch, J. Luis. 2004. "Granting and Renegotiating Concessions: Doing it Right." World Bank Institute of Development Studies: Washington, DC. http://dx.doi.org/10.1596/0-8213-5792-1

Guasch, J. Luis, Jean-Jacques Laffont and Stéphane Straub. 2007. "Concessions of Infrastructure in Latin America: Government-led Renegotiation." *Journal of Applied Econometrics* 22(7): 1267–94.

Ibrahim, Ibrahim and Frank Harrigan. 2012. "Qatar's Economy: Past, Present and Future." *QScience Connect.*

Marin, Philippe. 2009. *Public-Private Partnerships for Urban Water Utilities: A Review of Experiences in Developing Countries.* Washington, DC: World Bank Publications.

MEED Insight. 2012. "Qatar General Electricity and Water Corporation (Kahramaa)." Accessed December 15, 2013. http://www.meed.com/supplements/2012/qatar-projects/kahramaa/3123662.article

Poseidon Water. 2014. "The Carlsbad Desalination Project." Accessed January 15, 2015. http://carlsbaddesal.com/project-agreements

Qatar General Secretariat for Development. 2011. *Qatar National Development Strategy: 2011–2016.* http://www.gsdp.gov.qa/gsdp_vision/docs/NDS_EN.pdf

Qatar Information Exchange. 2014. "Population Structure." Accessed January 25, 2015. http://www.qix.gov.qa/portal/page/portal/qix/subject_area/General?subject_area=183

Sakhrani, Vivek. 2015. "Negotiated Collaboration: A Study in Flexible Infrastructure Design." PhD dissertation, MIT Institute for Data, Systems and Society, Cambridge, MA.

Voutchkov, Nikolay. 2010. *Desalination Cost Assessment and Management.* Bangkok: Water Treatment Academy, TechnoBiz Communications Co.

———. 2012. *Desalination Engineering: Planning and Design.* New York: McGraw-Hill Professional.

Weber, Barbara and Hans Wilhelm Alfen. 2010. *Infrastructure as an Asset Class: Investment Strategies, Project Finance and PPP*, West Sussex: John Wiley.

Wolfs, Maarten and Shane Woodroffe. 2002. "Structuring and Financing International BOO/BOT Desalination Projects." *The Journal of Structured Finance* 7(4): 19–24.

Wong, Melanie. 2013. "Flexible Design: An Innovative Approach for Planning Water Infrastructure Systems under Uncertainty." S.M. Thesis, MIT Technology and Policy Program, Cambridge, MA.

Wong-Turlington, Melanie, Richard de Neufville and Margaret Garcia. 2016. "Flexible Design of Water Infrastructure Systems." In: *Contingent Complexity and Prospects for Water Diplomacy: Understanding and Managing Risks and Opportunities for an Uncertain Water Future*. Eds. Shafiqul Islam and Kaveh Madani, London: Anthem Water Diplomacy Series.

World Bank. 2014. "World Development Indicators for Qatar." Accessed January 25, 2015. http://data.worldbank.org/country/qatar

———. 2013. "PPP Arrangements/Types of Public-Private Partnership Agreements." *Public-Private Partnership in Infrastructure Resource Center*. Washington, DC. Accessed January 25, 2015. http://ppp.worldbank.org/public-private-partnership/agreements.

Chapter 13

THE GRAND ETHIOPIAN RENAISSANCE DAM: CONFLICT AND WATER DIPLOMACY IN THE NILE BASIN

Ronny Berndtsson, Kaveh Madani,
Karin Aggestam and Dan-Erik Andersson

Abstract

The soon-to-be completed Grand Ethiopian Renaissance Dam (GERD) is abruptly modifying the hydropolitical map of the Nile Basin. Egypt, the historical hydro-hegemon that has preserved a strong position in the basin is challenged by Ethiopia, which seems to have decided to become a major African hydropower producer. Theoretical analysis indicates the risk for Egypt if it retaliates against Ethiopian intentions. However, it is obvious that retaliation does not lead to a sustainable future. Long-term sustainability includes collaboration between all riparians in the Nile Basin. Eventually, this will lead to greater prosperity for all involved countries and a more efficient use of the water but also a decrease in the flow of the Nile. Increasing the efficiency of water use means a larger volume of water for food production, better health of the population and sustainable economic development.

Introduction

In October 2014, the Ethiopian government announced that the Grand Ethiopian Renaissance Dam (GERD) was 40 percent complete. The dam will be the largest hydro-electric power plant and one of the largest reservoirs in Africa. The GERD will store a maximum of 74 km^3 of water corresponding to a bit less than the average annual inflow of the Nile to the Aswan high dam (a total storage of about 84 km^3). The dam surface area will be 1,680 km^2 at full supply level. The project has foreseen a power plant capacity of 6,000 MW and 15,692 GWh annual energy amount. The GERD is vital for energy production and a key factor for food production, economic development and poverty reduction in Ethiopia and the Nile Basin. However, the GERD is also a political statement that has already largely rewritten the hydropolitical map of the Nile Basin. The GERD has become a symbol of Ethiopian nationalism, or "renaissance" (*hidase* in Amharic). Historically, the Nile Basin has witnessed a rivalry between Egypt and

Ethiopia, on the one hand, but also uniting and cultural elements such as the relationships between the Coptic Church and the Orthodox Ethiopian Church on the other (e.g., Rubenson 2009). The Nile has also been the connecting link between Egypt and Ethiopia for thousands of years. Starting in the 3rd century, the Nile connected early Ethiopian Christianity with the Alexandrian Coptic Church, a link that continued uninterrupted for 1600 years (Ayele 1988, Erlich 2002).

In modern times, a series of international agreements regarding the Nile were made practically without the inclusion of Ethiopian interests. On May 7, 1929, an agreement between Egypt and Anglo-Egyptian Sudan gave the right to Egypt and Sudan to utilize annually 48 km³ and 4 km³ of the Nile water, respectively. Additionally, the flow of the Nile during January 20 to July 15 (the dry season) was reserved for Egypt, which also was given the right to monitor the Nile flow in upstream countries· The agreement gave Egypt complete control over the Nile during the dry season, when water is needed for irrigation. A second agreement was made in 1959 between the Sudan and Egypt, allocating the entire average annual flow of the Nile to be shared between them at 18.5 km³ and 55.5 km³, respectively. The annual water loss due to evaporation and other factors was agreed to be about 10 km³. The agreement gave Egypt the right to construct the Aswan High Dam, which has the capacity to store the entire annual Nile River flow. The creation of the Aswan High Dam in 1970 was to control floods, provide water for irrigation and generate hydroelectricity. Egyptians still celebrate the anniversary of its construction on January 9 every year. Emperor Haile Selassie was so offended by President Gamel Abdel Nasser's exclusion of Ethiopia from the 1959 Nile water agreement that he organized a complete separation of the Ethiopian Orthodox Church from the Orthodox Church in Alexandria, ending 1600 years of institutional marriage (WaterWorld 2015).

The Nile Basin Initiative (NBI) was established 1999 to promote a basin-wide cooperative framework. In fact, the NBI was mainly initiated with the active involvement of international donor agencies, and could be said to still survive only with their assistance (e.g., Swain 2011). Swain (2011) notes that even after more than a decade of existence, the NBI receives very little contribution from the basin states, exposing a lack of interest of these states in joint management of the shared river resources.

The case of GERD highlights a number of questions that have great importance beyond the dam itself, such as water sharing in transboundary rivers, collective claim-making processes and economic development in the region and globally. There are also important links and experiences that can be learned from the hydropolitics of the dam and evolving conflicts and collaborations at the interstate level. The Nile is one of the most important transboundary rivers in the Middle East, along with the Jordan, the Tigris and Euphrates. Thus, experiences in each of these river basins may be used to better understand and improve collaborative processes in other basins. Consequently, the objective of this paper is to use GERD as a case study and discuss the complex issue of large dams, specifically the GERD. We start with a general perspective on water scarcity in the riparian states of the Nile, then follow with a brief account of the history of the GERD. Finally, we outline a framework for collaborative measures involving GERD in the Nile Basin.

Nile Basin Water in Perspective

Most countries in the Nile Basin—as well as in the rest of the Middle East and North Africa (MENA)—are confronted by severe water shortage due to both climatic conditions and socioeconomic factors (e.g., ACSAD 1999). From an ecoclimatic point of view, most of the region extends across the semiarid, arid and hyperarid zones of the Sahel. The Sahel has been particularly affected by cycles of droughts and desertification during the past four decades. Socioeconomically, the region is characterized by a fast-increasing population, which has resulted in a sharp decline in the per capita availability of water. Under these conditions, water is the prime factor for a sustainable environment and, even more important, for economic development. Most MENA countries exploit more than 50 percent of its renewable water resources, some nearly 100 percent (notably, e.g., Egypt). The water availability per year and per capita will drop to less than 1,000 m^3 by 2025 in 11 MENA countries and to less than 500 m^3/year/habitant in as many as 8 countries, which is the level considered to be the shortage threshold (Margat and Vallée 1999). Agriculture is the largest consumer of water resources, with a strong trend toward developing irrigation. Urban development along the coasts increases both demand and the amount of fresh water discharged into the sea (e.g., Egypt). Tourists' needs amplify the demand for water in summer, as several countries around the Mediterranean are popular tourist destinations, and Ethiopia aspires to develop sustainable ecotourism. Linked to the forecasts of water scarcity are growing estimates of food dependence in the same countries. In the Nile Basin countries, key challenges are the rapid population increase and the need to supply new generations with food and energy.

The shrinking per capita water availability, according to the above, is unfortunately further aggravated by climate change. In general, MENA countries experience an average annual precipitation of less than about 100 mm in 65 percent of the region, 100 to 300 mm in 15 percent of the region, and more than 300 mm in the remaining area (AbuZeid and Abdel-Meguid 2006). On average, the precipitation is partitioned into about 80 percent evaporative losses and about 20 percent runoff that can be considered as renewable water. In the case of climate change, it is expected that average precipitation in the MENA will possibly decrease by more than 20 percent by 2050. Climate change will at the same time increase temperatures and thus also increase the evaporative forces, leaving behind just about 10 percent renewable water, as an average, which is schematized in Figure 13.1 (e.g., Wasimi 2010). By comparison, it can be mentioned that the renewable percentage of water from the Nile Basin at the outflow point to the Mediterranean is only 1.7 percent (Figure 13.2; Awulachew et al. 2010). The forecasts by GCMs for the upper basin tributary, the Blue Nile, which represents about 60 percent of the water supply at the High Aswan Dam, are somewhat mixed. Most GCMs forecast an increase in precipitation in the long range (50–80 years) over the upper Nile Basin. Thus, climate change would have negligible effect on low flow conditions of the river. Seasonal mean flow volume, however, may increase by more than double and +30 percent to +40 percent for the Belg (small rainy season) and Kiremit (main rainy season) periods, respectively (e.g., Dile et al. 2013). It should be noted that the above average values do

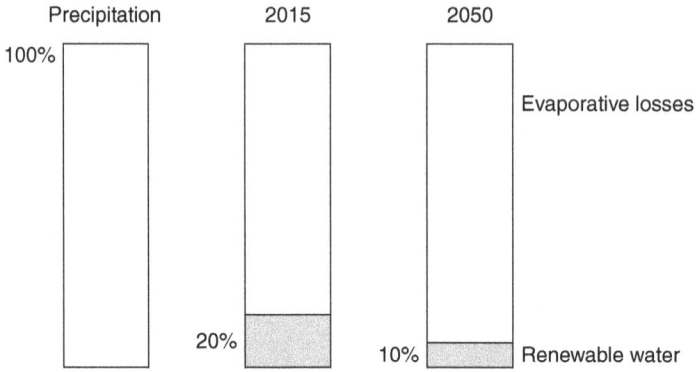

Figure 13.1 Schematic of total average precipitation over the MENA and the partitioning in to evaporative losses and renewable water in 2015 as compared to 2050.

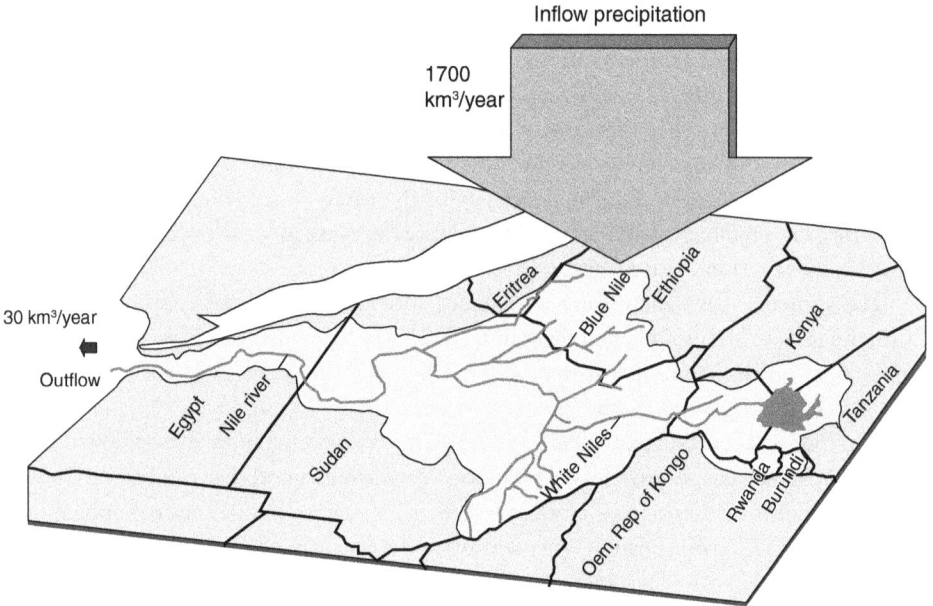

Figure 13.2 Schematic representation of total inflow of water to the Nile River Basin by precipitation as compared to total outflow to the Mediterranean. The ratio outflow to inflow represents 1.7 percent. The map does not correctly reflect the political disputable boundaries and South Sudan as new country.

not say anything about the variation across the MENA and, similarly, do not consider regional differences in the Nile Basin.

In the case of climate change, the generally renewable water amounting to about 10 percent of precipitation is still only a potential amount. A major portion of this water

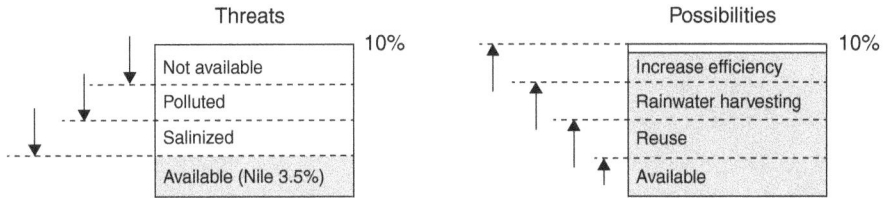

Figure 13.3 Threats and possibilities to MENA renewable water in view of global warming.

will not be available due to logistical reasons and difficulties in transporting water from uninhabited areas to populated areas, as seen from Figure 13.3. Also, much of this water may be polluted due to industrial and domestic waste and wastewater. A commonly occurring problem in the MENA as well as in the Nile Basin countries is salinization. Some water cannot be used due to the salinity of water and soil. Finally, only a fraction of the renewable water will in fact be available for drinking, irrigation or industrial use. For the Nile, the available part of this water is about 3.5 percent at the level of the Aswan Dam. The reduced amounts of water pose different threats to the water use.

However, various management strategies can increase the available renewable water volume. The most obvious is to reuse the water. Industrially used water needs to be treated before being discharged. Industries can as well reuse the water internally. Domestic wastewater is a valuable resource that can be reused in agriculture after primary treatment. Rainwater and floodwater harvesting is an ancient technique still utilized in many regions of the Nile Basin to improve agricultural yield. Finally, improved efficiency, especially, in agricultural water use can save large water volumes. All these measures increase the water availability and represent sound water management.

The Grand Ethiopian Renaissance Dam (GERD)

The history of GERD goes back to the 1950s, when the construction site was identified by the US Bureau of Reclamation (Water-Technology.net 2015). A pre-feasibility study of a border dam concluded under the EN Power Trade Study (ENTRO and EN Subsidiary Action Project) is the GERD dam initially appraised with smaller size. The Ethiopian government resurveyed the site in October 2009 and August 2010. In March 2011, a day after the GERD project was made public, a US$4.8 billion contract, without competitive bidding, was given to Salini Costruttori (contracting dam construction company). About 30 percent of the funding for the dam has been secured from China (*The Economist*, 2011), and the remainder is funded by the Government of Ethiopia through the selling of bonds and donations, both within Ethiopia and internationally (Water Technology. net 2015). The first two generators are expected to become operational after 44 months of construction. Construction on the dam is expected to be completed in 2017. Egypt has opposed the dam, claiming it will reduce the water volume from the Nile. The late Ethiopian prime minister, Meles Zenawi, claimed the dam would not reduce water availability downstream and would also regulate water for irrigation. In May 2011, it was announced that Ethiopia would share blueprints for the dam with Egypt so the down-stream impact could be examined. In July 2011, Sudan was divided into two countries. Earlier, Egypt had threatened war if Ethiopia tried to block the Nile flow and Ethiopia had responded that no country could prevent it from using the Nile water (the Harmon doctrine). Egypt, in turn, declared that it would not give up its share of the Nile. After the sudden death of Prime Minister Zenawi in 2012, however, a joint Egyptian, Sudanese and Ethiopian committee was formed to evaluate the downstream effects of the dam. Ethiopia, Egypt and Sudan agreed to form an international panel of experts to review the GERD's social and environmental impacts on downstream nations. The 10-member panel submitted its first report to the governments in June 2013. The panel found a need for more details on the impact assessment of the GERD (International Rivers 2014). In January 2014, after a series of high-level meetings between the three governments, dis-cussions broke down. In April 2014 Ethiopia's prime minister invited Egypt and Sudan to another round of talks over the dam. In August 2014, a tripartite ministerial-level meeting agreed to set up a meeting of the Tripartite National Committee (TNC) over the dam (Ethiopian Ministry of Foreign Affairs 2014). The first TNC meeting took place in September 2014 in Ethiopia (All Africa 2014). At present, about 40 percent of GERD is completed (Ahram 2014). The first stage of the dam was expected to be operational from June 2015 and will produce 700 MW of electricity. For further detail on the hydropolitics of the Nile Basin see, for example, Al-Atawy (1996) or Waterbury (2002).

Approaches to Analyzing and Synthesizing GERD

There are many different approaches to analyzing and synthesizing the effects of GERD to better understand the hydropolitics of the Nile Basin. Integrated water resources management (IWRM) has been used during the most recent decades to solve water-resources-related planning and management. However, it is more and more argued that the monolithic and complicated structure of IWRM does not work well for com-plex international river basins with several management hierarchies and strong hydro-hegemony (e.g., Biswas 2010). A weakness of IWRM is that it tends to treat water policy and related conflicts from a commodity perspective, largely overlooking the human and social dimensions of the process. Recently, Abdelhady et al. (2015) used the opposing notions of nationalism and hydrosolidarity (see Lundqvist 1999, Falkenmark 2005) at three different scales—everyday politics, state policies, interstate and global politics—to analyze aspects of the new hydropolitical map of the Nile Basin. They argued that nationalism and national interests are not necessarily negative standpoints but, instead, that there may be a meeting point where regional and national interests join with hydro-solidarity principles. The power asymmetry in the Nile Basin has been obvious, with Egyptian hydro-hegemony characterized by an inequitable share of the Nile flow in favor of Egypt (and Sudan) and at the expense of upstream riparians (Carles 2006). The construction of GERD because of a continuously economically stronger and politically more stable Ethiopia is changing the power balance in the Nile Basin (Cascão 2009). Through the GERD and also other, smaller dam construction, Ethiopia is contesting the hydro-hegemony of Egypt by not including Egypt as a downstream discussion partner.

Recently, however, by formation of the TNT as described above, Ethiopia, Sudan and Egypt appeared to frame a starting collaboration by technical development (see, e.g., Aggestam and Sundell-Eklund 2014). In this case, the underlying functionalism through technical collaboration may generate positive effects in a broader political sense. Madani et al. (2011) used a game theory approach to analyze and improve the understanding of water-use conflicts in the Nile Basin. They found that stability of the possible outcome is dependent on the preferences of Egypt, and not very sensitive to Ethiopia's preferences. In addition, a stable outcome between the nations in the Nile Basin included retaliation by Egypt. Although war over water has been a much-discussed issue over recent decades, it is more and more recognized that direct water wars are less likely (e.g., Wolf 1998, Amery 2002, Brown and Crawford 2009), although some authors have found strong causal evidence linking climatic events to human conflict across all major regions of the world (Burke et al. 2009, Hsiang et al. 2013). Thus, water conflict is perhaps more likely as embedded in a larger complex within climate change and food production.

Efficient Use of Nile Water

Ethiopia has enforced the construction of GERD despite protests, mainly from Egypt. Thus, it appears that Ethiopia is signaling its intentions to become an African power (e.g., *The Guardian* 2014). This means that the historical hegemony of Egypt is challenged. Exactly how Egypt will react to this is a matter for the future. The total cost of GERD is expected to be about $4.1bn. As Ethiopia has received limited international funding, the government is appealing to the population to buy treasury bonds. For civil servants this is mandatory. Companies have urged their personnel to give up a month's wages to support GERD (*The Guardian* 2014). Thus, it is obvious that the Ethiopian government is using nationalistic feelings to collect the necessary capital to build GERD. In any case, after its completion, it is clear that GERD will have rewritten the hydropolitics of the Nile Basin and given Ethiopia a much stronger negotiating position.

The use of Nile water in Egypt at present stands on average at about 55.5 km^3 per year, and Sudan at 18.5 km^3 per year (MWRI 2010). GERD will store a maximum of 74 km^3 of water, although the average storage is probably about 63 km^3 (Salini Impregilo 2015). Thus, the scale of the GERD is such that the effects downstream could be substantial. After the completion of GERD, downstream countries will be dependent on the management of the dam for their supply of Nile water. Obviously, this means first Egypt and secondly Sudan. The only way to minimize negative effects downstream of GERD is to build a strategy for efficient water management. This requires detailed collaborative action from first Ethiopia, Egypt, and Sudan. As indicated in previous chapters, this includes both ex-ante and ex-post actions. Typical ex-ante measures include treatment for reuse, recycling, and improving irrigation efficiency. For the Cairo area, this embraces the Bahr El Baqar Drain and Lake Manzala. Lake Manzala is one of the largest lakes and wetlands of Egypt. It is important for fish production and ecosystem services. Lake Manzala is, however, subject to various threats, among them agricultural drainage, municipal sewage water and industrial wastewater. About 3.7 km^3 per year of water (mostly from urban wastewater, urban and agricultural drainage) flow annually into Lake Manzala from nine major drains and canals. The most important of these are Faraskur, Al Sarw, Baghous, Abu Garida and Bahr El Baqar. The Bahr El Baqar is considered one of the most polluted drains in Egypt. It receives the greatest part of Cairo's wastewater (about 3 km^3 per year) and carries it into Lake Manzala through a very densely populated area of the Eastern Delta, passing through Qalubyia, Sharkia, Ismailia and Port Said governorates. Because of untreated industrial wastewater, the population around the Bahr El Baqar Drain and Lake Manzala is experiencing severe health problems. Improving domestic and industrial wastewater from the Cairo area would allow for reuse of a possible water volume of 3 km^3 per year (Hamed et al. 2013; Table 13.1). Further, improving irrigation efficiency, recycling water and better use of water-efficient crops could save as much as 21 km^3 per year (El-Quosy et al. 1999, Hamza and Mason 2004, Selim 2012). These measures, however, would mean economic and organizational commitment of the Egyptian authorities to deal with current pollution problems and wasteful irrigation.

GERD ex-post action includes jointly managing the High Aswan Dam and GERD to minimize losses caused by evaporation. Studies have indicated a possible reduction of current evaporation of more than 10 km^3 per year (e.g., Mulat and Moges 2014). For this to occur, Egypt and Ethiopia need to collaborate on an intimate basis. Further use of the Nile water may be to utilize the remaining about 30 km^3 per year outflow to the Mediterranean. This is a substantial amount of water that can be used for extending irrigation areas. The use of this water must, however, be balanced against the negative environmental effects of decreasing the outflow to the Mediterranean. Among these are increased risks for coastal erosion and inundation of low-lying coastal areas in view of sea level rise in connection to climate change.

It is important that water-saving measures be undertaken in not only Egypt but also in Ethiopia and, eventually, all Nile Basin countries. The GERD will not be the last large dam structure in the Nile Basin. Dam building without the combination of treatment of polluted water and improving agricultural water efficiency will just lead to a larger waste of water. Agricultural water management interventions in the Nile

Table 13.1 GERD ex-ante and ex-post actions to save water in the Nile Basin.

Saving by action	Ex-ante GERD	Ex-post GERD
3 km^3 per year	Improving water quality for safe re-use in the Cairo area	
21 km^3 per year	Improving irrigation efficiency, recycling water, and better use of water efficient crops in Egypt	
10 km^3 per year		Evaporation minimization by joint management of High Aswan Dam and GERD
30 km^3 per year		Extended use of remaining Nile River outflow to the Mediterranean

Basin can largely improve agricultural production and productivity. Awulachew et al. (2012) found that agricultural water management technology intervention combined with soil fertility and seed improvement may increase productivity up to threefold. This includes rainwater harvesting at different scales for increasing agricultural productivity. Planned irrigation areas in the Nile Basin amount to 10.6 million ha, as compared to the current area of 5.5 million ha. If executed with the present wasteful water application rates and without reservoir management the total water withdrawal rate would be 127 km^3 per year as compared to the Nile mean flow rate of about 84.5 km^3 per year. This means that, to a major extent, demand will surpass supply. In any case, Nile River flow will continue to decrease gradually in the future because of development of irrigation agriculture. Consequently, there is no turning back. The only way is forward with extensive collaboration between major Nile water users. Trondalen (2008) notes that unless the countries involved cooperate, the consequences will be devastating. Lack of clean water for the populations will result not only in severe human suffering but could potentially also have severe geopolitical consequences. Similarly, Grey and Sadoff (2007) show that achieving basic water security by both harnessing the productive potential of water and limiting its destructive impact is a continued societal priority.

Summary and Discussion

The construction of GERD is rewriting the hydropolitical map of the Nile Basin. Egypt is gradually losing its hydro-hegemony to Ethiopia and in the long-term eventually to other developing countries in the Nile Basin. The risk of Egyptian retaliation is obvious. However, we argue that this is not likely, as retaliation does not lead to a sustainable future. Sustainability includes collaboration and, even if it gradually will lead to decreasing the flow of the Nile, it will bring about a more efficient use of water. Increasing the efficiency of water use means a larger volume of water for food production, better health of the population and sustainable economic development.

Acknowledgments

Funding from the MECW project at the Center for Middle Eastern Studies, and the Hydrosolidarity in the Nile Basin project at the Pufendorf Institute for Advanced Studies, Lund University, is gratefully acknowledged.

References

Abdelhady, Dalia, Karin Aggestam, Dan-Erik Andersson, Olof Beckman, Ronny Berndtsson, Karin Broberg Palmgren, Kaveh Madani, Umut Ozkirimli, Kenneth M. Persson and Petter Pilesjö. 2015. "The Nile and the Grand Ethiopian Renaissance Dam: Hydrosolidarity vs. nationalism." *Journal of Contemporary Water Resources Education* 155: 73–82.

AbuZeid, Khaled and Afaf Abdel-Meguid. 2006. *Water Conflict and Conflict Management Mechanisms in the Middle East and North Africa*. Cairo: Centre for Environment and Development for the Arab Region and Europe, CEDARE.

ACSAD, 1999. *Water Resources Division; Objectives, Functions and Activities.* Damascus, Syria: The Arab Center for the Studies of Arid Zones and Dry Lands.

Aggestam, Karin and Anna Sundell-Eklund. 2014. "Situating Water in Peacebuilding: Revisiting the Middle East Peace Process." *Water International* 39: 10–22.

Ahram. 2014. "40% of Grand Ethiopian Renaissance Dam Completed: Ethiopian President," *Ahram Online* Friday October 3, 2014. Accessed at http://english.ahram.org.eg/NewsContent/1/64/112299/Egypt/Politics-/-of-Grand-Ethiopian-Renaissance-Dam-completed-Ethi.aspx.

Al-Atawy, Mohamed-Hatem 1996. *Nilopolitics: A Hydrological Regime 1870–1990.* Cairo: American University in Cairo Press, 62.

All Africa. 2014. *Ethiopia: The First Meeting of the Tripartite National Committee On the Grand Ethiopian Renaissance Dam Concludes.* September 22, 2014. Retrieved January 17, 2015.

Amery, Hussein A. 2002. "Water Wars in the Middle East: A Looming Threat." *The Geographical Journal* 168: 313–23.

Awulachew, Seleshi, Lisa-Marie Rebelo and David Molden. 2010. "The Nile Basin: Tapping the Unmet Agricultural Potential of Nile Waters." *Water International* 35: 623–54.

Awulachew, Seleshi B., Solomon S. Demissie, Fitsum Hagos, Teklu Erkossa and Don Peden. 2012. "*Water Management Intervention Analysis in the Nile Basin.*" In: Awulachew, Seleshi Bekele; Smakhtin, Vladimir; Molden, David; Peden D., eds, *The Nile River Basin: Water, Agriculture, Governance and Livelihoods.* Abingdon: Routledge–Earthscan. 292–311.

Ayele, Negussay. 1986. *The Blue Nile and Hydropolitics among Egypt, Ethiopia, Sudan, the Nile and the Blue Nile in Perspective.* Presented at the Ninth International Congress of Ethiopian Studies, Moscow, August 26–29, 1986.

Biswas, Asit. 2010. "Cooperation or Conflict in Transboundary Water Management: Case Study of South Asia." *Hydrological Sciences Journal* 56: 662–70.

Brown, Oli and Alec Crawford. 2009. *Rising Temperature, Rising Tensions: Climate Change and the Risk of Violent Conflict in the Middle East.* Winnipeg: International Institute for Sustainable Development (IISD).

Burke, Marshall B., Edward Miguel, Shanker Satyanath, John A. Dykema and David B. Lobell. 2009. "Warming Increases the Risk of Civil War in Africa." *Proceedings of the National Academy of Sciences* 106: 20670–4.

Carles, Alexis. 2006. *Power Asymmetry and Conflict over Water Resources in the Nile River Basin: The Egyptian Hydro-hegemony.* Dissertation submitted as a part of the MA in Environment and Development Degree in Geography, at King's College London.

Cascão, Ana E. 2009. "Changing Power Relations in the Nile River Basin: Unilateralism vs. Cooperation?" *Water Alternatives* 2: 245–68.

Dile, Yihun, Ronny Berndtsson and Shimelis Setegn. 2013. "Hydrological Response to Climate Change for Gilgel Abay River, in the Lake Tana Basin: Upper Blue Nile Basin of Ethiopia." *PLoS ONE* 8(10): e79296. doi:10.1371/journal.pone.0079296.

El-Quosy, Dia and Tarek A. Ahmed. 1999. "Water Saving Techniques, Lessons Learned from Irrigation of Agricultural Land in Egypt." In: Proceedings Seventh Nile Conference, Cairo, March 15–19, EGY-18: 1–18.

Erlich Haggai. 2002. *The Cross and the River, Ethiopia, Egypt, and the Nile.* London: Lynne Rienner Publishers, Inc.

Ethiopian Ministry of Foreign Affairs. Accessed: October 24, 2014

Falkenmark, Malin. 2005. "*Towards Hydrosolidarity: Ample Opportunities for Human Ingenuity.*" Fifteen Year Message from the Stockholm Water Symposia. Stockholm, Stockholm International Water Institute.

Grey, David and Claudia Sadoff. 2007. "Sink or Swim? Water Security for Growth and Development." *Water Policy* 9: 545–71.

Hamed, Yasser A., Tamer S. Abdelmoneim, Mohamed H. El-Kiki, Mohamed A. Hassan and Ronny Berndtsson. 2013. "Assessment of Heavy Metals Pollution and Microbial Contamination in Water, Sediments, and Fish of Lake Manzala, Egypt." *Life Science Journal* 10: 86–99.

Hamza, Waleed and Simon Mason. 2004. *Water Availability and Food Security Challenges in Egypt*. Internernational Forum on Food Security under Water Scarcity in the Middle East: Problems and Solutions. Como, Italy.

Hsiang, Solomon, Marshall Burke and Edward Miguel. 2013. "Quantifying the Influence of Climate on Human Conflict." *Science* 341: 6151, doi: 10.1126/science.1235367.

International Rivers. 2014. *The Grand Ethiopian Renaissance Dam Fact Sheet*. January 24, 2014. Accessed February 27, 2015 at http://www.internationalrivers.org/resources/the-grand-ethiopian-renaissance-dam-fact-sheet-8213.

Lundqvist, Jan. 1999. "Towards Upstream/Downstream Hydrosolidarity." *Water International* 24: 275–77.

Madani, Kaveh, David Rheinheimer, Laila Elimam and Christina Connell-Buck. 2011. *A Game Theory Approach to Understanding the Nile River Basin Conflict*. Festschrift, Lars Bengtsson – A Water Resource, Report no. 3235, edited by K. M. Persson, 97–114.

Margat Jean and Domitille Vallée. 1999. *Executive Summary of the Mediterranean Regional Vision Blue Plan MEDTAC*. Stockholm: Août.

Mulat, Asegdew G. and Semu A. Moges. 2014. "Assessment of the Impact of the Grand Ethiopian Renaissance Dam on the Performance of the High Aswan Dam." *Journal of Water Resource and Protection* 6: 583–98.

MWRI. 2010. *Water Challenges in Egypt*. Cairo: Ministry of Water Resources and Irrigation.

Rubenson, Samuel. 2009. *The European Impact on Christian-Muslim Relations in the Middle East during the Nineteenth Century. The Ethiopian Example*. In: *The Fuzzy Logic of Encounter: New Perspectives on Cultural Contact*. Münster: Waxmann.

Salini Impregilo. 2015. *Grand Ethiopian Renaissance Dam Project*. Accessed August 12, 2015 at: http://www.salini-impregilo.com/en/projects/in-progress/dams-hydroelectric-plants-hydraulic-works/grand-ethiopian-renaissance-dam-project.html.

Selim, Aboulila Tarek. 2012. *Evaluation of Modern Irrigation Techniques with Brackish Water*. PhD Thesis, Report 1056, Department of Water Resources Engineering, Lund University, Lund, Sweden.

Swain, Ashok. 2011. "Challenges for Water Sharing in the Nile Basin: Changing Geo-politics and Changing Climate." *Hydrological Sciences Journal* 56: 687–702. doi: 10.1080/02626667.2011.577037.

The Economist. 2011. "The River Nile, a Dam Nuisance—Egypt and Ethiopia Quarrel over Water." April 20, 2011.

The Guardian. 2014. "Ethiopia's Nile Dam Project Signals Its Intention to Become an African Power." July 14, 2014. Accessed August 12, 2015 at: http://www.theguardian.com/global-development/2014/jul/14/ethiopia-grand-renaissance-dam-egypt.

Trondalen, Jon Martin. 2008. *Water and Peace for the People: Possible Solutions to Water Disputes in the Middle East*. Water and Conflict Resolution Series – UNESCO-IHP, 9782-3-104086-3.

Wasimi, Saleh. 2010. "Climate Change in the Middle East and North Africa (MENA) Region and Implications for Water Resources Project Planning and Management." *International Journal of Climate Change Strategies and Management* 2: 297–320.

Waterbury, John. 2002. *The Nile Basin: National Determinants of Collective Action*. New Haven: Yale University Press.

Water-Technology.net. 2015. *Grand Ethiopian Renaissance Dam Project*, Benishangul-Gumuz, Ethiopia. Available at: http://www.water-technology.net/projects/grand-ethiopian-renaissance-dam-africa/.

WaterWorld. 2015. "Grand Designs North Africa: Impact of Ethiopia's Renaissance Dam." Accessed August 12, 2015 at: http://www.waterworld.com/articles/wwi/print/volume-29/issue-1/regional-spotlight/ethiopia-impact-of-renaissance-dam/grand-designs-north-africa-impact-of-ethiopia-s-renaissance-dam.html.

Wolf, Aaron T. 1998. "Conflict and Cooperation along International Waterways." *Water Policy* 1: 252–65.

Chapter Fourteen

ENGAGING STAKEHOLDERS FOR WATER DIPLOMACY: LESSONS FOR INTEGRATED WATER RESOURCES MANAGEMENT

Bhadranie Thoradeniya and Basant Maheshwari

Abstract

Stakeholder engagement and participatory approaches for water diplomacy are increasingly used for developing trust, consensus and communication while stimulating the water reform process. This is particularly important to addressing future water challenges and improving water security.

The effectiveness of stakeholder engagement in water resources management depends on a number of factors, including the method employed, inclusiveness of the engagement process, local knowledge and time made available for the stakeholder dialogue to develop.

There is a plethora of literature available on stakeholder engagement, both within and outside of water resources management discipline, and there are some important lessons that may be applicable in developing water diplomacy in a range of situations.

In this chapter, we identify key features of stakeholder engagement and water diplomacy from an array of disciplines as to their suitability for integrated water resources management.

The concepts and experiences examined suggest that involvement of stakeholders, especially in the early stages of an engagement and diplomacy process, has the advantage of pursuing participatory decision making and facilitating social learning. The stakeholder engagement process increases the likelihood of meeting the needs and priorities of local communities and improves the overall process of water diplomacy.

Overall, successful engagement processes for water diplomacy help in acceptability of difficult and complex decisions made through mutual trust with strong elements of transparency, accountability and legitimacy. Furthermore, the engagement process helps in an improved stakeholder relationship, adaptability and resilience decisions, dissemination of decisions to the wider community and the capacity building of the stakeholders.

Introduction

Traditional engineering and economic solutions are insufficient to resolve most complex water issues. Management of water as a resource is becoming increasingly critical in urban, peri-urban and rural areas. Watershed boundaries transcending political boundaries increase the complexity of issues. The major drive for this critical situation stems from two aspects; the experienced and predicted scarcity of water spatially and temporally all over the world (Faust et al. 2013) and the increasing demand for available water, both in quantity and quality, among multiple stakeholders (Pinto and Maheshwari 2010). Additionally, the anthropogenic activities degrading water quality have ironically impacted the very existence of humans (Erdogan 2013; Grover and Krantzberg 2015, 185–86). The sustainability in all aspects of water, from source to user, has thus drawn important and urgent attention from various use sectors around the world with the significant environmental and political issues created (Erdogan 2013; Gudowsky and Bechtold 2013; Webb et al. 2009). Sustainable water systems are defined as "water systems that are managed to satisfy changing demands placed on them (both human and environmental) now and into the future, whilst maintaining ecological and environmental integrity of water systems" (Pearson et al. 2010).

Stakeholder engagement and participatory approaches for decision making are increasingly considered in various sectors, including water, to overcome alienation, foster communication and stimulate the reform process (Larson and Williams 2009). As such, striking a balance between the traditional top-down and emerging bottom-up approaches is a part of the water diplomacy process so important to address future water challenges and improve water security in the longer term. Among the many attempts to address the numerous issues related to water resources, there is sufficient evidence that participatory or bottom-up approaches have gained growing recognition in decision making, strategic policy formulation and operational management as opposed to conventional top-down planning, which was mostly inefficient, unsuccessful in implementation, and unsustainable (Erdogan 2013; Faust et al. 2013; Gutrich et al. 2005; Zorrilla et al. 2010). The most important feature in a participatory approach to decision making is the conscious effort that is made to include and engage stakeholders in an attempt to find a holistic solution to the issue and validate the solution with stakeholders.

The two main objectives of this chapter are: (a) to identify key features of stakeholder engagement and water diplomacy from an array of disciplines as to their suitability for integrated water resources management (IWRM); and (b) to critically examine past stakeholder engagement models, frameworks and techniques and draw out some important lessons for the engagement in the context of water diplomacy.

Engagement for Water Diplomacy and Its Expectations

Engagement is a process that relies on social learning incorporating inclusion and interaction with stakeholders. Pearson et al. (2010) define social learning as "a dynamic process, which enables individuals to engage in new ways of thinking together to address difficult decision problems such as the unsustainable use of water," while HarmonyCOP

(2005, 2) summarize social learning for decision making as "Learning together to manage together." Reed (2008) claims that stakeholder engagement is increasingly sought in environmental decision making, which is equally complex and dynamic in nature and requires flexibility and transparency with diversity of knowledge and values. However, it is interesting to note the occasional contradictions, such as the claim by Christoff (2005 as cited in Schwecke 2009) that public concern for the environment had dropped from 75 percent to 57 percent in Australia within 12 years despite increasing reports of environmental problems.

The increased attention to environment protection and democratic governance are two key factors that emphasize the need to hear the lost voices in multi-sector technical solutions (Reed 2008). Carmona et al. (2013), citing literature justifies stakeholder engagement using three major rationales: "Democratic rationale: the public should be involved in decisions that affect them[.] Substantive rationale: citizens can provide scientists with their specialised knowledge, for better understanding of facts and value[.] Pragmatic rationale: an involved and educated public is more likely to support implementation of resulting policies." Despite this understanding, Webb et al. (2009) claim that much of the community work results in advocacy for people rather than activities by those affected leading to practical and political action on their own terms. It is understood with equal importance that for a solution to be sustainable all parties involved in the issue—those who impact and are impacted—should be part of the solution. Therefore, engagement is essential in instances where decisions are made on competing resources. Importantly, the process needs to facilitate mutual learning and development (Warburton et al. 2007) with dual knowledge transfer, social (experiential) knowledge of lay stakeholders and the scientific and technical (cognitive) knowledge of the officials who are usually the change initiators (Gudowsky and Bechtold 2013; Glicken 2000). Carson (2010) argues that with access to expertise and information the citizens (lay stakeholders) can make remarkably good collective judgements by weighing strengths and weaknesses of arguments.

Decision making through engagement in water resources under competitive demand requires due consideration for the ability of engaging stakeholders and advancing water diplomacy. This becomes the third feasibility: to evaluate engineering projects, in addition to the generally considered technical (scientific capacity to solve the issue) and economic (economic gains and losses of the decision) feasibilities. Chess and Gibson (2001) discuss these dimensions—scientific feasibility (the nature of issue and the scientific capacity to solve them); motivational feasibility (value or economic considerations of the solution); and social feasibility—as three attributes intrinsic in solving watershed issues. Social feasibility is the ability to involve stakeholders in a meaningful process to include their input, which would ideally occur through voluntary participation or be facilitated through an existing statutory infrastructure (Warburton et al. 2007). In the watersheds that lack social feasibility, government agencies need to build social capacity (Erdogan 2013). In this sense, stakeholder engagement provides for capacity building.

This could also be understood as building social capital. Social capital, a concept of a century ago (Maak 2007, 332), has been widely used in corporate business situations. Maak (2007), citing Woolcock and Narayan (2000), Adler and Kwon (2002) and Putnam (1993) explain social capital as those features such as networks, relationships, norms and trust that enable people to act collectively. In fact, the leadership of stakeholder engagement is expected to build social capital which, in turn, results in information flow among the stakeholders. It is argued here that the water diplomacy benefits from the confidence and ties thus built during the engagement process to help with informed decision making.

In support of the above argument we cite Liu et al. (2008) on the failure of computer models (scientific feasibility) on IWRM applications in Ethiopia and state: "Although such models can be applied for policy analysis, they typically need extensive calibration and cannot simulate the intricacies that farmers have to deal with on a day by day basis." On the same issue, Nyssen et al. (2006) emphasize the necessity of site-specific calibration and validation when using such models, which is actually promoting indigenous knowledge.

When data are incomplete or of poor quality, other integrated and participatory approaches will likely to be more effective as management tools than technical approaches such as sophisticated computer models. The case study of Amhara, Ethiopia, is an example for the failure of overdependence on scientific knowledge in the developing world, where Liu et al. (2008) discuss the failure of at least four such models. This case study speaks of the success of participatory approaches through enabling local potential instead of sophisticated technical approaches.

Water users, non-governmental organizations, researchers and education providers are often not directly connected with government agencies involved in decision making, but they can play an important role in implementation of policy decisions and trust-building efforts (Susskind and Islam 2012; Krantzberg et al. 2014). For this reason, public participation for water diplomacy is incorporated in many different forms in the planning processes of initiatives to deliver information and gain public support and trust (Hämäläinen et al. 2001). Nevertheless, communications and consultations between stakeholders of an issue in a fair and respectful manner do not necessarily mean that there is an interest in fulfilling each other's desires (Greenwood 2007). We emphasize that stakeholder engagement for water resources management decisions should not stem from a feeling of business responsibility or with a business-as-usual attitude, but should involve all the complex relationships with a genuine interest to achieve sustainable decisions. The engagement and water diplomacy should enable mutually beneficial relationships and not be a deceptive control mechanism (Greenwood 2007). Public engagement should not be undertaken when the decisions are already made, and there is no space to change, no intention to include outcomes of the engagement process or as a decision-delaying tactic whereby the outcomes are not recognized in the decision making (Warburton et al. 2007).

Issues related to transboundary rivers and the management of such river basins are often quite complex, and in such situations stakeholder engagement plays a prominent role over the institutionalized technical solutions, because of the emotional involvement and behavior of stakeholders involved in achieving solutions. Among the many examples worldwide, the Boundary Waters Treaty (BWT) signed in 1909 between Canada and the United States and the International Joint Committee (IJC) established under the provisions of BWT are worthy of special mention for their lessons learned, both in terms of their success and failure (Grover and Krantzberg 2015), as well as for the realiza-tions of the stakeholder engagement needs and continuous innovative attempts for such engagement within the ecosystem approach (Jetoo and Krantzberg 2014; Krantzberg et al. 2014).

A number of outcomes are expected of a successful engagement process for water diplomacy; apart from the decisions themselves, development of stakeholder trust—needed for the acceptance of the decisions—is the key outcome expected, according to the majority of literature. Thus, a successful stakeholder engagement process should develop stakeholders' mutual trust where transparency and accountability play a major role. To this effect, Measham et al. (2009) state that transparency, determination and abil-ity to adapt are the desired characteristics identified from a government perspective for successful engagement processes, while success factors from the community perspective are identified as being independent, respecting the landscape, getting on with the job and avoiding burnout.

In addition, effective stakeholder relationships in diplomacy build an approach that appears to be resilient and adaptive to future decisions (Johnson et al. 2013; Pearson et al. 2010). Further, the involvement of stakeholders, especially in the early stages of an engagement process, has the advantage of easy dissemination of the participatory deci-sion (Keown et al. 2008) because the process facilitates social learning and increases the likelihood that needs and priorities of local communities are met. Voinov and Gaddis (2008) suggest that community stakeholders can better deliver the findings and recom-mendations of an engagement process to the decision makers in government than can the scientists, who may be perceived as external to the issue and the locality. It is also suggested that the presentations to the wider community, other stakeholders and media should be made by members of the stakeholder group engaged in the process since they are more respected and can better handle the impacts of policy decisions on local com-munity decisions (Keown et al. 2008).

The dimensions considered for decision making would increase with non-technical information entering into the process (Johnson 2013; Gudowsky and Bechtold 2013). Additionally, the extent of effective participation of stakeholders in water resources man-agement is determined by: the methods employed in engaging stakeholders, inclusion of diverse groups, group size, incorporation of local knowledge and expertise, and the time available for the process to develop (Voinov and Gaddis 2008), because "the quality of decision is strongly dependent on the quality of the process that leads to it" (Reed 2008).

An important aspect of a successful stakeholder engagement is the early identifica-tion and proper handling of the factors that could hinder the successful completion of the process. The key factor for the success of an engagement process is clarity of goals, one of ten factors identified in a recent survey carried out among 215 stakeholders, both within and outside of the water sector (OECD 2015, 169). Unclear goals lead to stake-holder fatigue as well as to breach of trust, which is the most important characteristic of a process.

The majority of experiences caution against five major areas, including clarity of goals, as reasons for the failure of the engagement processes. The other reasons for failure relate to (a) poor communication affecting smooth knowledge transfer between the lay stakeholders and scientific or institutionalized stakeholders; (b) compromising the dignity of some stakeholders and creating bias toward one or more stakeholders; (c) accuracy and availability of relevant data and information vital for participatory decision making; (d) insufficient time, resulting in a rush through the engagement process; and (e) poor planning or insufficient financial and human resources to facilitate effective engagement.

An Organization for Economic Co-operation and Development (OECD) study (2016, 172) further identifies six principles for stakeholder engagement: (a) inclusiveness and equity; (b) clarity of goals, transparency and accountability; (c) capacity and information; (d) efficiency and effectiveness; (e) institutionalization, structuring and integration; and (f) adaptiveness followed by check lists and indicators provided for each of the principles (OECD 2016, 174–83). Nonetheless, checklist items and indicators need to be carefully selected to suit the specific engagement process. Further, identification of activities required for attaining the checklist items and indicators is critical for the success of the process.

In summary, this chapter argues in support of the important contribution of stakeholder engagement toward water diplomacy while cautioning about the risk factors. Figure 14.1 illustrates relational ties between the steps in achieving the goals of an initiative.

The complexity of water issues arise from the needs of environmental (ecosystem) protection and the democratic governance. Water diplomacy is essential for finding solutions for such issues, and it is strengthened by the bottom-up approach where the stakeholder capacity is built. In other words, the local stakeholders (both institutional and community types) become capable of advocating diplomacy over unsuccessful top-down approaches that often encompass alien and highly technical solutions. In a nutshell, stakeholder engagements are inclusive and interactive social learning endeavors without deception and can help build social capital through trust, transparency and accountability. The informed social capital is expected to possess the capacity of resilience and adaptability for present and future changes required in water resources management.

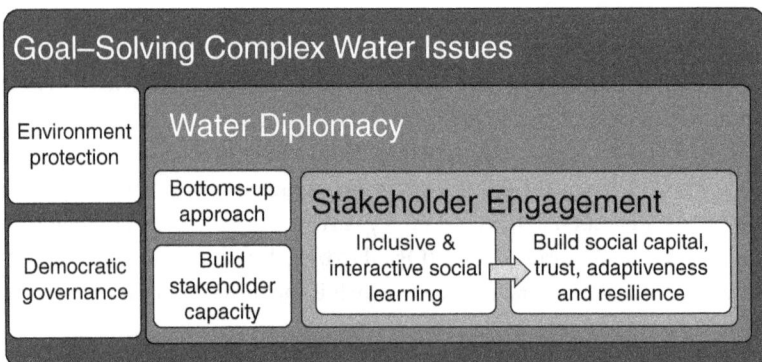

Figure 14.1 Relational ties in solving complex water issues.

Concepts Leading to Practice: Frameworks and Models

Notwithstanding the significant growth of knowledge since the early works of Freeman (1984), stakeholder engagement is said to be under-theorized in the corporate and management sectors, where Greenwood (2007) claims that "many accounts of stakeholder activities focus on the attributes of organisations or the attributes of stakeholders rather than on the attributes of the relationship between organisations and stakeholders."

We consider the above argument to be valid in the water resources management sphere since management of water resources involves complex decision making where such decisions must be agreed by the multi-use sectors and multi-stakeholder interests. Additionally, inconsistencies and knowledge gaps exist within the bureaucratic and political platforms in decision making. Comprehensive, yet easy-to-use frameworks for engagement will help managers make informed and strategic decisions (Mulley et al. 1997) in water diplomacy. Literature presents a number of frameworks spanning a wide spectrum of focus.

Literature from the corporate sector can be regarded as the origin of most concepts on stakeholder engagement. The stakeholder approach has been dealt with in corporate literature and constitutes two different schools of thought: first the thinking based on early works of Freeman (1984), who explicitly regarded the stakeholder approach as instrumental (a tool) in strategic management and, second, the thinking on the ethical and moral biases to consider it as a normative approach within the core functions of corporate strategy. Accordingly, the traditional stance of *stakeholder* engagement was buffering ("aimed at containing the effects of stakeholders on the firm") while later it changed to bridging ("forming strategic partnerships") (Sinclair 2011).

Engagement levels, stakeholder identification, stakeholder analysis and stakeholder salience have been studied from a number of viewpoints and theoretical traditions resulting in a multitude of engagement frameworks and models, both in the corporate and public sectors. Prominent theoretical concepts, frameworks and models from the literature are discussed here to elicit the factors contributing to the success or limiting the progress of an engagement process.

Continuum of Engagement Levels

Engagement could take place at different levels, from almost zero involvement of the stakeholder (beneficiary) to projects managed and funded by community organizations. Of the many models used for engagement, the three types presented by the American Bar Association's Standing Committee on Environmental Law for political and business decision making show the changes from the top-down to the bottom-up approaches. Peoples' participation in a process invited by the government, on terms defined by the government and in accordance with the government needs a paternalistic model in contrast to conflict or confrontational models where public participation is mainly through litigation. Demonstrating with examples, Glicken (2000) states these were the two models of historic participation in the United States. The third type is the consensus-building model whereby every affected group participates, often using self-designated representatives.

In relation to US federal social programs, Arnstein (1969) first proposed a typology of engagement and diplomacy with eight participation levels, using a laddering arrangement as a metaphor to illustrate the increasing stakeholder power in public participation (Figure 14.2). The eight levels from the bottom where citizens do not participate at all are manipulation, therapy, informing, placation, partnership, delegated power and citizen control (where fully empowered citizens are in control). This categorization has been referred to in many later works, for example, Canter (1996), Reed (2008) and the United Nations Centre for Human Settlement (UNCHS) (2001), which prove the claim by Arnstein that the categorization is suitable for application in many different fields.

The bottom two levels, manipulation and therapy, illustrate the insincere efforts by planners as a substitute for genuine participation. The real objective of the planners at this stage is not to enable people to participate but to enable the power holders to persuade the weaker public stakeholders. The third to fifth levels show the increasing level of tokenism, where planners allow the masses to hear and have a voice. However, at this stage there is no assurance that the will of public stakeholders will be heeded. The power of decision making still lies with the power holders. Levels six to eight denote where the citizens/public are involved in the decision-making process. The citizens at this stage will be able to enter partnerships enabling them to negotiate and to engage in trade-offs with traditional power holders (Canter, 1996). Arnstein (1969) dismissed the lower rungs in definitive terms for their dishonest and arrogant nature while advising counter arguments for higher levels also not to be taken lightly based on critical analyses of the federal programs of the time.

The Forum on Stakeholder Confidence (FSC, 2004) presents a similar concept with five levels of stakeholder participation (Figure 14.2): (a) Inform/Educate: This is the level when factual information is needed to describe a policy, program or process; a decision has been already made and the public need to know the results of a process; when there is a need for acceptance of a proposal before a decision may be made; or when the issue is relatively simple. (b) Gather information/views: The purpose is to listen and gather information; policy decisions are being made and there may not be firm commitment to do anything with the collected information. (c) Discuss or involve: This two-way information exchange is used when the stakeholders are likely to be affected by the outcome, and when there is an opportunity to influence the final outcome; input may shape policy directions and program delivery. (d) Engaged: When it is necessary for stakeholders to talk to each other regarding complex value-laden decisions, options generated together will be respected. (e) Partner: When the institutions want to empower the stakeholders to manage the process.

The International Association of Public Participation (IAPP) has developed a parallel spectrum (Figure 14.2); Inform, Consult, Involve, Collaborate and Empower (Warburton et al. 2007). All three spectrums of engagement by Arnstein, FSC and IAPP imply a hierarchy of levels with the higher levels providing for a more honest process and empowerment of stakeholders.

Contrasting the popular "laddering" systems, Davidson (1998) presents a "wheel of engagement (empowerment)" with Information, Consultation, articipation and Empowerment in the four segments of a circle in the context of council planning.

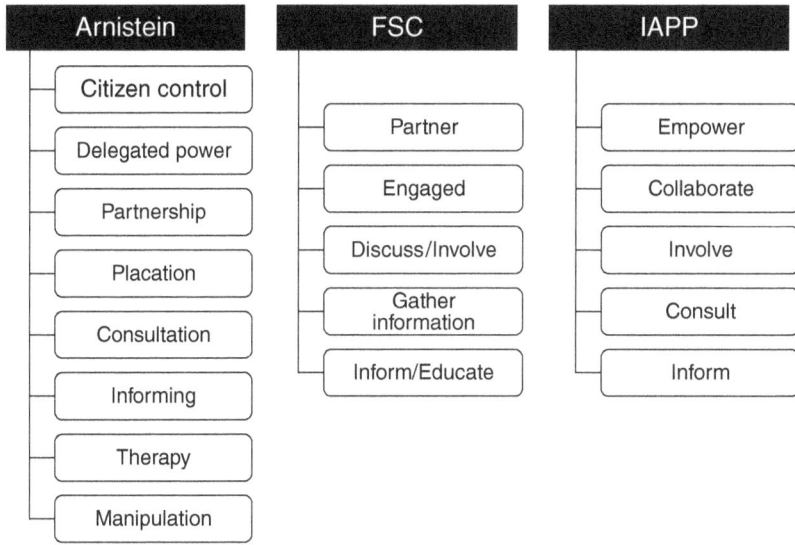

Figure 14.2 Comparison of three spectrums of stakeholder consultation levels.

Subdivisions within each quadrant facilitate minor variations of stakeholder power. The advantage of the "wheel" over the "ladder" is that a conscious collaborative decision on the objective of the engagement initiative sanctions the use of appropriate strategies and techniques that would otherwise be considered as low-level engagement. Further, choice of engagement level is limited by time, scale and resources available for the process (Warburton et al. 2007). Thus, we claim that the hybrid between the "ladder" and "wheel" could be advantageous in the engagement initiatives of water diplomacy since the strategies required for decision making differ in levels of objective and knowledge involved and, as such, the different contexts are likely to require different levels of engagement (Reed 2008; Warburton et al. 2007).

Stakeholder Analysis and Salience

Claiming that the public and non-profit sector literature is weak in systematic identification and analysis of stakeholders, Bryson (2004, 28–32) presents in detail five stakeholder identification and analysis techniques, which can be used to plan engagement processes in public-sector initiatives. He claims that the first three are particularly helpful in identifying stakeholders, and any one of them can be the direct staring point of the stakeholder identification process in water diplomacy. They are: (a) choosing stakeholder analysis participants; (b) basic stakeholder analysis technique and (c) power versus interest grids—after Eden and Ackermann.

Salience of the stakeholders is another aspect that needs attention in planning an engagement process, bearing in mind that as a project grows the salience could shift. The engagement process should, therefore, reflect the dynamic nature of salience (Johnson et al. 2013). Comprehensive identification analysis and ranking in terms of salience and capacity to influence are the key steps to recognizing stakeholders. Sinclair (2011) uses a four-tiered representation to illustrate levels of (salient) corporate stakeholder engagement, where high-level stakeholder input is useful in the joint planning on matters of strategic significance. It is proposed that this can be further adapted with three basic types of engagement processes for water-management initiatives (Figure 14.3). The salience of stakeholders varies among the types and, at times, within a type.

In corporate management, Mitchell et al. (1997) use three relational attributes of stakeholders to generate typology of stakeholders; (a)Their power to influence; (b) the legitimacy of their relationship; and (c) the urgency of their claims with the firm (Table 14.1). Salience will usually be increased with the number of attributes possessed: low with "latent stakeholders," moderate with "expectant stakeholders," and high with "definitive stakeholders." This theory has been successfully adopted by researchers in both the public and private sectors (Johnson et al. 2013; Erdogan 2013; Wamalwa

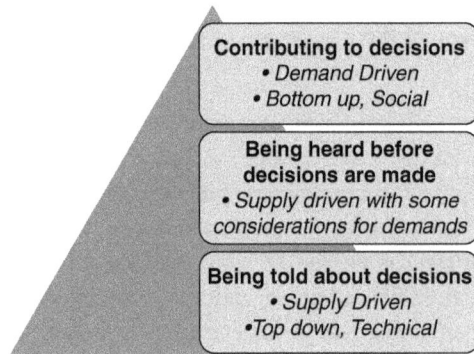

Figure 14.3 Three types of engagement processes in water management initiatives. *Source*: Adapted from Sinclair, 2011.

Table 14.1 Stakeholder typology.

Stakeholder Typology		Attribute		
		Power	Legitimacy	Urgency
Latent	Dormant	√	-	-
	Discretionary	-	√	-
	Demanding	-	-	√
Expectant	Dominant	√	√	-
	Dangerous	√	-	√
	Dependent	-	√	√
Definitive	Definitive	√	√	√

Source: Adapted from Mitchell 1997.

2009) for stakeholder identification and determining their salience. The advantage of the theory is its dynamic nature enabling managers to recognize situational uniqueness.

Frameworks and Models on Specific Features

Responsible Treatment of Stakeholders

Greenwood (2007, 322) presents a model that categorizes stakeholder involvement in eight segments that could merge and overlap, depending on the level of engagement process and the responsible treatment of stakeholders. He argues that the prevalence of corporate irresponsibility occurs when there is high engagement but low responsible treatment of the stakeholder—in other words when deception and manipulation are present. This argument and the model are helpful to view water management initiatives by government bodies when incorporating the public as stakeholders in an engagement process for water diplomacy.

Sample Surveys versus Action Conversations

Prior to making a decision on an issue, it is important to understand communities' views and attitudes toward the initiative as well as their awareness of the impacts of such decisions (Schwecke et al. 2009; Pinto and Maheshwari 2010; Faust et al. 2013). The usual practice is to employ large (in thousands) sample surveys that will return the attitude (acceptance/rejection) as well as the reasons for such an attitude (Schwecke et al. 2009; Pinto and Maheshwari 2010). In contrast, Webb et al. (2009) argue, based on their work, that an action conversation process-based framework is better for exploring broader community perceptions about water issues and future sustainability. Further, the process can be advantageously used to build a base of support for practical action within a community by engaging interested community stakeholders as full-time participants.

Engagement for Adaptive and Integrated Management

Carr et al. (2009) identified social learning and engagement as the drivers of adaptive and integrated approaches. Pearson et al. (2010) present a framework for the urban water management sector where Adaptive Management and Integrated Water Management are the two key principles for sustainable decision making at both strategic and operational levels.

Impact of Trust on Participation Strategy

Public trust is an important aspect in engagement initiatives for stakeholder decision making. Focht and Trachtenberg (2005, 86–102) developed and tested a framework for a participation strategy based on trust. They suggest that two judgments of trust, official trust and social trust, influence the willingness of stakeholders to show deference or be vigilant in the policy process. Accordingly, four stakeholder participation roles are identified: subdued, cooperative, enhanced and defensive. These require four stake-holder participation strategies: confirmation, facilitation, consultation and negotiation.

Techniques and Modelling Tools in Engagement

Scientific knowledge on catchments (watersheds) has been progressively built up over a long period of time. As a result there are numerous techniques, tools, procedures, methods, etc., based on different scientific knowledge (e.g., statistics, IT, etc.) to help make decisions in water resources management.

Techniques for Engaging Stakeholders

Choosing the appropriate technique for the involvement of stakeholders at the required level is a fundamental criterion for the success of an engagement process (Collier and Berman 2002; OECD 2004; FSC 2004). Different techniques can be grouped into four categories: (a) lower-level involvement techniques; (b) higher-level involvement techniques; (c) techniques for rural development work; (d) actions of facilitators. The first two categories broadly divide the involvement continuum into two segments. This categorization is especially for the purpose of presentation and does not prevent using the techniques differently when appropriate.

Lower-Level Involvement Techniques

These techniques especially suit engagement processes spanning from the inform/educate level to the discuss level and are mainly the development of appropriate public information materials for communication and techniques of consultations. The methods include but are not limited to advertising and social marketing, community mapping, fact sheets/backgrounder, focus groups, open house, bilateral meetings with stakeholders, polling, public hearing and seminars, questionnaires and surveys (for details see Minister of Public Works and Government Services Canada 2000).

Higher-Level Involvement Techniques

At the higher level of engagement spectrum from discuss/involve, and engage to partner, the techniques used to facilitate involvement of stakeholders include public hearings, deliberative polling, focus groups, citizen advisory groups, consultative groups, nominal group processes, multi-actor policy workshops, Charette, the Delphi method, round table, citizen task forces, study circles, scenario workshops, referendum, consensus conference, citizen juries and citizen panels (for details see FSC 2004).

Participatory Rural Appraisal (PRA) Techniques

In the rural development sector, a group of techniques evolved in the late 1970s under Rapid Rural Appraisal (RRA) to bridge the gap between limited information that can be elicited in surveys and the realities of rural deprivations, which are often missed (Chambers 1993). The PRA approach emphasizes local knowledge and encourages local people to work as teams. It encompasses a group of tools especially selected for their effectiveness in rural areas when it comes to encouraging stakeholders to share and analyze information through co-operation, team building and empowerment. These methods are also known as community-based approaches (World Bank 1996). Some of the PRA tools are: mapping and modelling, seasonal calendars (diagrams), institutional or Venn diagrams, timelines and trend and change analysis, wealth ranking and matrix scoring and ranking (Chambers 2002; Kumar 2002; Gosling and Edwards 1995).

Action Research Method

The action research framework is another development in community engagement facilitation where the role of a facilitator is to listen to local people and ensure their voices are heard by the decision makers. This approach has been proven for wider community engagement and its ability to identify the cultural elements that can inform further community education, leadership, ownership and capacity building for genuine and realistic community engagement (Webb et al. 2009). Grassroots communities usually possess broad awareness of issues associated with water; however, the solutions thrust upon them by professionals may be seen as complicated because of the use of too-high technological language (Craps and Prins 2004). Practical activities and visual communication methods have proved advantageous in engaging non-professional stakeholders as well as to develop alternative practical ideas for action. Action research is a cost-effective method which requires minimal training for participants (Webb et al. 2009).

The minister of Public Works and Government Services in Canada (2000) advocating the five levels of engagement of FSC (2004) illustrates the overlapping areas of involvement techniques. It highlights that the involvement techniques have no clearly demarcated boundaries between the levels, and most techniques of involvement could span different levels of involvement.

Modelling Tools for Engaging Stakeholders

Abstract representation of natural events and their causes is usually referred to as a model. The "simulation" models, which are built based on scientific inquiry and with assumptions to simplify natural systems and their relationships, allow the users to improve the insight of intricate real-life systems (Guitrich et al. 2005), in spite of their well-recorded deficiencies. On the other hand, the reasons for the limitations of "mental" models built by lay stakeholders are attributed to their qualitative nature, linear relationships, inability to incorporate lags in time and space and treating systems in isolation from their surroundings (Guitrich et al. 2005). Despite the above limitations, lay knowledge is increasingly accepted as essential in decision making.

Citing ten references for their important characteristics, Voinov and Gaddis (2008, 203) have categorized modelling tools into six groups: indices, statistical, spatial, process (temporal), lumped dynamic and distributed dynamic models. Nevertheless, the literature reports rejection of such models if based on scientific findings alone for the reason of non-incorporation of community knowledge, perspectives and values (Liu et al. 2008; Nyssen et al. 2006; Voinov and Gaddis 2008). This failure stems from the model assumptions associated with system uncertainties, which increase with the increasing complexity of issues. Conversely, engagement processes in the model-development stage allow discussions within which lay stakeholders are able to understand the unavoidable limitations due to a number of reasons, such as assumptions and the quality of available data: therefore, the extent of model reliability (Voinov and Gaddis 2008). Further, Voinov and Gaddis (2008) advocate that "Model selections should be determined based on the goals of the participants, the availability of data, the project deadlines and funding limitations rather than being determined by scientists' preferred modelling platform and methodology."

Five modelling concepts generally used in water management studies are briefly described in the following sections, and some of their applications are presented later.

System Analysis

Water, like any other natural resource, needs a whole-of-system approach based on the management of ecological, physical, social, cultural and other aspects of the environment (Larson and Williams 2009). Schwecke et al. (2009) suggest that technological fixes to a single issue in a system often only transfer the problem to another part of the system. Thus, in water resources management, finding solutions by using systems analysis enables all issues to be viewed at once and also facilitates dealing with all new impacts resulting from the proposed initiative (Mitchell 1990). Feasibility and the acceptability of the solution are determined by the important influence that stakeholders have on the system.

Group Decision Support Systems (GDSS)

GDSS combine communication, computer and decision technologies to support problem formulation and solutions in group meetings (DeSanctis and Gallupe 1987). The literature provides two definitions for GDSS: first, by Hurber (1984 cited in Gray 1987), "A GDSS consists of a set of software, hardware, and language components and procedures that support a group of people engaged in a decision-related meeting"; and, second, by DeSanctis and Gallupe (1985 cited in Gray 1987), "An interactive, computer-based system which facilitates solution of unstructured problems by a set of decision makers working together as a group." First emerging in the 1980s, GDSS are a combination of tools found on the Web and elsewhere to help generate, clarify, organize, reduce and evaluate ideas from groups, as well as tools to conduct surveys and rank criteria or alternatives to reduce process losses (all interaction within a group that slow down the process of making a decision) including disorganized activities, dominant members and social pressures (Koenen 2013). The specific decision-support tools and methods to be used depend on the nature of the specific case.

Though, one could expect GDSS to be very suitable for decision making in water resources management, its application in that management has been very rare in spite of the rapid technology enhancement. Koenen (2013) attributes the low-adoption of GDSS to fear of change, a need for more preparatory work for meetings, a fear of losing dialogue in meetings, the key role of the facilitator, over-enthusiasm leading to inappropriate use, a fear of disturbing the creative thinking in brainstorming (switch between thinking and using a computer) and the transparency of the system, which might be threatening to groups with power. In addition, the high technology may not be suitable for most stakeholders in the water sector engagement processes. Incorporation of GDSS is supported for its advantages, however, with taking maximum care to avoid pitfalls being documented. In the event of using GDSS for final decision making, with mathematical models to incorporate uncertainty, a preliminary testing system through role playing has been identified as mandatory for the processes to not be software-driven.

Multi-Criteria Decision Making (MCDM) Models

A single decision on water resource management impacts a number of sectors, and the decisions based on identified multiple criteria have been promoted in the literature. Multi-Criteria Analysis (MCA) analysis helps decision making when conflicting objectives are present, where the MCDM models have the advantage of accurately representing several objectives simultaneously (Grimble and Wellard 1997, 184). Of the many tools available for identifying decision making criteria, Hämäläinen et al. (2001) have used the "Value Tree Analysis" in two stages—first with the stakeholders obtaining their preliminary priorities to structure the problem by checking for any insignificant or missing criteria and, second, to prioritize the alternative decisions.

Bayesian Belief Network (BN)

Bayesian networks are perceived as effective tools in engagement as they enable dialogical learning with discussions and negotiations (Zorrilla et al. 2010; Portoghese et al. 2013). Based on Bayes's probability theory, the BN represents cause-and-effect relationships by using probability to convey the uncertainty between the input parameters and their dependent variables. Hence, they are argued to be especially suitable for group model building because of the difficulty in quantifying the exact values (uncertainty) in natural systems. The BN are claimed to be flexible for incorporating common understanding of stakeholders and can represent systems with changing environments in time and space to allow quantitative assessments of outcomes (Carmona et al. 2013).

Binary Probit Analysis

The particular strength of the method is claimed as a relatively low time-and-resources requirement to identify stakeholder groups through data collected on select behavioral characteristics of a sample and without the requirement of physical presence of key stakeholders for discussions and large time commitments.

Lessons from the Stakeholder Engagement Process

The report examining the natural-resource management organizations in Australia's Lake Eyre Basin shows how even the general factors of governance that impact engagement play out differently in remote areas requiring innovative and creative responses (Larson and Williams, 2009). Consequently, this section briefly considers salient points of global empowerment of the engagement process in watershed management from different viewpoints.

Two decades ago, a seven-country study by Mitchell (1990, 215) identified lack of processes and mechanisms as an explicit weakness in every country studied, and recommended development of better processes and mechanisms as a measure of integration of water resources management. This study also identified bottom-up approaches by the United States and Canada, while England was showing a shift away from the bottom-up approach. On the other hand, the case studied in Japan illustrated interest-based management.

Since then, stakeholder involvement in the decision making process has been increasingly advocated by the governments, practitioners and researchers from the problem-structuring stage to the group-consensus-seeking stage followed by a stage of seeking public acceptance for policy (Hämäläinen et al. 2001; Keown et al. 2008). Water resources management most frequently involves public projects. Public acceptance of a project is usually sought after making plausible decisions with some stakeholders representing the main interests. Therefore, the design of engagement processes to obtain such outcomes is important as the evaluated set of alternatives is presented to a larger audience and makes a good basis for public acceptance (Hämäläinen et al. 2001). The importance of process, continuous engagement, transparency and diverse stakeholder needs and communication preferences are clearly shown in the case of Tidal Power Development Project in Maine, in northeast United States (Johnson et al. 2013).

Hämäläinen et al. (2001) state that studying the overall value dimensions is important when first approaching a problem. The value dimensions include, for example, economical factors such as power production, flood damages, and ecological factors (e.g., changes in the spawning grounds of fish). The case of the Upper Guadiana Basin study (Zorrilla et al. 2010) is an example of such practical applications whereby stakeholders have been allowed to discuss the water management issues from a general perspective: agro-economic, institutional and hydrologic aspects. Grassroots stakeholders are able to make rational decisions when exposed to social, environmental and economic values in addition to the technical aspects of development initiatives (Thoradeniya and Ranasinghe 2012).

Integrated approaches rely heavily on legislation to provide legitimacy for decisions. Mitchell (1990, 211) confirms the above while arguing such legislations combined with political commitment results in steady progress of the integrated approach, either toward or away. UNCHS (2001), too, identifies political will and dedication as the most important factors in participatory approaches.

Wamalwa (2009), studying Kenya's Mara Watershed, identified a "disparity of views and perspectives, egoistic tendencies of tribal leaders, inadequate financial plans, lack of effective coordination mechanisms, ineffective multi stakeholder process, unwillingness of local community (if their interests are not addressed), lack of legitimacy for the institutional structures, lack of clearly spelt out roles among stakeholders" as barriers to integration.

Hämäläinen et al. (2001) emphasize the possibility of Evolutionary Systems Design by incorporating strong interactive participation in multi-criteria decision methods. In an interactive GDSS experiment incorporating "Value Tree Analysis" and "Method of Improving Directions," which allow learning and changes in the preferences for identifying Pareto-optimal alternatives for a river–lake regulation development project in Finland, they conclude that the framework supports interactive analysis of the problem and learning during all stages: problem structuring, identification of efficient alternatives, group consensus seeking and public acceptance seeking. The study also stresses the need for a preliminary testing system to decide on the most useful methods before final implementation, through role playing, which is also identified as a requirement for the processes not to be software-driven.

Zorrilla et al. (2010), in their study in Spain's Upper Guadiana Basin, state that the ability of BN to explicitly express the uncertainties is a particular strength in stakeholder engagement. The study has dealt with two key parameters (groundwater level and the local agrarian economy) and a relatively small number (20) of representative participants. It is claimed that the participatory process was structured in an attempt to develop a BN. Further, the use of BN in water planning could result in increased endorsement of strategies by the stakeholders. Another supportive argument, presented by Portoghese et al. (2013), is that the conflicting issues of water management could be a result of different knowledge frames, interests and beliefs of stakeholders, and BN are able to model such situations.

Nonetheless, insufficient integration of quantitative data (e.g., climatic data, groundwater levels, etc.) could lead to more qualitative bias outcomes, and stakeholders who are used to quantitative assessment may see the methodology as weak (Portoghese 2013). Further drawbacks of the BN include time-intensive training required to master the tool and its unsuitability in the dynamic systems due to the high volume of computational work required to solve probabilistic relations (Carmona et al. 2013; Castelletti and Soncini-sessa, 2007 as cited in Zorrella et al. 2010). Thus, its usage is limited by the initiators' knowledge in the theory of probability and their computer ability. Similarly, it is important to bear in mind that the accuracy of BN applications depends on the quality of input on dynamic stakeholder behavior where their thoughts and needs have to be converted as inputs.

In a case study on early decision making on a water system in the Sacramento–San Joaquin Delta, Faust et al. (2013) claims that water systems, although typically analyzed as monolithic, display System-of-System (SoS) traits according to the five characteristics: operational independence, managerial independence, evolutionary development, emergent behavior and geographic distribution of subsystems. The SoS has been used to complement binary probit analysis. In this instance, SoS analysis identifies stakeholders who need to be considered in the decision-making process, which then becomes an input to the stakeholder engagement process (which had been a questionnaire designed to obtain responses) for the binary probit analysis.

Building on the argument that the varied perceptions, beliefs and attitudes of stake-holders are based on their demographic and behavioral characteristics (Burgin et al. 2010), Faust et al. (2013) propose binary probit analysis to identify opposing and sup-porting stakeholder groups at the early stages of development initiatives.

What Can We Do to Achieve Effective Stakeholder Engagement?

The objective of engaging stakeholders and initiating water diplomacy in a decision-making process is to achieve well-accepted, Pareto-optimal solutions to address difficult-decision problems through social learning. The two principles, adaptive management and integrated management, are the drivers of social learning and engagement in sus-tainable decision making in water resources management. The experiences reviewed show that key requirements for the success of an engagement process are: inclusion and interaction of stakeholders, flexibility and transparency with diversity of knowledge and values, knowledge transfer among stakeholders and continuous engagement.

Engagement process for higher levels of diplomacy, that is, where the stakeholder contribution is made for decision making, will usually include three phases: problem structuring, identification of alternative solutions and decision making (there is also a phase of public-acceptance seeking). Establishment early in the process of the broader objective of the engagement initiative is important to avoid later misconceptions.

Winning the trust of stakeholders for the decisions made on IWRM is a main objective of the engagement process, which becomes second only to the decision itself. As shown by Focht and Trachtenberg (2005), the strategies of participation for winning stakeholder trust needs to be carefully chosen by understanding the attitudes of stakeholders.

Engagement processes should be initiated with genuine interest to see sustainable decisions and not be deceptive control mechanisms. Arnstein (1969) dismissed the lower two rungs of engagement, and both FSC and IAPP have not included those levels in their engagement continuum, clearly showing that engagement does not prevail at the lower levels. On the other hand, government and government agencies initiating stakeholder engagement should be vigilant to exercise responsible treatment of stake-holders whenever a high level of engagement is present to ensure that deception and manipulation will not set in during the process of engagement and breach the trust of stakeholders. Further, IWRM initiatives should hold stakeholder engagement as a normative approach and include appropriate types of engagement initiatives for diplo-macy in engineering projects when developing competing water resources, which will enhance their social acceptability. This will require a study to ascertain social feasibility to be incorporated in the watershed development projects and building the capacity (if necessary), which otherwise could become a major factor that restrains engagement processes.

Compared to corporate-sector engagement initiatives, time, site and issue perspec-tives play specific important roles in water management initiatives. In contrast, struc-tured engagement is a well-identified need in the stakeholder theories in the corporate sector, whereas engagements in water resources management lack such well-structured processes. Instead, they are usually initiated with a core group, and the process expands outward to major stakeholders and the public.

Nonetheless, it is argued here that the concept of Sinclair (2011), where the engagement framework progressively applied from core groups of stakeholders (corporate in this case) to other major stakeholder groups (communities, customers and employees) in the corporate sector is effective for water resources management as well. An engagement process could be required at any of the three stages shown in Figure 14.3, especially because of the space variations from international and national to local that is commonly experienced in water management. The salience of stakeholders could usually shift within a progressive engagement process.

Exposure of overall value dimensions encompassing technical, social, environmental and economic values of the resource uses, and allowing deliberations on a holistic approach, enrich the engagement process, which enables stakeholders to better understand the limitations and make conscious decisions.

Success of stakeholder engagement in IWRM heavily relies on the legitimacy of the process and the political will, which go hand in hand most of the time. Insufficiency of legislation for integration, noted by Mitchell (1990) over two decades ago, still prevails.

A well-developed array of processes and mechanisms for engagement is available in the current literature. Specific mention is due for the works of the World Bank (1996), the Minister of Public Works and Government Services Canada (2000) and OECD (2004) in the form of handbooks. However, the need for creative tools and techniques that suit specific situations, such as the case of Lake Eyre Basin, warrant further innovations and research.

It appears that the philosophy of continuum of engagement levels remain after four decades with only minor modification. The majority of the literature advocates the laddering arrangement compared to a circular arrangement in describing the continuum of engagement levels as it obviously depicts progressively increasing stakeholder power. Nonetheless, the three different types (situations) of engagement processes in water diplomacy need different levels of engagement and, in that context, a circular continuum of levels is more applicable (Figure 14.4).

In contrast the engagement techniques have shown remarkable development from early initiator-leading methods to action-research methods wherein the role of initiator reduces to a facilitator. The growing emphasis in the techniques is to hear the voice of the grassroots, lay stakeholders despite their urban or rural locality. The literature evidence also shows there is continuing research, for example the aspect of communication, for the improvement of the engagement processes (Gudowsky and Bechtold 2013).

The choice of mathematical, IT and other models as tools for engagement needs careful consideration of the stakeholder knowledge profile, transparency, availability and quality of data, financial and technical resources. As has been suggested by many researchers, the knowledge of the grassroots community stakeholder should never be subordinated to sophisticated computer-driven software and theoretical models for the simple reason that water management decisions are time-, site- and issue-specific.

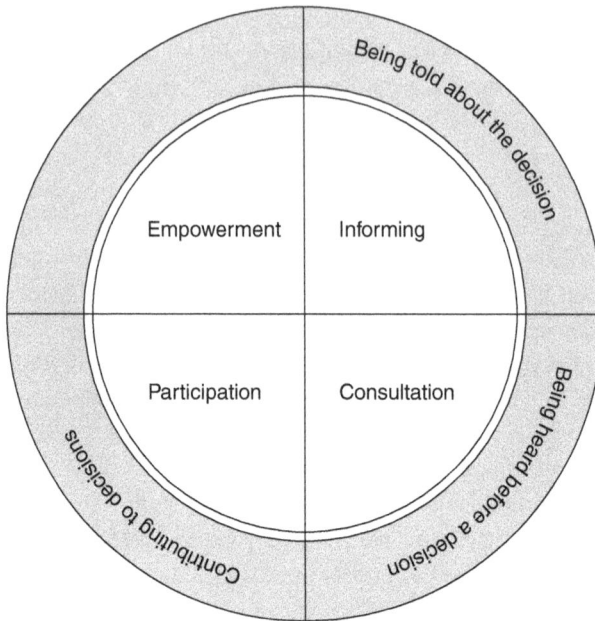

Figure 14.4 Types of IWRM engagement with engagement continuum.

Conclusion

Water being an important component of the environment and an essential commodity of humanity, its management has drawn the attention of the activists and researchers of environmental protection and democratic governance. In the cause of environmental conservation and safeguarding public interests, social learning through stakeholder engagement has gained a prominent position in decision making with regard to water resources management. The subject of stakeholder engagement has been studied, researched and practiced in abundance; resulting in a plethora of conceptual and practical outcomes in an array of subject areas, including but not limited to business, management, social and environment. The aim of this chapter is to review such literature and identify theoretical and practical aspects that would suit decision making in water resources management.

Democratic governance, diplomacy and increased attention to environmental protection have very firmly established the participatory bottom-up approaches over the former top-down approaches in decision making of IWRM. Stakeholder engagement and diplomacy in the context of water resources management is further justified with democratic, substantive and pragmatic rationales.

It is evident that the stakeholder engagement and water diplomacy process are still evolving as the theories of engagement in business and social spheres. Essential factors inherent in the engagement processes were identified through such past experiences.

Successful engagement processes for water diplomacy have threefold advantages: acceptability of difficult complex decisions made through the development of trust, which is achieved through transparency, accountability and legitimacy; adaptability and resilience for future decisions made through stakeholder relationships and better dissemination of decisions to local stakeholders. In addition, capacity building would also occur where there is insufficient social feasibility.

Theoretical models, analytical models and GDSS tools are all contributing to improved outcomes of an engagement process for diplomacy. This chapter does not attempt to critique individual models or tools that have been addressed in the literature, but it cautions the reader that the selection of those tools needs careful consideration of the factors: the issue that requires a decision; resources availability (time, specialists, equipment and funds); and the type of knowledge transfers that are anticipated, among others.

References

Arnstein, Sherry R. 1969. "A Ladder of Citizen Participation." *Journal of American Institute of Planners* 35(4): 216–24.

Bryson, John M. 2004. "What to Do When Stakeholders Matter: Stakeholder Identification and Analysis Techniques." *Public Management Review* 6(1): 121–53.

Burgin, Shelley, Daniel Williamson and Basant Maheshwari. 2010. "Natural Spaces—How Do They Influence Stewardship Attitudes and Actions of University Students?" *International Journal of Environmental Studies* 67(1): 63–78. doi: 10.1080/0020723090273892

Canter, Larry W. 1996. *Environmental Impact Assessment*. Boston: McGraw-Hill.

Carmona, Gema, Consuelo Varela-Ortega and John Bromley. 2013. "Participatory Modeling to Support Decision Making in Water Management under Uncertainty: Two Comparative Case Studies in the Gudiana River Basin, Spain." *Journal of Environmental Management* 128: 400–12

Carr, Dafna, Arlene Howells, Melissa Chang, Nadir Hirji and Ann English. 2009. "An Integrated Approach to Stakeholder Engagement." *Healthcare Quarterly* 12 (special issue).

Carson, Lyn. 2010. "Growing Up Politically: Conducting National Conversation on Climate Change." *Australian Policy On Line*. Accessed May 14, 2014 http://apo.org.au/commentary/growing-politically-conducting-national-conversation-climate-change.

Chambers, Robert. 1993. *Challenging the Professions: Frontiers for Rural Development*. West Yorkshire: Intermediate Technology Publications.

———. 2002. *"Participatory Workshops: A Source Book of 21 Sets of Ideas and Activities."* Florence, KY: Earthscan Publications.

Chess, Caron and Ginger Gibson. 2001. "Watersheds Are Not Equal: Exploring the Feasibility of Watershed Management." *Journal of the American Water Resources Association* 37(4).

Collier, C. and Greenstreet Berman. 2002. "Community Stakeholder Involvement" CIRI-6349A.

Craps, Marc and Silvia Prins. 2004. *Participation and Social Learning in the Development Planning of a Flemish River Valley: Case Study Report Produced under Work Package 5*. K. U. Leuven: Centre for Organizational and Personnel Psychology (COPP).

Davidson, Scot. 1998. "Community Planning, Spinning the Wheel." *Planning*, April 1998.

DeSanctis, Gerardine and R. Brent Gallupe. 1987. "A Foundation for the Study of Group Decision Support Systems." *Management Science* 33(5): 589–609.

Erdogan, Reyhan. 2013. "Stakeholder Involvement in Sustainable Watershed Management." In: *Advances in Landscape Architecture*. Rijeka: INTECH.

Faust, Kasey, Dulcy M. Abraham and Dan De Laurentis. 2013. "Assessment of Stakeholder Perceptions in Water Infrastructure Projects Using Systems-of-Systems and Binary Probit Analyses: A Case Study." *Journal of Environmental Management* 128: 866–76.

Focht, Will and Zev Trachtenberg. 2005. *Swimming Upstream: Collaborative Approaches to Watershed Management*, edited by Sabatier Paul A, Will Focht, Mark Lubell, Zev Trachtenberg, Arnold Vedlitz and Marty Matlock. Cambridge, MA: Massachusetts Institute of Technology Press.

Freeman, R. Edward. 1984. *Strategic Management: A Stakeholder Approach*. Boston: Pitman.

FSC. 2004. *Stakeholder Involvement Techniques, Forum on Stakeholder Confidence*. Organisation for Economic Co-operation and Development (OECD), www.nea.fr/html/rwm/docs/2004/rwm-fsc2004-7.pdf

Glicken, Jessica. 2000. "Getting Stakeholder Participation 'Right': A Discussion of Participatory Processes and Possible Pitfalls." *Environmental Science and Policy* 3: 305–10.

Gosling, Louisa and Mike Edwards. 1995. "Toolkit: A Practical Guide to Assessment, Monitoring, Review and Evaluation." *Development Manual 5*. London: Save the Children.

Gray, Pual. 1987. "Group Decision Support Systems." *Decision Support Systems* 3: 233–42.

Greenwood, Michelle. 2007. "Stakeholder Engagement: Beyond the Myth of Corporate Responsibility." *Journal of Business Ethics* 74: 315–27.

Grimble, Robin and Kate. Wellard. 1997. "Stakeholder Methodologies in Natural Resource Management: A Review of Principles, Contexts, Experiences and Opportunities." *Agricultural Systems* 55(2): 173–93.

Grover, Velma I. and Gail Krantzberg. 2015. "Transboundary Water Management: Lessons Learnt from North America." *Water International* 40(1): 183–98.

Gudowsky, Niklas and Bechtold Ulrike. 2013. "The Role of Information in Public Participation." *Journal of Public Deliberation* 9(1).

Gutrich, John, Deanna Donovan, Melissa Finucane, Will Focht, Fred Hitzhusen, Supachit Manopimoke, David McCauley, Bryan Norton, Paul Sabatier, Jim Salzman and Virza Sasmitawidjaja. 2005. "Science in the Public Process of Ecosystem Management: Lessons from Hawaii, Southeast Asia, Africa and the US Mainland." *Journal of Environmental Management* 76: 197–209.

Hämäläinen, Raimo, Eero Kettunen, Harri Ehtamo and Mika Marttunen. 2001. "Evaluating a Framework for Muti-Stakeholder Decision Support in Water Resources Management." *Group Decision and Negotiation* 10: 331–53.

HarmonyCOP. 2005. *Learning Together to Manage Together*. Osnabrück: University of Osnabrück, Institute of Environmental System Research.

Jetoo, Savitri and Gail Krantzberg. 2014. "Donning Our Thinking Hats for the Development of the Great Lakes Nearshore Governance Framework." In: *Journal of Great Lakes Research (in press)*. http://dx.doi.org/10.1016/j.jglr.2014.01.015

Johnson, Teresa R., Jessica S. Jansujwicz and Gayle Zydlewski. 2013. "Tidal Power Development in Maine: Stakeholder Identification and Perceptions of Engagement, Estuaries and Coasts." doi:10.1007/s12237-013-9703-3, Coastal and Estuarine Research Foundation.

Krantzberg, Gail. 2014. "Community Engagement Is Critical to Achieve a 'Thriving and Prosperous' Future for the Great Lakes–St. Lawrence River Basin." *Journal of Great Lakes Research* (in press), http://dx.doi.org/10.1016/j.jglr.2014.11.015

Keown, Kiera, Dwayne Van Eerd and Emma Irvin. 2008. "Stakeholder Engagement Opportunities in Systematic Reviews: Knowledge Transfer for Policy and Practice." *Journal of Continuing Education in the Health Professions* 28(2): 67–72.

Koenen, Sebastiaan. 2013. "Why Does Almost Nobody Use Group Decision Support Systems?" Paper presented at the 18th Twente Student Conference on IT, University of Twente, Enschede, The Netherlands.

Kumar, Somesh. 2002. *Methods for Community Participation: A Complete Guide for Practitioners*. New Delhi: Vistaar Publications.

Larson, Silva and Liana J. Williams. 2009. "Monitoring the Success of Stakeholder Engagement: Literature Review." In: *People, Communities and Economies of the Lake Eyre Basin: DKCRC Research Report 45*, edited by Tom G. Measham and Lynn Brake, 251–98. Alice Springs: Desert Knowledge Cooperative Research Centre.

Liu, Benjamin M., Yitayew Abebe, Oloro V. McHugh, Amy S. Collick, Brhane Gebrekidan and Tammo S. Steenhuis. 2008. "Overcoming Limited Information through Participatory Watershed Management: Case Study in Amhara, Ethiopia." *Physics and Chemistry of the Earth* 33: 13–21.

Maak, Thomas. 2007. "Responsible Leadership, Stakeholder Engagement and the Emergence of Social Capital." *Journal of Business Ethics* 74: 329–43. doi:10.1007/s10551-007-9510-5

Measham, Thomas G., Cathy Robinson, Carol Richards, Silva Larson, Mark Stafford Smith and T. Smith. 2009. "Tools for Successful NRM in the Lake Eyre Basin: Achieving Effective Engagement." In: *People, Communities and Economies of the Lake Eyre Basin: DKCRC Research Report 45*, edited by Tom G. Measham and Lynn Brake, 125–70. Alice Springs: Desert Knowledge Cooperative Research Centre.

Minister of Public Works and Government Services Canada. 2000. *Health Canada Policy Toolkit for Public Involvement in Decision Making*. Ottawa: Corporate Consulation Secretariat, Health Policy and Communications Branch

Mitchell, Bruce. 1990. *Integrated Water Management: International Experiences and Perspectives*. London: Belhaven Press.

Mitchell, Ronald K., Bradley R. Agle and Donna J. Wood. 1997. "Towards a Theory of Stakeholder Identification and Salience: Defining the Principle of Who and What Counts." *Academy of Management* 22(4).

Mulley, Paul, Bruce Simmons and Basant Maheshwari. 2007. "Sustaining Public Open Spaces through Water Recycling for Irrigation: Developing Decision Support Tools and Framework." *International Journal of Water* 3(4): 397–406.

Nyssen, Jan, Nigussie Haregeweyn, Katrien Descheemaeker, Desta Gebremichael, Karen Vancampenhout, Jean Poesenc, Mitiku Haile, Jan Moeyersons, Wouter Buytaert, Jozef Naudts, Jozef Deckers and Gerard Govers. 2006. "Comment on 'Modelling the Effect of Soil and Water Conservation Practices in Tigray, Ethiopia.'" Agriculture Ecosystems & Environment 105 (2005) 29–40. *Agriculture, Ecosystems & Environment* 114(2–4): 407–11.

OECD. 2004. *Stakeholder Involvement Techniques, Short Guide and Annotated Bibliography*. Paris: Nuclear Energy Agency, Organization for Economic Co-operation and Development.

———. 2015. *Stakeholder Engagement for Inclusive Water Governance: OECD Studies on Water*. Paris: OECD Publishing. http://dx.doi.org/10.1787/9789264231122-en

Pearson, Leonie J., Anthea Coggan, Wendy Proctor and Timothy F. Smith. 2010. "A Sustainable Decision Support Framework for Urban Water Management." *Water Resource Management* 24: 363–76.

Pinto, Uthpala and Basant L. Maheshwari. 2010. "Reuse of Greywater for Irrigation around Homes in Australia: Understanding Community Views, Issues and Practices." *Urban Water Journal* 7(2): 141–53. doi:10.1080/15730620903447639

Portoghese, Ivan, Daniela D'Agostino, Raffaele Giordano, Alessandra Scardigno, Ciro Apollonio and Michele Vurro. 2013. "An Integrated Modelling Tool to Evaluate the Acceptability of Irrigation Constraint Measures for Groundwater Protection." *Environmental Modelling & Software* 46: 90–103.

Reed, Mark S. 2008. "Stakeholder Participation for Environmental Management: A Literature Review." *Biological Conservation* 141: 2417–31.

Schwecke, Melanie, Bruce Simmons, Basant Maheshwari and Gavin Ramsay. 2009. "Community Awareness of the Use of Alternative Water Sources for Irrigation of Golf Courses: A Case Study of a Selected Site within the Sydney Metropolitan Area." *International Journal of Management and Decision Making* 10(3/4): 247–57.

Sinclair, Marie-Louise. 2011. "Developing a Model for Effective Stakeholder Engagement Management." *Asia Pacific Public Relations Journal* 12(1): 1–20.

Susskind, Lawrence and Shafiqul Islam. 2012. "Water Diplomacy: Creating Value and Building Trust in Transboundary Water Negotiations." *Science & Diplomacy* 1(3). http://www.sciencediplomacy.org/perspective/2012/water-diplomacy.

Thoradeniya, Bhadranie, Malik Ranasinghe and Nalin T. Sohan Wijesekera. 2012. "A Decision Support Tool for Stakeholder Involvement in Sustainable Water Resources Development." *Engineer, Journal of the Institution of Engineers, Sri Lanka* 45(4).

UNCHS. 2001. *27 Participation to Partnership: Lessons from UMP City Consultations, Urban Management Programme.* Nairobi: UNCHS (Habitat).

Voinov, Alexey and Erica J. Brown Gaddis. 2008. "Lessons for Successful Participatory Watershed Modeling: A Perspective from Modeling Practitioners." *Ecological Modelling* 216: 197–207.

Wamalwa, Isaac Wafula. 2009. "Prospects and Limitations of Integrated Watershed Management in Kenya: A Case Study of Mara Watershed." M.Sc. Thesis, Lund University, Sweden. http://www.oceandocs.org/handle/1834/7327?locale-attribute=fr

Warburton, Diane, Richard Wilson and Elspeth Rainbow. 2007. *Making a Difference: A Guide to Evaluating Public Participation in Central Government.* London: Involve and Shared Practice.

Webb, Tony, Shelly Burgin and Basant Maheswari. 2009. "Action Research for Sustainable Water Futures in Western Sydney: Reaching Beyond Traditional Stakeholder Engagement to Understand Community Stakeholder Language and Its Implications for Action." *Systemic Practice and Action Research.* 22: 1–14, doi:10.1007/s11213-008-9102-z

World Bank. 1996. *The World Bank Participation Sourcebook.* Washington, D.C.: The International Bank.

Zorrilla, Pedro, Gema Carmona García, Africa de la Hera, Consuelo Varela-Ortega, Pedro Martinez-Santos, John Bromley and Hans Jorgen. Henriksen. 2010. "Evaluation of Bayesian Networks in Participatory Water Resources Management, Upper Guadiana Basin, Spain." *Ecology and Society* 15(3):12. URL: http://www.ecologyandsociety.org/vol15/iss3/art12/

Chapter Fifteen

STRATEGIC INSIGHTS FOR CALIFORNIA'S DELTA CONFLICT

Mohammad R. Moazezi, Kaveh Madani and Keith W. Hipel

Abstract

Graph Model for Conflict Resolution (GMCR) is a systematic approach to handling strategic conflicts. GMCR is categorized as a non-quantitative game-theoretic model that provides a systematic framework to study conflicts and analyze behavioral characteristics of decision makers (DMs) in conflicts with respect to their preferences. The Sacramento–San Joaquin Delta—the major water supply source in California and a unique ecosystem—is on the verge of collapse because of an ongoing conflict over how to manage the Delta. In this chapter, GMCR is used to model this continuing conflict to better understand the options, preferences and strategies of the major decision makers participating in the Delta conflict. Coalition analysis is also used to identify the possible subset of decision makers with the motivation and opportunity to form coalitions and to reveal how creating a coalition can influence the outcomes of the conflict-resolution process. We anticipate our study provides strategic insight into the conflict for all parties and California policy makers. Furthermore, our study shows that, since a full-cooperation approach among the players hardly exists, external force to lead all players toward cooperation could be helpful.

Introduction

California's Sacramento–San Joaquin Delta is the largest estuary on North America's western coast. It is located in Northern California where it forms the eastern portion of the San Francisco Estuary, at the junction of the Sacramento and San Joaquin Rivers. The Delta has always played an important role in the wealth and economy of California, meeting agricultural and recreational needs as well as providing land and major water supplies for its residents (Lund et al. 2007). Its unique and complex ecosystem has made it a desirable home for more than 750 animal and plant species (McClurg 2010). The lives of Californians from the Bay Area and Silicon Valley to the San Joaquin Valley, the Central Coast and Southern California, including metropolitan areas such as Los Angeles and San Diego, directly depend on the Delta's water resources, its water conveyance infrastructures and its complex management (Hanak et al. 2012). The State Water Project (SWP) and the federal Central Valley Project (CVP) are two important water-delivery projects that export more than six billion cubic meters of water per year by their pumping facilities from the central and southern Delta to Southern California (Lund et al. 2007; McClurg 2010; Newlin et al. 2002).

Originally, the Delta was a marshland. In the late nineteenth century, to preserve existing farmlands, expand the farming capacity, and control floods, hundreds of kilometers of levees were constructed around the Delta and throughout its waterways (Suddeth et al. 2010). Today, the Delta consists of a network of channels and land areas hosting multiple developed cities and industrial agriculture lands that are protected by the extended levee system. These levees confine water flow to the riverbeds and thereby make it feasible to have major infrastructure such as water pumping plants, aqueducts, highways, railroads, shipping channels, natural gas pipelines and high-voltage transmission lines (McClurg 2010; Lund et al. 2007). California's growing population, expected to reach 45 million in 2020 and 92 million in 2100 (Lund 2004), will increase water and environmental demands, making the Delta's role more significant in California's future, especially in the face of climate change (Medellin-Azuara et al. 2008; Connell-Buck et al. 2011; Mirchi et al. 2013).

Management of the fragile Delta with its limited resources has always been associated with conflict, and the situation continues to exacerbate as a result of increasing demand and external factors such as droughts, floods, climatic changes, earthquakes and sea-level rise. For decades, the Delta has been faced with major problems. However, because of opposing interests of its various competing decision makers, no comprehensive and long-lasting solution has been developed to date. As a result, the Delta is in a serious crisis, with undesirable managerial, institutional, structural and hydro-environmental aspects.

Past studies have focused on characterizing the Delta crisis, trying to understand the problem and proposing solutions for it. Lund et al. (2007, 2008b) provided a detailed review of the history and various aspects of the crisis and suggested some alternatives for sustainable management of the Delta. This was followed by a variety of studies trying to select the best option out of the suggested alternatives considering the reported performance uncertainties under different decision criteria (Lund et al. 2008a; Madani and Lund 2011; Shalikarian et al. 2011; Mokhtari et al. 2012; Rastgoftar et al. 2012; Madani et al. 2014). Other studies focused on evaluation of the institutional and management aspects of the Delta problem. Hanemann and Dyckman (2009) focused on the CALFED institution and discussed why it failed to function effectively in resolving the Delta conflict. They concluded that the Delta problem is a zero-sum game with no feasible win-win solution. Madani and Lund (2012) challenged this conclusion, claiming that the Delta problem cannot be a zero-sum game because of non-linearity of the utility functions of the involved parties. They suggested that win-win solutions could be developed by taking advantage of issue linkage and "strategic loss" (Madani 2011) opportunities. Nevertheless, they agreed that the Delta stakeholders may still perceive the problem as a zero-sum game and prefer non-cooperation over win-win cooperative solutions. Based on the evolving/dynamic game structures concept (Madani 2010) and simple non-cooperative game theory stability definitions, they suggested that the Delta problem has evolved over time, going through different game structures. In their view, the Delta game, which initially started as a cooperative game, has experienced a Prisoner's Dilemma structure for many years. Under this structure parties preferred non-cooperation in fear of free rides by other parties, resulting in the failure of cooperative institutions such as CALFED. Nevertheless, they postulated why the chicken game would be the terminal stage of the Delta conflict in which one party must bear the cost of solving the Delta problem individually, considering the high risk of significant losses under non-cooperation. They suggested that the State of California might become the "chicken" of the game, and end up compensating for the losses, unless it tries to influence the game and force the parties to take action before it is too late.

While the previous studies focused on the institutional aspects of the Delta problem or on finding the best alternative to solve the Delta water-export problem, they did not comprehensively consider the strategic interactions of the Delta parties. Thus, the objective of this chapter is to provide strategic insights into the problem through a systematic study of the Delta conflict at the strategic level by formal modelling of this complex conflict in terms of decision makers, their options, objective preferences and the stability of feasible conflict outcomes. The Graph Model for Conflict Resolution (Fang et al. 1993; Fraser and Hipel 1984; Kilgour and Hipel 2005) is used as a method for modeling and analyzing the Delta conflict because of GMCR's unique capabilities in studying real-world interactive conflicts (Fang et al. 1993). GMCR constitutes a significant expansion of the conflict analysis methodology of Fraser and Hipel (1984), which is itself an improvement of Howard's Metagame approach (1971).

This chapter is organized in five sections. Section 2 describes the various aspects of California's Delta problem and discusses the main factors that have led to the conflict's status quo. Section 3 describes the GMCR method, and Section 4 applies this method to analyze the Delta conflict. Section 5 presents conclusions and discusses the major policy implications of the analysis.

California's Delta Crisis

The current Delta crisis is a product of a set of physical and institutional drivers (Lund et al. 2008a). The major factors that have contributed to this crisis are briefly described here.

Degrading Levees

The Delta's islands are mainly used for agriculture. These islands are protected by 1,770 kilometers of levees and have been subjected to profound land subsidence. This anthropogenic phenomenon is mainly because of using local material to elevate the barrier level in the initial levee construction. Decomposition of peat soil as a result of soil burning for weed control, and drainage activities for agricultural purposes along with natural erosion can be considered as main factors of lowering the Delta's lands, especially in the western and central Delta (Lund et al. 2007). Land subsidence has continued for many years and, consequently, in some places the land surface is estimated to be several meters below sea level (Lund et al. 2007), which has been rising because of climate change (Pfeffer et al. 2008). Sea-level rise and land subsidence have led to increased water pressure gradients and the threat of levee failure because of soil liquefaction. In addition, water-inflow changes and earthquakes can escalate the risk of levee failure and island flooding (Suddeth et al. 2010).

Since most levees were constructed with relatively primitive technology and on unstable foundations, they cannot properly resist natural disasters. This has become a major issue of concern, especially after the devastating Hurricane Katrina struck the Gulf Coast in 2005 with catastrophic consequences. Therefore, the feeling of insecurity about the Delta's flood protection system has undeniably increased in hazardous zones, especially in the western Delta, the central Delta and the Bay Area, where residents are in danger of losing lives and billions of dollars (Lund et al. 2007; 2008a; Suddeth et al. 2010).

Delta Salinity and Deteriorating Water Quality

The Delta's fresh water inflows have declined as a result of droughts, extensive water exports, upstream diversions and the failure of fragile levees. Since the Delta is the conjunction point for seawater and fresh water, a decreasing fresh water inflow has disturbed the balance between saline and fresh water, leading to saltwater intrusion, mainly in the central Delta. Moreover, contaminants resulting from the drainage of pesticide and nitrates from farms are deteriorating the Delta's water quality (Lund et al. 2008a; 2012). Water-quality issues have serious impacts on interests of the major decision makers of the Delta because of the tangible reduction in agricultural products, change in the taste of drinking water, increase in health risks and a range of harmful effects on the Delta's invaluable species and ecosystem (Lund et al. 2007).

Delta Ecosystem Degradation

The Delta is a unique resource and a desirable habitat for a diverse ecosystem, including 22 different fish species. In recent years, some of the Delta species such as delta smelt, Central Valley steelhead and winter-run Chinook salmon have been officially listed as endangered and threatened species that could become extinct (California Department of Fish and Wildlife, 2014) because of a range of natural and mostly human-caused phenomena. The human impacts include pollution, declining Delta inflows because of CVP and SWP operations, salinity, invasion of alien species, entrainment of fish at pumping facilities and lack of suitable spawning beds (Moyle et al. 2010; Lund et al. 2007; Madani and Lund 2011). These problems have caused a series of socioecological concerns which, in turn, have formed waves of political and legal pressure to reduce—through legislation and litigation—the total volume and frequency of water diversions (Lund et al. 2008a; Newlin et al. 2002; Norgaard et al. 2009; Tanaka et al. 2011).

Water Resources Availability and Management

The Delta is California's major water resource, providing two-thirds of its water supply. On average, half of the state's annual streamflow occurs in the Delta (Lund et al.

2007). Many California water users who live outside the Delta and rely on this water hub believe that they absolutely have the same right as in-Delta users to take advantage of the Delta as an invaluable asset of the state. They expect the state and federal agencies to facilitate their water exploitation, putting them in opposition to the in-Delta users' interests. Nevertheless, the Delta has a limited water supply that is simply not sufficient to satisfy all the urban, agricultural and environmental demands (Medellin-Azuara et al. 2011).

Questioning the reliability of Delta water transfers and discussing alternative water-transfer solutions create intense conflicts between decision makers with inherently opposing interests (Hanemann and Dyckman 2009). Despite all the controversies, four major alternative policies have been considered in recent years to deal with water exports in the long term: (a) continue exports via pumping from the Delta as usual; (b) construct a peripheral canal (or a tunnel) and convey upstream water around (or below the surface of) the Delta; (c) combine the current water-pumping method with a peripheral canal (hereafter, the dual conveyance strategy); (d) stop water exports completely to maximize the ecosystem benefits (Lund et al. 2008a).

The Delta's crisis suffers from management institutions and expectations that are undeniably divergent (Hanemann and Dyckman 2009; Madani and Lund 2012). Because of the absence of an authority with strong institutional support, and also the lack of a statewide vision, the controversies over how to manage the Delta resources among agricultural and urban water users, environmentalists and exporters, and Northern and Southern California have continued for six decades, and voluntary consensus seems unachievable (Madani and Lund 2012; Hanemann and Dyckman 2009). It is necessary to consider CALFED, here, as the major organized effort to develop a consensus-based solution to the Delta problem. The CALFED framework emerged in the mid-1990s in order to avoid a collision course between the state and federal governments, and also to enhance functionality and comprehensibility of the decisions that might affect not only Californians but also anyone whose interests are associated with the Delta's issues. CALFED consisted of 12 federal and 11 state agencies (Chung et al. 2002) and was supposed to be in charge of mitigating the crisis and fixing problems, as well as developing appropriate strategies and a comprehensive management plan (Hanemann and Dyckman 2009; Lund et al. 2007). However, in spite of some successes, the CALFED program failed in its mission (Hanemann and Dyckman 2009; Lund et al. 2007; Madani and Lund 2012) for many reasons, but mainly because of the lack of key factors that could enable CALFED to develop the required substance for success of environmental negotiations (Bruce and Madani 2014), such as consistent power and authority, uninterrupted political support, strong financial support, external incentives and decision makers' trust as to its impartiality.

Method: Graph Model for Conflict Resolution

Natural-resources management is overwhelmed by strategic conflicts at all interaction levels, from regional to international. Strategic conflicts are defined as decision situations involving two or more independent decision makers with individual choices and individual preferences that may conflict or coincide (Peng et al. 1997). Game theory provides a systematic framework to study conflicts and to analyze behavioral character-istics of DMs in conflicts with respect to their preferences (Madani 2010). Game theory is divided into two basic branches: quantitative and non-quantitative. Among the non-quantitative approaches, there is a wide variety of techniques for handling strategic con-flicts (Madani and Hipel 2011), including Metagame Analysis (Howard 1971), Conflict Analysis (Fraser and Hipel 1984), Drama Theory (Howard 1999) and Graph Model for Conflict Resolution (GMCR) (Fang et al. 1993).

GMCR is a flexible, comprehensive and easy-to-use method for modeling and analyzing strategic conflicts (Kilgour and Hipel 2005). It provides a holistic tool for the systematic study of real-world disputes and consideration of the strategic aspects of multiple-participant, multiple-objective decision making problems to identify reachable and stable solutions. Major advantages of GMCR for game theory model-ing include: the ability to handle any finite number of decision makers and options/outcomes; keeping track of both reversible and irreversible moves; inclusion of com-mon moves for players in the game; determining the strength (likelihood) of equilib-ria (possible outcomes); tracing the possible evolution path of the conflict; systematic tracking of all feasible movements during the game; and the ability of coalition analysis and determination of equilibria under different coalitions (Fang et al. 1993). Because of its inherent capabilities, GMCR has been widely used for analyzing a range of real-world strategic conflicts, including conflicts over water and environ-mental resources (Fang et al. 1997; Hipel et al. 1997; Kilgour et al. 2001; Li et al. 2004; Kilgour and Hipe 2005; Obeidi et al. 2006; Nadalal and Hipel 2007; Madani and Hipel 2007; Inohara and Hipel 2008; Madani et al. 2011; Madani 2013). For a detailed discussion of different characteristics of GMCR, readers are referred to Fang et al (1993).

Studying a conflict using the GMCR framework includes two major phases: mod-eling and analysis. The key components for constructing a conflict model based on GMCR are: (a) the set of DMs; (b) the set of options for each DM; (c) the set of feasible *states* (all feasible combinations of DM's option selections); (d) the set of allow-able states transitions, which shows the possible moves between states for each DM; and (e) the set of DMs' relative preferences over the feasible states (Kilgour and Hipel 2005). Typically, because of uncertainties and lack of information at the initial stage of the conflict, the preferences of the DMs may be intransitive. But, as the conflict evolves and the DMs gain more information about the decisions and preferences of other players, it is possible to rank all states of each DM from the most preferred to the least preferred to form transitive preferences. Figure 15.1 depicts the key components of GMCR method.

The analysis phase starts once the essential elements of a conflict model are recog-nized. It includes two steps: stability analysis and interpretation. A stability analysis is carried out using a range of stability definitions. Stability analysis involves using a set of rules for calculating whether a DM would prefer to stay at a state or move away from it unilaterally. In this way, stability analysis provides a comprehensive procedure for

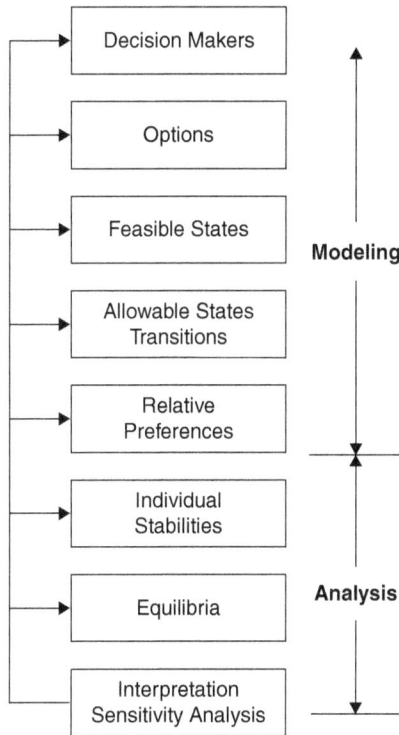

Figure 15.1 Overview of GMCR structure (Fang et al. 1993).

answering what-if questions regarding a stability of each possible outcome. The stability definitions, or solution concepts, describe different styles of human behavior under the assumption of rationality, reflecting the DMs' foresight level, risk acceptance and conservatism, strategic readiness and knowledge of opponents' preferences (Fang et al. 1993; Madani and Hipel, 2011; Madani, 2013). By definition, a given state is stable for a DM if it is not advantageous for the DM to move away from it. A state that is stable for every DM under the same solution concept is an equilibrium or a possible resolution of the conflict. If a state is an equilibrium under a range of stability definitions, it is considered as a strong equilibrium or a likely resolution of the conflict (Madani and Hipel, 2007).

The major stability definitions normally used within the GMCR framework include Nash stability (R) (Nash, 1951), General Meta-rationality (GMR) (Howard, 1971), Symmetric Meta-rationality (SMR) (Howard, 1971), Sequential Stability (SEQ) (Fraser and Hipel, 1984), Limited Move Stability (L_h) (Fang et al., 1993; Zagare, 1984), and Non-Myopic Stability (NM) (Brams and Wittman, 1981). Table 15.1 summarizes the main characteristics of these solution concepts, discussed by Fang et al. (1993) and Madani and Hipel (2011).

Table 15.1 Behavioral characteristics of the main stability definitions in GMCR (Fang et al. 1993; Madani and Hipel 2011).

Stability Definition	Foresight		Knowledge of Preferences	Willingness for Disimprovement	How does a focal DM anticipate the other DMs will react to his improvement?
	Level	Number of Moves Ahead			
Nash Stability (R)	Low	1	Own	Never	None
General Metarationality (GMR)	Medium	2	Own	Only for sanctioning	May sanction focal DM's improvement at any cost
Symmetric Metarationality (SMR)	Medium	3	Own	Only for sanctioning	May sanction focal DM's improvement at any cost
Sequential Stability (SEQ)	Medium	2	All	Never	May sanction focal DM's improvement, but only using their own improvements
Limited Move Stability (L_h)	Variable	$h \geq 1$	All	Strategic disimprovement for future increased gain	Symmetric, i.e. others optimize for themselves just like the focal DM
Non-myopic Stability (NM)	High	Unlimited	All	Strategic disimprovement for future increased gain	Symmetric, i.e. others optimize for themselves just like the focal DM

Delta Conflict Model

Major Players

In this analysis, the Delta's diverse DMs are categorized into four major player groups based on similar interests, shared powers, and common goals. The major Delta DM groups are: water exporters, environmentalists, upstream and in-Delta water users and the State of California.

Water Exporters

Water exporters transfer fresh water from the Bay-Delta (mostly from the North) to elsewhere (mostly to the South) for use in metropolitan and agricultural areas. Users in Southern California, the Bay Area and the San Joaquin Valley rely on these water supplies. The total water export by the two water export facilities, CVP and SWP, is between 6,200 and 7,400 million cubic meters per year, which is equal to about 15 percent of the natural flows into the Delta watershed and 25 percent of inflows after reduction through other upstream diversions (Lund et al., 2007). The CVP has 20 reservoirs that are mostly used to supply agricultural demands The SWP has 22 reservoirs, with over 70 percent of their total storage capacity dedicated to cities and industries (Chung et al., 2002). Recently, attention has been given to the southwestern edge of the Delta where the CVPs and SWPs pump intakes are located, with the maximum capacity of 425 cubic meters per second. Beside these two major exporters, there are some minor exporters, such as the Contra Costa Water District (CCWD), the federally owned Contra Costa Canal and hundreds of small diversions used by the Delta farmers (Hanemann and Dyckman, 2009; Lund et al., 2008a). In our analysis here, we only consider the operators of the CVP and SWP as the water exporters.

The water quality and reliability of the Delta's fresh water supplies are of significant importance for this DM group (Lund et al., 2007). Traditionally, they have been seeking to extract more water at the lowest price. But, today, more than anything else they want sustainable deployment without frequent restriction or interruption.

Environmentalists

Environmentalists include groups and organizations concerned with the Delta's ecosystem and key species. They are interested in having healthier native fish species populations as well as urban and agricultural water efficiency. This player group has made a major effort in educating the public and policymakers about the causes and consequences of the Delta's environmental problems, and in helping the Delta improve its critical condition and quality standards to become a desirable habitat for its entire terrestrial and aquatic species, especially the endangered species, as it used to be.

This DM group prefers lower water export levels and more restriction on water diversions. Water quality control, as well as fish population viability and increase are the other main concerns of environmentalists.

Upstream and In-Delta Water Users

This group includes all Delta water users, except for those who rely on imported water from the Delta. Hereafter, the term *water users* includes those users who live in the Delta area, regardless of whether they are urban users or farmers. The main reason for excluding the Delta water importers from this group is that their interests and preferences are mostly in line with those of the water exporters. The water importers, such as Southern Californians, play their role in the process of decision making by pressuring the water exporters. Therefore, they are not included in the analysis as a separate DM group, under the assumption that they have no direct control over the conflict. The water users group includes both agricultural users and urban users. This is because, although their water demands are temporally divergent, both sectors are mainly concerned with sufficiency and acceptability of Delta water quantity and quality to satisfy their water demands.

The history of the Delta's conflict (Hanemann and Dyckman, 2009; Lund et al., 2007; Madani and Lund, 2012) reveals the important role of the water users in this long-standing dispute. One piece of evidence that supports this claim is an incident in 1982, when SWP's efforts over several years to construct the Peripheral Canal were defeated in a referendum by a coalition of water users and environmentalists, despite the fact that it was supported politically and financially by the state and the US Bureau of Reclamation (Lund et al., 2007).

The water users seek reliable fresh water supplies and a levee system. Salinity intrusion and degraded water quality, fragile levees, expansion of water exports and diversions are some of the major concerns of this group.

The State of California

The State of California is in charge of sustainable development of the Delta and the whole state. The major costly effort of the state to find a solution to the Delta crisis through a cooperative mechanism, CALFED, was a failure. California is one of the major players in the Delta problem and is subject to potential major losses, unless it takes some serious action to resolve the Delta conflict (Madani and Lund, 2012).

Having a safe society with a prosperous economy and developing a consensus-based solution for the Delta crisis by involving different DMs who have strong opposing interests are the main objectives of this focal DM.

The Players' Options

The options or courses of action for each of the main DMs in the Delta conflict are described below.

Water Exporters

The water exporters are assumed to have three mutually exclusive options: continuing as usual, constructing a new water export system, and stopping exports.

Continue as Usual: Taking this option means that water exporters continue to pump and export Delta fresh water regardless of the consequences. Although this approach has no new or extra financial burdens for the focal DM, it exacerbates the Delta's situation in terms of water quality, reliability and the ecosystem. This option makes the Delta more unstable and vulnerable, which can lead to sociopolitical unrest, with a high risk of economic loss for the state. This option might be eventually exposed to strong sanctions by the environmentalists, who can take legal action against pumping operations to lessen entrainment and to protect fish and wildlife populations. Under this option, the water-delivery capability of the water exporters is also under threat by probable failure of the fragile levees. Water-delivery interruptions along with severe reduction in water quality can have negative effects on the water market and the water exporters' revenues (Jenkins et al., 2001).

Construct a New Water Export System: This action would be a significant step forward to attaining a dramatically improved situation in the Delta. As mentioned earlier, the Peripheral Canal option, first proposed by the water exporters in the 1960s, was rejected by California voters in 1982, mainly because they found it incompatible with their interests. Recently, the construction of a new water conveyance system has been a subject of discussion on technical solutions for the Delta crisis. The purposes of building a new system for exporting water have been fundamentally changed relative to the Peripheral Canal of the 1980s and have tended to be more sustainable rather than being greedy.

This action involves construction of a new water conveyance system. This new system, regardless of being a peripheral canal, a dual conveyance system, or any similar set of transfer canals or tunnels, is supposed to provide fresh water of good quality, and to protect water supplies and their serviceability from the risk of a levee failure and also from frequent interruptions because of tense conflicts over the Delta's ecosystem.

Stop Exporting Water: This environmental-friendly option is the most, extreme alternative for the water exporters. Historically, the water exporters have made substantial investments in implementation and operation of the water-export facilities. Hence, reducing their water shares or, in the worst case, stopping water exports means a significant economic failure for them. On the other hand, the water users in Southern California, the Bay Area and the San Joaquin Valley rely on this resource, and the exporters should satisfy their demands. Furthermore, one of the main financial sources for the northern agriculture users is leasing their excess water to their southern counterparts. Therefore, cutting off water exports might have considerable consequences for California's economy and society, while it could serve the environment more than any other option.

Environmentalists

Since one of the significant aspects of the crisis is the health and population of the endangered fish species in the Delta, the decisions of the environmentalists as a group concerned with this issue have a great impact on the conflict. Historically, the Delta has experienced three waves of environmental concerns related to emerging the SWP, the Peripheral Canal proposal and unrestricted water extractions during extended drought conditions. The environmentalists want to make sure that enough water is guaranteed for fish species (Lund et al., 2008a). They can take legal action, including filing new lawsuits against the water exporters or can pursue pressuring them through various political channels. During the Delta conflict, this strategy has been effective from time to time in bringing other key DMs to negotiations for compromise on environmental issues (Lund et al., 2007).

Water Users

The water users have two basic options at their disposal: levee improvement and sharing reduction.

Levee Improvement: The Delta's fragile levee system is a common concern for all DMs in the Delta conflict. The citizens of developed cities and the farmers, surrounded by the levee system, are concerned about their lives, safety and properties. They are interested in having a cost-effective fix and rehabilitation of this poorly constructed protective system. This issue is also a major concern for the water exporters and the State of California since their major infrastructures are located in this area and are subject to the risk of failure. Likewise, the environmentalists want immediate improvement of the levee system to address salinity intrusion.

Although improving the levee system is an issue of concern for all parties, they prefer to have a free ride and let other players, especially the wealthier players (the water exporters), bear the costs of this improvement. This attitude leads to parties' gambling with assets and resources by not contributing their fair share and delaying addressing the problem while gaining trivial benefits in return. This is perhaps a momentous opportunity for the State of California to take advantage of its leadership role to resolve the issue.

Reducing Water Exports: Reduction of water exports is the least-preferred option for the water users. As a strategy to improve the water quality and ecosystem, this option could be chosen voluntarily or by compulsion (under pressure from the legal system). However, this option is costly for both the focal DM and the State of California because of its adverse impact on agricultural productivity. This option may also contain political costs for the state and bring turmoil to the community.

California

The State of California has a range of reward and punishment policies as options to force the DMs to develop consensus. The most notable alternatives among them are putting pressure on water exporters and providing financial incentives to water users.

Pressure on the Water Exporters: California can exert pressure on water exporters to construct a modified water export system. A solution can be achieved if the water exporters agree with the new water-transfer option and agree to bear the major expense of implementing this option. The history of the conflict suggests that it is unlikely that the exporters will voluntarily agree to help in implementing this option, which seems unnecessary in their view.

Provide Financial Incentives: Immediate improvement of the levee system seems an unlikely outcome of the conflict, as levee improvement is not at the same priority level for all players. However, California can provide financial incentives to convince the water users to improve the levees and prevent further postponement of rehabilitation of the fragile levees system.

Table 15.2 summarizes the players, their options and the status quo of the Delta conflict. Each DM controls one or more options that it can choose to exercise or not. Here, "Y" means selected and "N" means not selected. As can be seen, in status quo, water exporters continue business as usual, do not construct a water conveyance system and do not stop exporting water. The environmentalists do not take any serious legal action against the water exporters. The water users do not want to pay for levee improvement and do not reduce their water shares either. The State of California does not intervene in the conflict and lets the DMs find the solution voluntarily.

Table 15.2 The DMs, their options, and status quo of the Delta Conflict.

Decision makers	Options	Status quo
Water exporters	1. Continue as usual (status quo)	Y
	2. Construct a new water export system	N
	3. Stop exports	N
Environmentalist	4. Legal action	N
Water users	5. Levee improvement	N
	6. Reduction	N
California	7. Pressure	N
	8. Financial incentive	N

Preparing Input Information for GMCR II

The information pre-processing stage includes scenario generation, scenario elimination by removing infeasible states, and ordinal preference specification using GMCR II (Fang et al., 1997)—the decision-support tool used in this study to model the Delta conflict within the GMCR framework.

Generation and Elimination Scenario

In principle, all possible combinations of players' options represent the possible scenarios of the problem. However, not all of these combinations are feasible in practice. Thus, there is a need to remove the infeasible scenarios or states before carrying out stability analysis for identifying the likely outcomes of the conflict.

With eight options, $2^8=256$ outcomes are mathematically possible in the Delta conflict before eliminating infeasible states. A large number of states are logically infeasible for certain players. For example, the water exporters' options are mutually exclusive in this dispute. This player-group also has to select at least one option among its available options, i.e. it cannot say no (N) to all options. Some other states might be preferentially infeasible (Fraser and Hipel 1984) for some players because of option-dependence and the players' strategic view of the game. For example, the water exporters are not expected to stop exports without any sanction (e.g., court order). Removing all infeasible states leaves 80 feasible states for the stability analysis.

Specifying DM Preferences

Once the feasible states are identified, players' preferences among these states should be specified. The ordinal preference information for each player can be determined within GMCR II using three methods: direct ranking, option weighting and option prioritizing. The direct ranking method involves pair-wise comparisons among the feasible options in order to explicitly express the preferences. This ranking method is simple and useful only for small-size problems (Peng et al. 1997). For larger games, option-weighting and option-prioritizing methods are more efficient and practical. In option weighting, for each player a numeric positive or negative weight is assigned to each option to show its degree of desirability for that player. In option prioritizing, a set of preference statements is listed according to importance for each player. Preference statements are combinations of non-conditional, conditional, bi-conditional and logical terms (Peng et al. 1997). In many problems, neither of the methods can fully order the feasible states according to the preferences of the players, and an additional adjustment step is needed in order to ensure the preference rankings are accurately represented in the model. Therefore, direct ranking is used as a final complementary step to revise rankings.

Here, the option-prioritizing method was used to develop the initial ranking of the states based on the DMs' preferences as illustrated in Table 15.3. This table provides the players' preference statements from the most important at the top to the least important at the bottom. Option prioritizing was followed by direct ranking to achieve better accuracy in preference specification. Numbers in the first column of Table 15.3 refer to the option numbers listed in Table 15.2, and the symbols "-", "IF" and "IFF" represent "not," "if" and "if and only if," respectively.

Table 15.3 Preference statements of the DMs in the Delta conflict.

Preference Statement	Description
Option Prioritization for "Water Exporters"	
–4	Water Exporters mostly prefer that the Environmentalist do not take any legal action against exporting.
–3	Next, they prefer not to stop water exports.
–7	Then, they would not prefer to be under pressure by the State of California.
2 IF 4	Water Exporters prefer to construct a new water conveyance system if the Environmentalists take legal action against them.
1	Water Exporters prefer to continue water exports as usual.
5	Water Exporters prefer that Water Users improve the levees for pumping plants reliability and pipelines safety.
Option Prioritization for "Environmentalists"	
3	The Environmentalists mostly prefer that the Water Exporters stop exports from the Delta (either with external force or without).
2 IF –3	Next, they prefer that the Exporters construct a new water conveyance system if they do not stop exporting water.
7	They would prefer that the State of California also exerts pressure on the Water Exporters for constructing a new water conveyance system.
6	The Environmentalists prefer share reduction by the Water Users.
–4	Taking no legal action is the least preferred option of the Environmentalists.
Option Prioritization for "Water Users"	
–6	Water Users mostly prefer not to reduce their water shares.
5	Next, they are willing to improve the levee system.

Table 15.3 Continued

Preference Statement	Description
8	The next preference of the Water Users is that California provides financial support to them.
3	They prefer that the Water Exporters stop their water extraction.
2 IF –3	They would like to construct a new water conveyance system if the Water Exporters do not stop water transfers.
–4	Water Users prefer that the Environmentalists would not take any legal action.

Option Prioritization for "California"

–8	California would not like to provide financial incentives for levee improvement.
2	California prefers that the Exporters construct a new water conveyance system.
–7	California does not prefer to pressure the Water Exporters.
5	California prefers that Water Users improve the levee systems.
–1	California prefers that the Water Exporters do not continue with water exports as usual.
7 IFF 1	California prefers putting pressure on Water Exporters if and only if they decide to continue exporting water as usual.
–3	California prefers that the Water Exporters do not stop the exports.
8 IFF –5	California is willing to provide financial incentives to Water Users if and only if they refuse to improve the levees voluntarily.
4 IFF 1	California prefers that the Environmentalists take legal action if and only if the Water Exporters decide to continue doing business as usual.

The following general observations about the final ranking of the 80 feasible states of the Delta conflict are noteworthy:

- The most desirable state for the water exporters is the status quo, and the least preferred state for them is the one that includes stopping water exports under the state's pressure and in response to legal action by the environmentalists. The exact opposite is true for the environmentalists—that is, the ideal state for the environmentalists is the worst scenario for the water exporters. Table 15.4 shows the ten most and ten least preferred states for the water exporters, suggested by GMCR II after introducing option prioritization to the model.
- For the water users, the most preferred state is the one including levee improvements without any reduction in water shares along with the state's financial support. The worst state for the water users occurs when water-use reduction is added to the status quo.
- The State of California prefers resolution of the conflict on a voluntarily basis without its involvement. Being aware of the current risky situation of the Delta, California prefers constructing a new water conveyance system by the water exporters and improvement of the levee system by the water users without any direct intervention. The worst

Table 15.4 The ten most and least preferred states of the water exporters in the Delta conflict.

EXPORTER'S PREFERENCES		10 MOST PREFERRED STATES									
DMS	Options										
Exporters	1- Continue	Y	Y	Y	Y	Y	Y	Y	Y	N	N
	2- Construct	N	N	N	N	N	N	N	N	Y	Y
	3- Stop	N	N	N	N	N	N	N	N	N	N
Environmentalists	4- Legal	N	N	N	N	N	N	N	N	N	N
Users	5- Levees improvement	Y	Y	Y	Y	N	N	N	N	Y	Y
	6- Reduction	Y	N	Y	N	N	N	Y	Y	Y	N
California	7- Pressure	N	N	N	N	Y	Y	Y	Y	N	N
	8- Financial help	Y	Y	N	N	N	Y	Y	N	N	Y

EXPORTERS' PREFERENCES		10 LEAST PREFERRED STATES									
DMS	Options										
Exporters	1- Continue	N	N	N	N	N	N	N	N	N	N
	2- Construct	N	N	N	N	N	N	N	N	N	N
	3- Stop	Y	Y	Y	Y	Y	Y	Y	Y	Y	Y
Environmentalists	4- Legal	Y	Y	Y	Y	Y	Y	Y	Y	Y	Y
Users	5- Levees improvement	N	N	Y	Y	Y	Y	N	N	N	N
	6- Reduction	Y	N	Y	N	N	Y	Y	N	Y	N
California	7- Pressure	Y	Y	N	N	Y	Y	N	N	Y	Y
	8- Financial help	N	N	N	N	Y	Y	Y	N	N	Y

situation for California would be when water exports are stopped as a result of legal action by the environmentalists and under California's pressure while water users do not improve levees despite the provided financial incentives.

Stability Analysis

GMCR II was used for stability analysis using the six stability definitions illustrated in Table 15.1. The equilibria of the game (stable solutions for all parties and likely resolutions of the conflict) are given in Table 15.5. Based on this table, the game's equilibria can be divided into two groups. The stronger equilibrium group includes states 8, 28, 48 and 68, which are stable for all DMs under all six solution concepts considered (Nash or Rational (R), GMR, SMR, SEQ, L(2) and NM). The second equilibrium group (weak equilibria) includes states 7, 27, 47, and 67, which are not Nash-stable, but are in equilibrium under the other solution concepts.

None of the eight equilibria of the game involves stopping water exports by the water exporters and sharing reduction by the water users. This clearly indicates the significance of keeping the same level of access to the Delta's water for these two groups of stakeholders and underlies the fact that they will not accept any resolution that involves reduction in their water use. On the other hand, all equilibria include improvement of the levee system by the water users. Given the vulnerability of the water users to failure of the

Table 15.5 Stability analysis results.

DMs	Options	States							
		7	8	27	28	47	48	67	68
Water Exporters	1-Continue	N	Y	N	Y	N	Y	N	Y
	2-Construct	Y	N	Y	N	Y	N	Y	N
	3-Stop	N	N	N	N	N	N	N	N
Environmentalist	4-Legal action	N	Y	N	Y	N	Y	N	Y
Water Users	5-Levee improvement	Y	Y	Y	Y	Y	Y	Y	Y
	6-Reduction	N	N	N	N	N	N	N	N
California	7-Pressure	N	N	Y	Y	N	N	Y	Y
	8-Financial support	N	N	N	N	Y	Y	Y	Y
Stability Definition	R		√		√		√		√
	GMR	√	√	√	√	√	√	√	√
	SMR	√	√	√	√	√	√	√	√
	SEQ	√	√	√	√	√	√	√	√
	L(2)	√	√	√	√	√	√	√	√
	NM	√	√	√	√	√	√	√	√

√ shows that the column state is stable for all players (is an equilibrium) under the row stability definition.

levee system, this group may eventually become willing to improve levees in the absence of financial support of the state. Nevertheless, this group prefers financial support from California for improving the levee system.

California's level of intervention might depend on the various circumstances in all noted equilibria. Results show that a strong equilibrium (likely outcome) can develop even without the state's involvement (outcome 8). However, all other strong equilibria (outcomes 28, 48 and 68) involve some intervention by California. All strong equilibria of the game involve a legal action (or pressure) by the environmentalists, who use this strategy as leverage to force the water exporters to stop business as usual. Nevertheless, two of the strong equilibria (outcome 8 and 28) are not environment-friendly because, despite the legal action by the environmentalists, and even with pressure from the State of California (outcome 28), water exports continue as usual. This, indeed, is indicative of the importance of water exports, a strong position of water exporters in this problem and their tendency to act non-cooperatively. The two strong environment-friendly equilibria, however, include implementation of a new water conveyance system by the exporters.

In the weak equilibria or less-likely outcomes of the game (states 7, 27, 47 and 67), the major stakeholders are willing to solve the problem on a cooperative basis. These environment-friendly outcomes do not involve legal action by the environmentalists. On the other hand, levees are improved and a new water conveyance system is built to enhance the conditions of the Delta for all stakeholder groups. The State of California has an opportunity to wisely adjust its action to the decision situation by choosing the appropriate level of pressure on the water exporters and providing financial support for levee improvements. The lower strength of these equilibria, however, suggests that

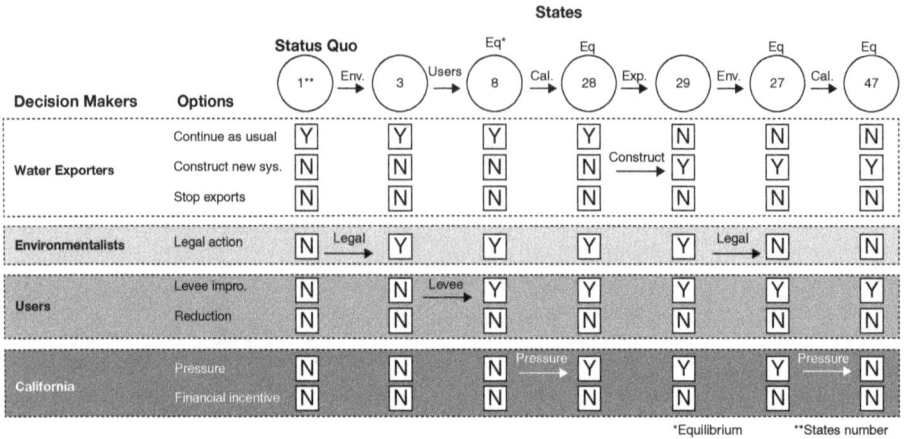

Figure 15.2 Possible evolution path of the Delta conflict from the status quo to the final cooperative equilibrium.

a cooperative resolution of the Delta conflict is less likely. It can be also concluded that California's intervention can play a major role in enforcing a cooperative outcome that serves the objectives of all parties in this conflict.

Figure 15.2 shows a possible evolution of the Delta conflict from the status quo through intermediate outcomes to the final resolution. Environmentalists cannot stand the business-as-usual situation and move to state from status quo through legal actions against (or pressuring) the water exporters. Water users evolve the status of the conflict by improving the levees. By putting pressure on water exporters, California can help to move the game from state 8 to state 28. Under the pressure of California and the environmentalists and to decrease tensions, water exporters need to step back from business as usual policy to exporting water through a new conveyance system, changing the game's state from 28 to 29. At this stage, environmentalists may decide to withdraw their legal pressure. Similarly, California has this chance to release the pressure and provide financial support to water users for improving the levee. Thus, the sequence of actions is expected to move the game from the status quo, state 1, to state 47.

Coalition analysis (Inohara and Hipel 2008) can be performed to further examine the strength of equilibria and investigate the opportunities for cooperation of a subset of DMs (involving at least two DMs) to improve outcomes that are not achievable if they act individually. Parties might form coalitions to move jointly to a better outcome that all DMs in the subset prefer to the status quo (win-win move). The new outcome might be threatening to other players, forcing them to take further action to change the state of the game. Accordingly, the objective of coalition analysis is to determine the vulnerable equilibria to coalition formation (Kilgour et al. 2001) based on group rationality as opposed to individual rationality. The result of the coalition analysis using GMCR II suggests that formation of a coalition between the environmentalists and the water exporters can change the Delta game from a non-cooperative one to a cooperative one. In the cooperative mode the water exporters commit to construction a new conveyance system to avoid unnecessarily political tension and environmentalists do not pursue any legal action to pressure the exporters. This would make the strong equilibria of the game (states 8, 28, 48 and 68) unstable and help with development of environment-friendly equilibria (outcomes 7, 27, 47 and 67). Nevertheless, a natural formation of the environmentalists–water exporters coalition is less likely unless California wants to catalyze the process through pressuring the exporters.

Conclusion

The Graph Model for Conflict Resolution (GMCR) was used in this study as a powerful game-theoretic methodology to strategically investigate California's Sacramento–San Joaquin Delta conflict. GMCR II was employed as a decision-support system to facilitate the analysis of the game based on various stability definitions, representing various behavioral characteristics of the decision makers. Modeling results suggest that: (a) any resolution to the conflict involves improvement of the Delta's fragile levee system, and that stopping water exports or reduction of water uses seem highly implausible; (b) the confrontation between the environmentalists and water exporters is expected to continue, as the latter group might not be willing to change its water-export strategies even under legal action by the environmentalists; and (c) the State of California needs to play a key role in resolving the conflict.

Through reward or punishment strategies, the State of California can encourage or force the parties to resolve the conflict on a cooperative basis and to develop environment-friendly outcomes. A cooperative resolution of the conflict involves construction of a new water-export facility. A coalition between the environmentalists and the water exporters makes the non-cooperative equilibria unstable and enforces a more environment-friendly water-export option. Considering the evolution path of the conflict, as well as results of the stability analysis and coalition analysis, outcome 47 is the most likely and promising resolution of the Delta conflict. Realization of this outcome includes construction of a new water conveyance system to change the current water exports method (without change in the withdrawal amount) as well as improvement of the levees system with financial support from the State of California.

In theory, full cooperation in search of a win-win situation should lead to a sustainable resolution in most conflicts. However, a desirable resolution would not emerge without a willingness to compromise and taking advantage of "strategic loss" opportunities (Madani 2011). In practice, perfect cooperation hardly exists because of various factors including conflicting values and interests; bounder rationality; a lack of trust; and communication among players. Thus, mediation and having a strong external force to push all players toward cooperation could be helpful.

References

Brams, Steven J. and Donald Wittman. 1981. "Nonmyopic Equilibria in 2 X 2 Games." *Conflict Management and Peace Science* 6: 39–62.

Bruce, Christopher and Kaveh Madani. 2014. "The Conditions for Successful Collaborative Negotiation over Water Policy: Substance versus Process," Working Paper 2014–36, Department of Economics, University of Calgary.

Chung, Francis, Katherine Kelly and Kamyay Guivetchi. 2002. "Averting a California Water Crisis." *Journal of Water Resources Planning and Management.* 128: 237–39.

Cornell-Buck, Christina R., Josué Medellin-Azuara, Jay R. Lund and Kaveh Madani. 2011. "Adapting California's Water System to Warm vs. Dry Climates." *Climatic Change* 109(1): 133–49.

California Department of Fish and Wildlife. 2014. "Endangered and Threatened Animal List." Available at: http://www.dfg.ca.gov/biogeodata/cnddb/pdfs/TEAnimals.pdf

Fang, Liping, Keith W. Hipel and D. Marc Kilgour. 1993. *Interactive Decision Making: The Graph Model for Conflict Resolution.* New York: Wiley.

Fang, Liping, Keith W. Hipel and D. Marc Kilgour. 1997. "Scenario Generation and Reduction in the Decision Support System GMCR II." In: IEEE International Conference on Systems, Man, and Cybernetics 1997. Computational Cybernetics and Simulation. IEEE, Orlando, Florida, 341–45.

Fraser, Niall M. and Keith W. Hipel, 1984. *Conflict Analysis: Models and Resolutions.* New York: North-Holland.

Hanak, Ellen, Jay Lund, Barton "Buzz" Thompson, W. Bowman Cutter, Brian Gray, David Houston, Richard Howitt, Katrina Jessoe, Gary Libecap, Josué Medellín-Azuara, Shelia Olmstead, Daniel Sumner, David Sunding, Brian Thomas and Robert Wilkinson. 2012. *Water and California Economy.* San Francisco: Public Policy Institute of California.

Hanemann, Michael and Catlin Dyckman. 2009. "The San Francisco Bay-Delta: A Failure of Decision-Making Capacity." *Environmental Science and Policy* (12): 710–25.

Hipel, Keith W., Marc D. Kilgour, Liping Fang and John Peng. 1997. "The Decision Support System GMCR in Environmental Conflict Management." *Applied Math and Computation.* 83, 117–52.

Howard, Nigel. 1971. *Paradoxes of Rationality: Theory of Metagames and Political Behavior.* Cambridge, MA: MIT.

Howard, Nigel. 1999. Confrontation Analysis: How to Win Operations Other than War. Washington, DC: Command and Control Research Program.

Inohara, Takehiro and Keith Hipel. 2008. "Coalition Analysis in the Graph Model for Conflict Resolution." *Systems Engineering* 343–59.

Jenkins, Marion W., Andrew A. Draper, Guilherme F. Marques, Randall S. Ritzema, Stacy K. Tanaka, Kenneth W. Kirby, Megan S. Fidell, Siwa Msangi, Richard E. Howitt and Jay R. Lund. 2001. *"Economic Valuation of California's Water Resources and Infrastructure."* In: ASCE Specialty Conference on Water Resources and the Environment, Orlando, FL.

Kilgour, D. Marc, Keith W. Hipel, Liping Fang and John Peng. 2001. "Coalition Analysis in Group Decision Support." *Group Decision and Negotiation* 10: 159–75.

Kilgour, D. Marc and Keith W. Hipel. 2005. "The Graph Model for Conflict Resolution: Past, Present, and Future." *Group Decision and Negotiation* 14: 441–60.

Li, Kevin W., D. Marc Kilgour and Keith W. Hipel. 2004. "Status Quo Analysis of the Flathead River Conflict." *Water Resources Research* 40, W05S03.

Lund, Jay R. 2004. "Re-assembling California's Water: How Water Management Will Change and Why." In: *World Water and Environmental Resources Congress 2003*: 1–6.

Lund, Jay R., Ellen Hanak, William Fleenor, Richard Howitt, Jeffrey Mount and Peter Moyle. 2007. *Envision Futures for the Sacramento–San Joaquin Delta.* San Francisco: Public Policy Institute of California.

Lund, Jay R., Ellen Hanak, William Fleenor, William Bennett, Richard Howitt, Jeffery Mount and Peter Moyle. 2008a. *Comparing Futures for the Sacramento–San Juaquin Delta.* San Francisco: Public Policy Institute of California.

Lund, Jay R., Ellen Hanak, William Fleenor, William Benett, Richard Howitt, Jeffery Mount and Peter Moyle. 2008b. *Decision Analysis of Delta Strategies.* Technical appendix J, comparing future for the Sacramento–San Joaquin Delta. San Francisco: Public Policy Institute of California.

Madani, Kaveh. 2010. "Game Theory and Water Resources." *Journal of Hydrology* 381: 225–38.

————. 2011. "Hydropower Licensing and Climate Change: Insights from Cooperative Game Theory." *Advances in Water Resources* 34: 174–83.

————. 2013. "Modeling International Climate Change Negotiations More Responsibly: Can Highly Simplified Game Theory Models Provide Reliable Policy Insights?" *Ecological Economics* 90, 68–76.

Madani, Kaveh and Keith W. Hipel. 2007. "Strategic Insights into the Jordan River Conflict." In: *World Environmental and Water Resources Congress 2007*. American Society of Civil Engineers, Reston, VA, 1–10.

Madani, Kaveh and Keith W. Hipel. 2011. "Non-cooperative Stability Definitions for Strategic Analysis of Generic Water Resources Conflicts." *Water Resources and Management* 25: 1949–77.

Madani, Kaveh and Jay R. Lund. 2011. "A Monte-Carlo Game Theoretic Approach for Multi-Criteria Decision Making under Uncertainty." *Advances in Water Resources.* 34: 607–16.

————. 2012. "California's Sacramento–San Joaquin Delta Conflict: From Cooperation to Chicken." *Journal of Water Resources Planning and Management.* 138: 90–99.

Madani, Kaveh, David Rheinheimer, Laila Elimam and Christina Connell-Buck. 2011. *"A Game Theory Approach to Understanding the Nile River Basin Conflict,"* 97–114. In: *A Water Resource Festschrift* in honor of Professor Lars Bengtsson. Lund, Sweden: Lund University.

Madani, Kaveh, Laura. Read and Laleh Shalikarian. 2014. "Voting under Uncertainty: A Stochastic Framework for Analyzing Group Decision Making Problems." *Journal of Water Resources Management* 28: 1839–56.

McClurg, Sue. 2010. *Layperson's Guide to the Delta.* Sacramento: Water Education Foundation.

Medellin-Azuara, Josué, Julien J. Harou, Marcelo Olivares, Kaveh Madani, Jay R. Lund, Richard Howitt, Stacy K. Tanaka, Marion Jenkins and Tingju Zhu. 2008. "Adaptability and Adaptations of California's Water Supply System to Dry Climate Warming." *Journal of Climate Change* 87: 75–90.

Medellin-Azuara, Josué, Ali Mirchi and Kaveh Madani. 2011. "Water Supply for Agricultural, Environmental, and Urban Uses in California's Borderlands." In: Contreras, L. *Agricultural Policies: New Developments.* New York: Nova Science Publishers, 201–12.

Mirchi, Ali, Kaveh Madani, Maurice Roos and David W. Watkins. 2013. "Climate Change Impacts on California's Water Resources." In: Kurt Schawbe, José Albiac, Jeffery Connor, Rashid Hassan and Liliana González. *Drought in Arid and Semi-arid Regions.* Dordrecht: Springer, 301–19.

Mokhtari, Soroush, Kaveh Madani and Ni-Bin Chang. 2012. *"Multi-Criteria Decision Making under Uncertainty: Application to the California's Sacramento-San Joaquin Delta Problem."* In: *Crossing Boundaries.* World Environmental and Water Resources Congress 2012, American Society of Civil Engineers, 2339–48.

Moyle, Peter B., William A. Bennett, Cliff Dahm, John R. Durand, Christopher Enright, William E. Fleenor, Wim Kimmerer and Jay R. Lund. 2010. *Changing Ecosystems: A Brief Ecological History of the Delta: Major Change in the Delta.* Report to the California State Water Resources Control Board, Sacramento.

Nandalal, K. D. W., and Keith W. Hipel. 2007. "Strategic Decision Support for Resolving Conflict over Water Sharing among Countries along the Syr Darya River in the Aral Sea Basin." *Journal of Water Resources Planning and Management* 133: 289–99.

Nash, John F. 1951. "Non-cooperative Games." *Annals of Mathematics* 54, 286–95.

Newlin, Brad, Marion Jenkins, Jay R. Lund and Richard Howitt. 2002. "Southern California Water Markets: Potential and Limitations." *Journal of Water Resources. Planning and Management* 128: 21–32.

Norgaard, Richard B., Giorgos Kallis and Michael Kiparsky. 2009. Collectively Engaging Complex Socio-ecological Systems: Re-envisioning Science, Governance, and the California Delta. *Environmental Science and Policy* 12, 644–52.

Obeidi, Amer, Keith W. Hipel and D. Marc Kilgour. 2006. "Turbulence in Miramichi Bay: The Burnt Church Conflict over Native Fishing Rights." *Journal of American Water Resources Association.*

Peng, Xiaoyong, Keith W. Hipel, D. Marc Kilgour and Liping Fang. 1997. "Representing Ordinal Preferences in the Decision Support System GMCR II." In: *IEEE International Conference on Systems, Man, and Cybernetics*. Orlando, Florida, 341–45.

Pfeffer, W. Tad, Joel T. Harper, and Shad O'Neel. 2008. "Kinematic Constraints on Glacier Contributions to 21st Century Sea-level Rise." *Science* 321: 1340–43.

Rastgoftar, Hossein, Sanaz Imen and Kaveh Madani. 2012. "*Stochastic Fuzzy Assessment for Managing Hydro-Environmental Systems under Uncertainty and Ambiguity.*" In: *Crossing Boundaries*. World Environmental and Water Resources Congress 2012, American Society of Civil Engineers, 2413–21.

Suddeth, Robyn, Jeff Mount and Jay R. Lund . 2010. "Levee Decisions and Sustainability for the Sacramento–San Joaquin Delta." *San Francisco Estuary and Watershed Science* 8(2): 1–23.

Tanaka, Stacy, Christina R. Connell-Buck, Kaveh Madani, Josué Medellin-Azuara, Jay R. Lund and Ellen Hanak . 2011. "Economic Costs and Adaptations for Alternative Regulations of California's Sacramento-San Joaquin Delta." *San Francisco Estuary and Watershed Science* 9(2): 1–28.

Zagare, Frank C. 1984. "Limited-move Equilibria in 2 x 2 Games." *Theory Decision* 16: 1–19.

CONTRIBUTORS

Aggestam, Karin
Peace & Conflict Studies, Lund University, Lund, Sweden

AlMisnad, Abdulla
Public Investment Management, Ministry of Finance, Doha, Qatar

Andersson, Dan-Erik
Center for Middle Eastern Studies, Lund University, Lund, Sweden

Berndtsson, Ronny
Center for Middle Eastern Studies, Lund University, Lund, Sweden, and Department of Water Resources Engineering, Lund University, Lund, Sweden

Bhatia, Udit
Sustainability & Data Sciences Laboratory, Civil & Environmental Engineering, Northeastern University, Boston, Massachusetts, USA

Carr, Joel A.
Department of Environmental Sciences, University of Virginia, Charlottesville, Virginia, USA

Choudhury, Enamul
Department of Urban Affairs and Geography, Wright State University, Ohio, USA

D'Odorico, Paolo
Department of Environmental Sciences, University of Virginia, Charlottesville, Virginia, USA
National Social-Environmental Synthesis Center, University of Maryland, Annapolis, Maryland, USA

Ganguly, Auroop R.
Sustainability & Data Sciences Laboratory, Civil & Environmental Engineering, Northeastern University, Boston, Massachusetts, USA

Garcia, Margaret
Department of Civil & Environmental Engineering, Tufts University, Medford, Massachusetts, USA

Hipel, Keith W.
Department of Systems Design Engineering, University of Waterloo, Waterloo, Canada

Huntjens, Patrick
The Hague Institute for Global Justice

Ibrahim, Yosif
Fairfax County Department of Public Works and Environmental Services, Virginia, USA

Islam, Shafiqul
Water Diplomacy Program, Civil and Environmental Engineering, Tufts University, Massachusetts, USA

Kumar, Devashish
Sustainability & Data Sciences Laboratory, Civil & Environmental Engineering, Northeastern University, Boston, Massachusetts, USA

Kodra, Evan
Sustainability & Data Sciences Laboratory, Civil & Environmental Engineering, Northeastern University, Boston, Massachusetts, USA

Madani, Kaveh
Centre for Environmental Policy, Imperial College London, London, UK

Maheshwari, Basant
School of Science & Health, University of Western Sydney, New South Wales, Australia

Marshall, Dena
Four Worlds, Oregon, USA

Mirchi, Ali
Department of Civil and Environmental Engineering, Michigan Technological University, Houghton, Michigan, USA

Moazezi, Mohammad R.
Department of Civil and Environmental Engineering, University of Waterloo, Waterloo, Canada

Mondal, Arpita
Divecha Center for Climate Change, Indian Institute of Science, Bangalore, India

Majumder, Maimuna
Engineering Systems Division at Massachusetts Institute of Technology, Cambridge, Massachusetts, USA, and HealthMap Computational Epidemiology Group at Boston's Children's Hospital of Harvard Medical School, USA

Mujumdar, P. P.
Divecha Center for Climate Change and Department of Civil Engineering, Indian Institute of Science, Bangalore, India

Neufville, Richard de
Institute for Data, Systems, and Society, Massachusetts Institute of Technology, Cambridge, Massachusetts, USA

Pohl, Benjamin
Adelphi, Berlin, Germany

Read, Laura
Department of Civil and Environmental Engineering, Tufts University, Medford, Massachusetts, USA

Salamé, Léna
UNESCO, Paris, France

Swain, Ashok
Uppsala University, Uppsala, Sweden

Thoradeniya, Bhadranie
Institute of Technology, University of Moratuwa, Moratuwa, Sri Lanka

Wolf, Aaron T.
Program in Water Conflict Management and Transformation, Oregon State University, Oregon, USA

Wong Turlington, Melanie
Arcadis, San Francisco, California, USA

Zarezadeh, Mahboubeh
Department of Water Resources Engineering, Tarbiat Modares University, Tehran, Iran

INDEX

www.ingramcontent.com/pod-product-compliance
Lightning Source LLC
Chambersburg PA
CBHW030638270326
41929CB00007B/120